LUNG
SURFACTANTS

LUNG BIOLOGY IN HEALTH AND DISEASE

Executive Editor

Claude Lenfant
Director, National Heart, Lung and Blood Institute
National Institutes of Health
Bethesda, Maryland

1. Immunologic and Infectious Reactions in the Lung, *edited by C. H. Kirkpatrick and H. Y. Reynolds*
2. The Biochemical Basis of Pulmonary Function, *edited by R. G. Crystal*
3. Bioengineering Aspects of the Lung, *edited by J. B. West*
4. Metabolic Functions of the Lung, *edited by Y. S. Bakhle and J. R. Vane*
5. Respiratory Defense Mechanisms (in two parts), *edited by J. D. Brain, D. F. Proctor, and L. M. Reid*
6. Development of the Lung, *edited by W. A. Hodson*
7. Lung Water and Solute Exchange, *edited by N. C. Staub*
8. Extrapulmonary Manifestations of Respiratory Disease, *edited by E. D. Robin*
9. Chronic Obstructive Pulmonary Disease, *edited by T. L. Petty*
10. Pathogenesis and Therapy of Lung Cancer, *edited by C. C. Harris*
11. Genetic Determinants of Pulmonary Disease, *edited by S. D. Litwin*
12. The Lung in the Transition Between Health and Disease, *edited by P. T. Macklem and S. Permutt*
13. Evolution of Respiratory Processes: A Comparative Approach, *edited by S. C. Wood and C. Lenfant*
14. Pulmonary Vascular Diseases, *edited by K. M. Moser*
15. Physiology and Pharmacology of the Airways, *edited by J. A. Nadel*
16. Diagnostic Techniques in Pulmonary Disease (in two parts), *edited by M. A. Sackner*
17. Regulation of Breathing (in two parts), *edited by T. F. Hornbein*
18. Occupational Lung Diseases: Research Approaches and Methods, *edited by H. Weill and M. Turner-Warwick*
19. Immunopharmacology of the Lung, *edited by H. H. Newball*
20. Sarcoidosis and Other Granulomatous Diseases of the Lung, *edited by B. L. Fanburg*
21. Sleep and Breathing, *edited by N. A. Saunders and C. E. Sullivan*
22. *Pneumocystis carinii* Pneumonia: Pathogenesis, Diagnosis, and Treatment, *edited by L. S. Young*
23. Pulmonary Nuclear Medicine: Techniques in Diagnosis of Lung Disease, *edited by H. L. Atkins*

ADDITIONAL VOLUMES IN PREPARATION

The opinions expressed in these volumes do not necessarily represent the views of the National Institutes of Health.

LUNG SURFACTANTS
BASIC SCIENCE AND CLINICAL APPLICATIONS

Robert H. Notter

University of Rochester School of Medicine
Rochester, New York

MARCEL DEKKER, INC.   NEW YORK · BASEL

ISBN: 0-8247-0401-0

This book is printed on acid-free paper.

Headquarters
Marcel Dekker, Inc.
270 Madison Avenue, New York, NY 10016
tel: 212-696-9000; fax: 212-685-4540

Eastern Hemisphere Distribution
Marcel Dekker AG
Hutgasse 4, Postfach 812, CH-4001 Basel, Switzerland
tel: 41-61-261-8482; fax: 41-61-261-8896

World Wide Web
http://www.dekker.com

The publisher offers discounts on this book when ordered in bulk quantities. For more information, write to Special Sales/Professional Marketing at the headquarters address above.

Current printing (last digit):
10 9 8 7 6 5 4 3 2 1

PRINTED IN THE UNITED STATES OF AMERICA

INTRODUCTION

Over the last 40 years or so, we have witnessed remarkable advances in medicine which are best demonstrated by a decline in the death rate for many acute and chronic conditions. This has translated into an increase in longevity measured in years.

Although not evenly distributed in the major urban areas of the United States or among the various ethnic groups of the country, improvements in health and survival have occurred in all age groups. Forty years ago, mortality in the newborn period was highest among premature infants. Among the many causes of death, respiratory distress syndrome (RDS) was the most prevalent one. Data from the late 1960s indicate that approximately 30% of premature infants had RDS, and its severity and frequency were closely related to the degree of prematurity. Of all these infants who had RDS, up to 40% died from this condition—about 30,000 to 35,000 each year.

Today we have an amazingly different situation. Even though the number of births per year has significantly increased over the last 30 years, and even though infants are born alive at much younger gestational ages today than then, the number of deaths from RDS is less than 1500 per year. What a remarkable success indeed!

The reason for this success is very simple: it is the result of a research effort that has ranged from very basic research to clinical investigation. A large part of this research focused on the lung surfactant—its identification, its structure, its regulation, and its role in maintaining the integrity of lung function. As we all know well, this led to the development of lung surfactant replacement therapy for use in those pathological situations resulting from its absence or limited presence. Respiratory distress syndrome in premature infants is the "prototype" of such conditions.

The story of lung surfactant is a wonderful example that justifies our commitment to biomedical research and illustrates the public health benefit derived from that research.

This book is, in fact, the story of lung surfactant. It departs from the usual presentations in the Lung Biology in Health and Disease series of monographs because it is a single-author volume. Dr. Notter brings to the series decades of work and expertise in lung surfactant. The resulting textbook will undoubtedly be an aid to the researchers in this area and to physicians who care for the patients. Ultimately, more patients, born before their time, will be the beneficiaries.

As the editor of this series of monographs, I thank Dr. Notter for the opportunity to include this volume in the series.

Claude Lenfant, M.D.
Bethesda, Maryland

PREFACE

This book grows out of the research of basic scientists and physicians whose work over the past five decades has taken us from the discovery of pulmonary surfactant through the development of life-saving exogenous surfactant therapy for premature infants. The text emphasizes the interrelationship between the basic science of lung surfactants and the development of effective clinical surfactant therapies for lung disease and injury. The history of lung surfactant research shows successes and failures that can be explained by reference to biophysical and physiological principles. Understanding the basic science of lung surfactants ensures optimal surfactant therapy in the future.

The book attempts to address the lack of didactic material accessible to physicians, physicians-in-training, and other medical personnel wishing to understand surfactants and their activity at a fundamental level. The integrated presentation of basic and clinical topics will also hopefully interest nonclinical biomedical and physical scientists involved in research on the lungs and lung surfactants. Although by no means comprehensive, the text includes introductory chapters on surfactants and surface films, phospholipids and their physical behavior, and experimental methods and materials used in lung surfactant research. An overview of the pathology and clinical features of lung surfactant-related diseases is also given. Primary emphasis, of course, is on the composition, biophysics, and physiology of lung surfactants, and on clinical surfactant therapies in the context of this basic science. Bringing together these multiple aspects into what is hopefully a cohesive whole was not without its challenges, and apologies are offered in advance for any deficiencies that remain.

Robert H. Notter

ACKNOWLEDGMENTS

The encouragement and perspectives of many colleagues, students, and friends were invaluable in the preparation of this book. Particular thanks is given to my faculty colleagues at the University of Rochester and SUNY Buffalo, including Drs. Zhengdong Wang, Bruce Holm, Jacob Finkelstein, Goran Enhorning, Edmund Egan, Dale Phelps, Michael Apostolakos, Richard Hyde, and Donald Shapiro (deceased). Insights gained through current or prior collaborations with other colleagues including Drs. Stephen Hall, Fred Possmayer, John Baatz, Jeffrey Whitsett, Joseph Turcotte, Douglas Willson, and Darryl Absolom (deceased) were also invaluable, as were the contributions of my graduate students and postdoctoral fellows, whose hard work over the years was integral to many of the results and perspectives presented. Mrs. Kim Butler provided great assistance with the manuscript and references, and Mrs. Jenny Smith was responsible for helping with all the figures. The forbearance of my wife and children, who accepted the stress and time associated with this endeavor, needs no elaboration.

A special acknowledgment is given to the faculty and staff of the Division of Neonatology at the University of Rochester, who over the years have contributed significantly to basic and clinical research helping to establish the efficacy of surfactant therapy for premature infants and extending it to patients with respiratory failure and acute lung injury. I am proud to have been associated with this basic and clinical research effort at the University of Rochester.

CONTENTS

LUNG
SURFACTANTS

1

Introduction

This book focuses on the basic science of endogenous and exogenous lung surfactants and its translation to clinical lung surfactant therapies. Initial chapters provide an introduction to surface tension, surfactants, surface films, phospholipid structure and biophysics, and experimental methods and materials used in studying lung surfactants. Following chapters cover the discovery of lung surfactant, the theoretical basis of its pulmonary activity, its functional composition and molecular biophysics, and the mechanisms by which it can become dysfunctional in lung disease and injury. The direct connection between the biophysical properties and physiological actions of lung surfactants is detailed, and biophysical-physiological correlations of lung surfactant activity and inhibition are provided. The final third of the book emphasizes applications involving surfactant replacement therapy for the neonatal respiratory distress syndrome (RDS), acute lung injury, and the acute respiratory distress syndrome (ARDS). Animal models of surfactant deficiency and acute lung injury are summarized, and their use in developing exogenous surfactant therapy for RDS and ARDS are illustrated in a variety of examples. The history and current status of clinical surfactant therapy for RDS and ARDS are then discussed in detail, along with current and future exogenous surfactants and their composition and activity. The necessary integration of basic and clinical research on lung surfactants is emphasized throughout coverage here. The style of presentation is designed for physicians and scientists with varying backgrounds in interfacial phenomena and biophysics. Each chapter (other than this Introduction) has an initial Overview and a final Summary covering main points and perspectives. Some chapter text is also printed in smaller type to designate supplementary material or explanation. The body of each chapter contains extensive literature citations, which are compiled in alphabetical order at the end of the book along with a glossary of common terms and abbreviations.

Understanding the biophysical basis of lung surfactant activity and inhibition is highly important for physicians and basic scientists interested in respiration. Normal respiration depends on the surface active function of endogenous surfactant, and severe pulmonary disease results if this activity is deficient or compromised. Endogenous and exogenous lung surfactants generate their direct physiological effects through specific biophysical properties rather than by acting on cell-based receptors or substrates. Knowledge of the component-specific interfacial biophysics of any lung surfactant material underlies rational assessment of its physiological activity and use in surfactant therapy for RDS and ARDS.

Although the functions of the mammalian lungs are straightforward conceptually, rigorous descriptions of the pulmonary system and respiration are complex and necessitate a multidisciplinary perspective. As in any living system, biochemistry, cell and molecular biology, and physiology are required for understanding lung development, growth, and function. In addition, morphometrics, analytical geometry, and biomechanics are needed to describe the complexities of lung structure and pressure-volume behavior. Medicine, pathology, and pharmacology are also necessary for applications involving lung diseases and their therapy. Of all the scientific disciplines relevant for the lung, perhaps the most surprising is interfacial phenomena. This discipline and the related areas of molecular biophysics, physical chemistry, and thermodynamics apply to pulmonary mechanics and function because of the importance of surface tension and surface active agents (*surfactants*) in respiration.

The alveolar airsacs in mammals are stabilized by pulmonary surfactant, a complex mixture composed primarily of phospholipids and specific proteins synthesized, stored, secreted, and recycled by type II alveolar epithelial cells. The lungs of mammals have a huge internal surface area of order 1 m^2/kg body weight at total lung capacity. This surface area in a 70 kg adult approximates that of a badminton court, and much of it is lined by a thin liquid film or "alveolar hypophase." Surface tension forces at the extensive air-hypophase interface are sufficiently large to dominate the quasi-static work of breathing. The pulmonary surfactant system plays crucial roles in respiratory physiology by moderating these surface tension forces. As detailed in subsequent chapters, surfactant secreted by alveolar type II epithelial cells adsorbs at the air-hypophase interface and lowers and varies surface tension as a function of alveolar size during breathing. This regulation of surface tension reduces the work of breathing while stabilizing alveoli against collapse and overdistension. It also leads to a smaller hydrostatic pressure driving force for edema fluid to move into the interstitium from the pulmonary capillaries. Functional lung surfactant is necessary for life, and its deficiency or inactivation is inevitably associated with clinically significant pulmonary disease.

RDS and ARDS are the two major surfactant-related pulmonary diseases worldwide. RDS, also called hyaline membrane disease (HMD), is a disease of prematurity. It is primarily found in infants <32 weeks gestation, although older premature infants are sometimes affected. RDS is caused by a deficiency of surfactant in the lungs of premature infants at birth, although elements of surfactant dysfunction and lung injury, along with multiple complications of prematurity and intensive care, can complicate its clinical course. In contrast to RDS, ARDS-related respiratory failure can affect patients of all ages, from full-term infants to adults. Lung surfactant dysfunction is just part of the complex pathology of ARDS, a severe lung injury syndrome arising from multiple causes. ARDS has a substantial mortality and morbidity despite sophisticated ventilatory management and intensive care, and a number of therapeutic approaches for it are being investigated in current research.

If endogenous lung surfactant is deficient or dysfunctional, exogenous surface active material can in principle be delivered to the alveoli as a substitute. This is the basic concept of exogenous surfactant therapy. Inherent in this concept is the presumption that the lungs will eventually establish or recover the ability to synthesize their own surfactant and maintain a normal pulmonary environment without further dysfunction. Although conceptually simple, effective exogenous surfactant therapy for RDS took many years to develop and depended on basic science understanding gained through extensive biophysical and animal model research. Exogenous surfactant therapy, along with advances in mechanical ventilation and neonatal intensive care, has had a dramatic impact on the outcome

---

Major Lung Surfactant–Associated Diseases

Neonatal respiratory distress syndrome (RDS)
    Also known as hyaline membrane disease.
    Primarily affects premature infants <32 weeks gestation.
    Initiating cause is surfactant deficiency.
    Lung injury and complications of prematurity can also be present.

Acute respiratory distress syndrome (ARDS)
    Formerly called the "adult" respiratory distress syndrome.
    Involves acute lung injury but may have multiorgan involvement.
    Has multiple etiologies and affects patients of all ages.
    Surfactant dysfunction is one aspect of a complex pathology.
    Surfactant deficiency may or may not be present.

Concept of Exogenous Surfactant Therapy

Surfactant replacement therapy involves the alveolar delivery of exogenous surfactants to replace or supplement endogenous surfactant that is deficient or has become dysfunctional.

Both lung surfactant deficiency and lung surfactant dysfunction should in principle respond to exogenous surfactant therapy.

Surfactant replacement therapy is a short-term measure only. Patients must eventually produce their own surfactant and restore a normal pulmonary environment.

---

and survival of premature infants. Surfactant therapy for RDS is currently being optimized and extended to acute lung injury and ARDS. Surfactant therapy for ARDS is still in the developmental stages, with mixed success particularly in adult patients. To be effective in ARDS, surfactant therapy needs to employ exogenous surfactants with maximal activity and inhibition resistance and may also require a "multimodal" approach that simultaneously targets other aspects of pathophysiology.

The development and optimization of exogenous surfactant therapy for RDS and ARDS depend fundamentally on basic science understanding. Controlled clinical trials are essential in establishing therapeutic efficacy but are less suited to examine detailed physiological responses and mechanisms of activity. Clinical trials by their intrinsic nature do not provide assessments at the same level of specificity possible in basic laboratory and animal research. This is particularly true for clinical studies in diseases like RDS and ARDS that involve populations of critically ill patients receiving invasive intensive care. Outcomes for premature infants receiving exogenous surfactants to prevent or treat RDS reflect multiple developmental and pathophysiologic variables only some of which relate to lung surfactant activity. Patient populations with ARDS are more heterogeneous and incorporate an even broader pathophysiology. The limited resolving power of clinical studies in RDS and ARDS makes it more difficult to demonstrate differences in efficacy and long term outcome among agents and treatment strategies. Basic research understanding, emphasized in coverage here, provides crucial mechanism-based data and correlations to complement and supplement clinical findings.

This book attempts to provide an integrated view of basic and clinical research that has led to the development of successful surfactant therapy for RDS and that is essential to optimize and extend it effectively to ARDS. Some of the kinds of questions addressed in subsequent chapters include:

> *What is surface tension and why is it important in the lungs?*
> *What are surface active agents (surfactants) and how do they act?*
> *What materials, methods, and animal models are used in surfactant research?*
> *What is endogenous lung surfactant and how was it discovered?*
> *What are the biophysical and physiological actions of endogenous lung surfactant and its individual lipid and protein components?*
> *What are the history and current status of exogenous surfactant therapy for RDS and ARDS?*
> *How do current clinical exogenous surfactants compare with endogenous surfactant and with each other?*
> *What new exogenous surfactants may be developed in the future?*

Answering such questions requires examining lung surfactants and their activity from perspectives beginning with biophysics and physical chemistry and ending with clinical medicine. This multidisciplinary approach is reflected in the chapter organization summarized below.

Importance of Basic Science Evaluations of Lung Surfactants

Basic research on endogenous and exogenous lung surfactants provides detailed, comprehensive evaluations of composition and biophysical activity.

The direct effects of surfactants on pulmonary function result from composition-dependent surface active properties that can be measured readily and comprehensively *in vitro*.

Basic research in animal models can similarly define in detail the physiological effects of lung surfactants on respiratory function and mechanics.

Clinical trials in critically ill patients with RDS or ARDS are essential for establishing therapeutic efficacy, but complementary basic research on mechanisms and activity is also required.

Summary of Chapter Organization

Chapters 2–4 cover basic concepts related to surface tension, surfactants, surface films, adsorption, dynamic surface tension lowering, phospholipid structure and biophysics, and physicochemical methods used to study lung surfactants. *Chapter 5* covers the range of materials used in lung surfactant research and shows how many of them are produced; it also outlines some of the complexities that make it essential to correlate composition, surface activity, and physiological effects in lung surfactant research. *Chapter 6* details the rationale for the existence of endogenous lung surfactant and summarizes its discovery, metabolism in type II pneumocytes, and biophysical and physiological actions in respiration. *Chapter 7* covers theories and experiments relating alveolar surface tension–area behavior to pulmonary pressure-volume mechanics, as well as analyses of lung surfactant surface behavior and contributions to alveolar stability. *Chapter 8* details the functional composition of endogenous lung surfactant and the biophysical roles and interactions of its individual lipid and protein components. *Chapter 9* describes the mechanisms and characteristics of lung surfactant dysfunction (inactivation) by a variety of different inhibitors and indicates how dysfunction can in many cases be reversed. The latter third of the book then emphasizes applications to surfactant replacement therapy. *Chapter 10* summarizes the pathophysiology and clinical features of RDS and ARDS and more briefly describes hereditary SP-B deficiency. *Chapter 11* covers surfactant replacement research in animal models of RDS and ARDS and its crucial importance in developing and optimizing surfactant therapy for humans. *Chapter 12* reviews the history and current status of clinical surfactant therapy for RDS in premature infants. *Chapter 13* describes surfactant therapy for ARDS and clinical lung injury, including the use of exogenous surfactants in combined-modality strategies directed against multiple aspects of lung injury. Finally, *Chapter 14* summarizes the composition and activity of exogenous surfactants in current clinical use worldwide and gives examples of research approaches and new synthetic exogenous surfactants under investigation for future use.

2

Introduction to Surface Tension and Surfactants

I. Overview

This chapter introduces the general concepts of surface tension, surface active agents (surfactants), surface films, and adsorption. The study of these topics utilizes principles of interfacial phenomena, physical and surface chemistry, and thermodynamics. Surfactants are important in a variety of physical processes and include materials ranging from simple soaps and detergents to complex biological molecules such as lipids and proteins. Coverage here is an introductory overview of interfacial phenomena and surfactant behavior. Surface tension and pressure are defined and discussed, as are various aspects of surfactant film behavior including surface molecular states, dynamic and equilibrium surface tension lowering, film collapse, squeeze-out, and respreading. The adsorption of surfactants is also described, and a brief introduction to surface thermodynamics and surface phase behavior is also provided. These concepts are more specifically applied to lung surfactants in subsequent chapters.

II. Interfaces

A wise man once said that "the color of the world one sees depends on the color of the glass one looks through" [986]. From the viewpoint of someone interested in surfactants and their behavior, the glass is colored by the behavior of interfaces. An interface is a boundary between phases. The entire universe can be divided into five types of interfaces: liquid-gas, liquid-liquid, liquid-solid, solid-solid, and gas-solid (Figure 2-1). Because molecules at or near an interface are subject to different interactions than those in the surrounding bulk phases, they exhibit special behavior. Unique forces arise in the interfacial region. These forces can

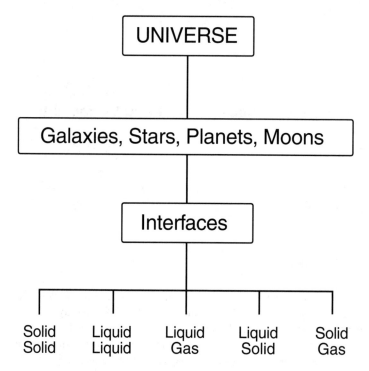

Figure 2-1 Types of interfaces in the universe. The entire universe can be divided into subsystems containing only five kinds of stable interfaces if matter is assumed to exist as three-dimensional solids, liquids, or gases. A stable gas-gas interface is not formed. (Adapted from Ref. 986.)

be modified by substances called surfactants, molecules that have an energetic preference for the interface and affect intermolecular forces there. The study of interfaces, interfacial forces, and surfactants is called *interfacial phenomena*. This scientific area, historically of most interest in engineering and physics, turns out to be highly relevant for the lungs. In particular, surface tension forces in the pulmonary alveoli, and their modification by surfactant molecules during breathing, are fundamentally important for respiratory function in humans and other air-breathing animals as described in subsequent chapters.

Note that at the molecular level, the interfacial region between two bulk phases in real systems is not a plane of zero thickness. Instead, there is an interfacial region across which properties vary between those in the two surrounding bulk phases (a phase is defined as a region of matter which is homogeneous in its properties) (Figure 2-2). In theory and practice, however, it is convenient to view the interfacial region as a

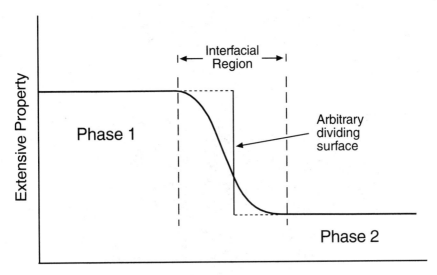

Figure 2-2 Real and idealized interfaces. Real interfaces comprise a region across which properties change from those in the surrounding bulk phases. However, an arbitrary dividing surface designating the interface as an abrupt step change from one phase to the other is a convenient conceptual fiction.

two-dimensional arbitrary dividing surface. Particularly for plane (flat) interfaces, many mathematical relations of surface thermodynamics can be formulated so as to be independent of the location of the dividing surface within the interfacial region [5, 10, 198, 203, 295, 991]. The position of the dividing surface can thus be chosen arbitrarily so as to simplify specific relationships such as those relating surface concentration, surface tension, and surface area (see Appendix to this chapter).

III. Surface Tension

Surface tension is the common name for the interfacial tension at a liquid-gas interface. In any liquid, the distance between molecules is orders of magnitude smaller than in a gas at atmospheric pressure[1] (Figure 2-3). A typical molecule in the bulk of the liquid is attracted equally in all directions by its nearest neighbors (molecule 1, arrows). This is not the case, however, for a molecule in the

---

1. The gas phase is about 1000 times less dense than the liquid phase. For example, one mole of an ideal gas at STP occupies 22.4 liters while one mole of water occupies about 18 ml.

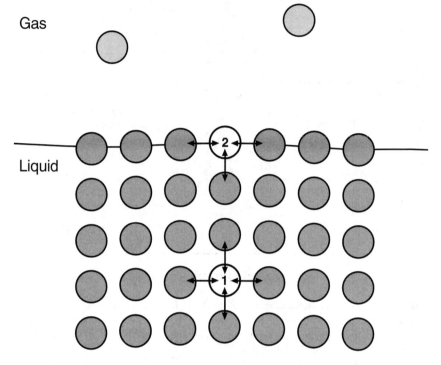

Figure 2-3 Molecular forces leading to surface tension. Attractive forces from nearest neighbors are illustrated for an idealized bulk liquid molecule ("1") and an interfacial region molecule ("2") contacting a gas. Because the gas phase is much more dilute than the liquid, gas molecules exert a negligible attraction on interfacial molecules. This leads to an unbalanced inward attractive force on interfacial molecules that causes the surface to seek a minimal area, generating surface tension.

interfacial region (molecule 2, arrows). Due to a relative lack of attraction from the dilute gas phase, there is an unbalanced attraction on surface molecules toward the bulk of the liquid. This causes the surface to minimize its area, resulting in the force called surface tension. The influence of surface tension forces is widely observable in nature, generating the preferred spherical shape of falling drops, the meniscus at the top of a glass of water, and the support for water bugs on ponds [5, 10, 295, 991]. Surface tension is a thermodynamic quantity related to the work needed to expand the surface area of the system (Figure 2-4). In the lung, a major part of the static work of breathing results from expanding the alveolar airsacs against surface tension forces. Surface tension forces also directly affect the stability and inflation uniformity of the alveolar network as detailed later (Chapters 6, 7).

Surface Tension and Work

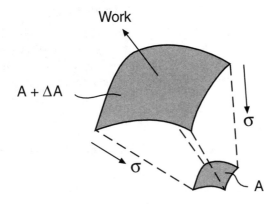

Increasing surface area A requires work to
overcome the opposing action of surface
tension σ in minimizing the extent of the surface

Figure 2-4   Surface tension is related to the work necessary to expand a surface. Surface
tension has units of force per length or work or energy per area. Work must be done against
surface tension in order to increase the surface area of a system. Surface tension is defined
thermodyamically as a change in free energy per unit surface area (see text).

Thermodynamic Definition and Conceptual Interpretation of Surface Tension.
The observed contraction of a liquid surface indicates that it has free energy that is
minimized by reducing surface area. Work must be done against net inward attractive
forces in order to bring a molecule from the bulk liquid to the surface. Surface tension
is effectively a measure of this work per unit of surface area. Surface tension can also
be viewed as interfacial free energy per unit of surface area. Thermodynamically,
surface tension (σ) is defined as the partial derivative of the Gibbs free energy G with
respect to surface area A at constant temperature, pressure, and number of moles
(see Appendix on surface thermodynamics at the end of the chapter) [5, 10, 198, 203,
295, 991]:

$$\sigma = (\partial G/\partial A)_{T,P,n} \tag{2-1}$$

An alternative definition of σ can also be derived in terms of the change in Helmholtz
free energy F with surface area. The Gibbs free energy and the Helmholtz free energy
are work-related thermodynamic functions. These thermodynamic definitions are con-

sistent with surface tension being viewed either as an interfacial energy per area or as a measure of the work associated with increasing the surface area of a system.

A. Units of Surface Tension and Surface Pressure

Surface tension is generally expressed in the numerically equivalent units of millinewtons per meter (mN/m) or dynes/cm. The units of surface pressure (see below) are the same as those of surface tension. Since work and energy have units of force times distance, the units of surface tension can be interpreted as work or energy/area as noted above. The work necessary to extend a surface having a tension of σ dynes/cm by 1 cm^2 of area is σ ergs (Figure 2-4).

B. Changes in Surface Tension with Temperature and Pressure

The surface tension of a liquid is a function of temperature and in general decreases as temperature increases. The surface tension of water, for example, decreases by 3.7% from 72.75 to 70.06 mN/m between room temperature (20°C) and body temperature (37°C) (Chapter 4). This temperature dependence is consistent with the conceptual interpretation of surface tension in terms of the work necessary to expand surface area. The kinetic energy of molecules and their tendency to move outward increase as temperature increases, reducing the work of expansion. Surface tension is also in principle a function of pressure, but in practice substantial variations with pressure do not occur until it is raised substantially (e.g., by many atmospheres). However, if gas pressure is increased to large enough values that the attraction of the gas phase for interfacial molecules increases significantly, surface tension is correspondingly reduced.

IV. Surface Active Agents: Definition and Behavior

Surfactant is a contraction for "surface active agent." The term is a general one and does not refer only to pulmonary surfactant. *A surfactant by definition is any molecule that by virtue of its structure has an energetic preference for an interfacial location.* As detailed later, the presence of surfactant molecules at a gas-liquid interface always tends to reduce surface tension. For many surfactants, including those found in the lungs, the interface of interest is the *air-water interface.* Soaps are one group of surface active compounds that have been used for centuries. Another historical example is the oil used by ancient Greeks to calm rough seas, a wave-damping action related to the action of oil in distributing motion across large areas of surface and in reducing surface tension at wave peaks relative to wave troughs. Surfactant films have been studied for centuries. In 1774, Benjamin Franklin layered a teaspoon of oil on Clapham Pond and followed the rate and extent of spreading [295]. Experiments and calculations by

Lord Rayleigh in 1890 showed conclusively that olive oil films spread on water were of the order of molecular dimensions in thickness [888]. This, along with pioneering research by Gibbs, Pockels, Langmuir, and many others in the late 1800's and early 1900's, formed the basis for quantitative understanding about how surface active molecules interact and behave at an interface (for review see Refs. 5, 10, 183, 198, 203, 295, 991).

Surfactants differ in molecular structure depending on the interfacial system in which they are surface active. Molecules that are surface active at the air-water interface all share the structural characteristic of being *amphipathic*. This means that they have a portion of their structure that is polar and another portion that is nonpolar. Since water is a polar solvent, the polar portion of surfactant molecules is designated as *hydrophilic* (water liking), while the nonpolar portion is *hydrophobic* (water hating). The air-water interface is an energetically preferred location for molecules with an amphipathic structure. At the interface, the polar headgroup can be in the polar water phase, while the nonpolar portion is in the air. This arrangement minimizes free energy, and surfactant molecules will occupy the interface preferentially if it is available to them. Surfactant molecules are often drawn schematically in idealized fashion with a single line for the nonpolar portion and a circle for the polar portion, despite the fact that many such compounds have a more complex structure (Figure 2-5).

Surfactants or Surface Active Agents

The term surfactant is a contraction for "surface active agent," designating molecules with an energetic preference to locate at an interface.

Molecules that are surface active at the air-water interface have polar (hydrophilic) and nonpolar (hydrophobic) regions of structure.

These surfactants prefer an interfacial location where the polar groups are in the water phase while the nonpolar chains extend into the air phase.

There are many kinds of surfactants, ranging from simple soaps, detergents,and oils to complex biological molecules like lipids and proteins as found in endogenous lung surfactant and in many exogenous lung surfactants.

Idealized Surfactant

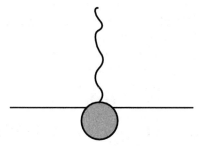

Surfactant molecules can be complex in structure,
but they are often drawn in simplified form
with a single hydrophobic region ($\sim\sim$)
and polar head group ($\bullet$)

Figure 2-5 Schematic of an idealized surfactant molecule. All compounds that are surface active at the air-water interface are amphipathic: they have polar and nonpolar regions of structure. An idealized surfactant molecule is often drawn with a polar head-group represented by a circle and a nonpolar tail or chain represented by a wavy line. This idealization is widely used for convenience, although many surfactants including proteins, polymers, and even lipids have multiple or complex polar and nonpolar regions of structure.

V. Surface Films (Surfactant Monolayers)

Surfactant molecules tend to arrange themselves at the interface in a thin *surface film*, which is the site where they affect surface tension. This surface film is typically composed of a single layer of surfactant molecules and is thus called a monomolecular film or *monolayer* (Figure 2-6). The liquid phase supporting the surfactant film is often referred to as the liquid *subphase*. A surfactant film always lowers surface tension below values found without surfactant present. To rationalize this, recall that surface tension is generated by the unbalanced attraction between interfacial molecules and those in the bulk liquid (Figure 2-3). The presence of surfactant molecules at the interface reduces this imbalance because the attractive forces between surfactant molecules and liquid molecules are less than the attractive forces of liquid molecules for each other. If this were not so, and surfactant molecules had greater attraction for the liquid than for each

---

Surfactant Films or Monolayers

Surfactant molecules occupy the interface as a thin surface film. If sufficient area is available, this film is monomolecular in thickness and is called a surfactant monolayer.

Surfactant films always act to lower surface tension from that found without surfactant present.

The amount by which surface tension is lowered depends on the specific surfactant and its concentration in the film. Surface tension lowering also depends on physical conditions such as temperature and cycling rate.

The quantitative behavior of surfactant films is typically defined in terms of a surface tension–area (σ-A) or surface pressure-area (π-A) isotherm.

---

other, they would necessarily go into solution. They would not, by definition, be surface active. The amount of surface tension lowering generated by a surfactant film is dependent on its concentration. The maximum reduction in surface tension occurs when the monolayer is in its most highly compressed state, i.e., where the packing of surfactant molecules is as close as possible and their monolayer surface concentration is maximal. The magnitude of surface tension lowering also

Surfactant molecules organize at the interface
in a one-molecule-thick film or
monolayer that acts to lower surface tension

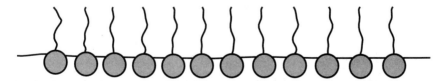

Figure 2-6 A surfactant monolayer at the air-water interface. Surfactant molecules form an interfacial film that minimizes their free energy. This one-molecule-thick film, called a monolayer, acts to lower and vary surface tension as surfactant concentration changes. See text for details.

depends on the type of surfactant molecules present, as well as on physical conditions such as temperature and the rate of film compression. The quantitative behavior of surfactant films is typically defined in terms of a surface tension–area (σ-A) or surface pressure-area (π-A) isotherm as described in following sections.

VI. Behavior of Surfactant Films at the Air-Water Interface

There is an extensive body of knowledge about the behavior and properties of surfactant films at the air-water interface. This section summarizes aspects of film behavior most relevant for pulmonary surfactants, with further details on surfactant films and their properties given in texts on surface chemistry and interfacial phenomena (e.g., [5, 10, 86, 183, 198, 295, 991]).

A. Surface Pressure and Its Definition

Changes in surface tension in a surfactant film are often described in terms of the *surface pressure* (π). Surface pressure is the amount by which surface tension is lowered by the surfactant film. Mathematically, surface pressure is defined as:

$$\pi = \sigma° - \sigma \qquad\qquad\qquad (2\text{-}2)$$

where $\sigma°$ is the surface tension of the pure liquid subphase without any surfactant and σ is the surface tension when the surfactant film is present. Although surface tension and surface pressure are measured in the same units, they vary in opposite directions. When surface tension is high, surface pressure is small, and vice versa. Numerical limits on the surface pressure π are between 0 and $\sigma°$. For water or physiologic saline, $\sigma°$ is about 70 mN/m at 37°C. This means that surfactant films on a subphase of water or saline generate values of π between 0 and 70 mN/m at 37°C. A value of $\pi = 0$ indicates that no reduction in surface tension has occurred, a condition that exists at very low surfactant concentrations. At the other extreme, a value of $\pi = \sigma°$ indicates that surface tension has been lowered all the way to zero by the surfactant film. The highest surface pressures (lowest surface tensions) generally occur in surfactant films when they are rapidly compressed to high surface concentration.

B. Molecular States in Surface Films

The organization of surfactant molecules in a film at the air-water interface has been described conceptually in terms of two-dimensional surface states with analogy to three-dimensional gases, liquids, and solids (e.g., [5, 10, 295]). Examples of two-dimensional surface states are *gaseous*, *expanded*, and *condensed* films (Figure 2-7). In the gaseous state, film molecules are dilute and do

Two-dimensional molecular states in interfacial films

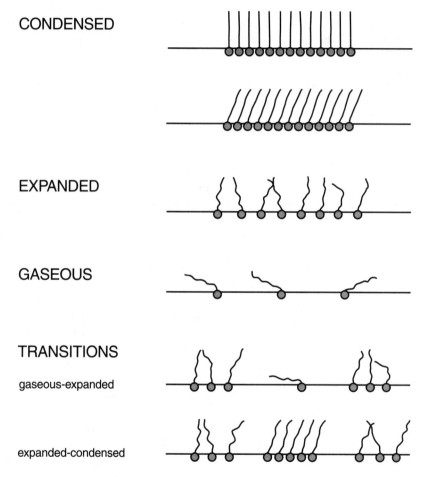

CONDENSED

EXPANDED

GASEOUS

TRANSITIONS

gaseous-expanded

expanded-condensed

Figure 2-7   Two-dimensional molecular states in interfacial films. Surfactant molecules in interfacial films are sometimes viewed as assuming different two-dimensional states depending on concentration and molecular characteristics. Shown are hypothetical molecular arrangements in the gaseous, expanded, and condensed (solid) states, as well as transitions between them.

not interact significantly, so that the film obeys a two-dimensional ideal gas law ($\pi A = kT$, where k is Boltzmann's constant). In the expanded state, also known as the liquid-expanded state, film molecules do interact in both their chain and headgroup regions. In expanded films, molecular packing is closer than in gaseous films and mobility is more constrained. However, film behavior is sufficiently compressible to be more analogous to a bulk liquid rather than a solid. Condensed films, on the other hand, are solid-like in behavior, with closer molecular packing, higher viscosity, and lower compressibility than expanded films. The fatty chains of surfactant molecules in condensed films can be oriented with varying degrees of tilt, with the most condensed films having vertical, fully extended chains. *Transitions* between the gaseous-expanded and expanded-condensed films can also be defined where both of the relevant states coexist at the interface (Figure 2-7).

C. Surface Pressure–Area (π-A) and Surface Tension–Area (σ-A) Isotherms

The quantitative surface tension lowering behavior of a surfactant film is typically reported in terms of isotherms of surface tension or pressure vs surface area. The π-A or σ-A isotherm for a surfactant film defines how it lowers and varies surface tension as a function of interfacial concentration at fixed temperature. The surface tension–area isotherm is measured with an instrument called a *surface (or film) balance* (Figure 2-8). This instrument allows a surfactant film to be compressed and expanded while surface tension or surface pressure is measured.[2] Surface pressure typically rises (surface tension falls) as the film is compressed from high to low area. For films of defined composition, area is often expressed as an inverse surface concentration in Å^2/molecule (1 $\text{Å} = 10^{-8}$ cm or 0.1 nm). Alternatively, in complex mixed films where molecular composition is uncertain, area can be expressed as a simple percentage of the maximum surface area.

Idealized π-A and σ-A isotherms for a compressed surfactant film are shown in Figures 2-9 and 2-10, respectively. These isotherms are rotated mirror images. The curves shown are composites for a hypothetical surfactant that passes through the gaseous, expanded, and condensed film states. An actual surfactant film may not exhibit all of the indicated states and transitions. The highest surface pressures (lowest surface tensions) tend to be generated by condensed or solid surface films with tight molecular packing. Expanded films can also generate substantial surface pressures, but maximum values during dynamic compression usually do not greatly exceed the equilibrium spreading limit. The study of film

2. Details on the experimental use of surface balances, particularly the Wilhelmy surface balance, are given in Chapter 4.

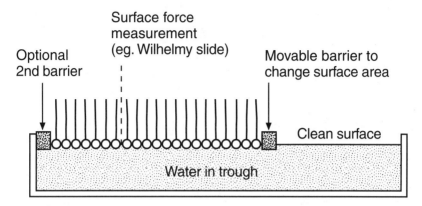

Figure 2-8 Conceptualized surface or film balance. The surface tension–area behavior of surfactant films is studied with instruments called surface or film balances. An interfacial film in the balance trough is compressed and expanded by a movable barrier while surface tension (or surface pressure) is measured by a force-sensing mechanism such as a Wilhelmy slide inserted into the interface. Alternatively, a second barrier such as a float attached to a torsion balance can be used to determine surface pressure as described originally by Langmuir [624]. Adapted from Ref. 301.

molecular states and surface phase transitions is theoretically complex and technically difficult. Concepts of idealized surface states such as those in Figures 2-7, 2-9, and 2-10 were introduced in the first half of the 1900's and are being refined using newer and more sophisticated instruments and theories. Nonetheless, these idealized film states remain useful conceptually for correlating the behavior and properties of different surfactants.

D. Dynamic versus Equilibrium Surface Tension Lowering in Surfactant Films

Many surfactant films generate significantly different surface pressures at the same area depending on how rapidly they are compressed. For films exhibiting such rate dependence, surface pressures tend to increase (surface tensions decrease) as cycling rate increases. Surface pressures measured during rapid compression reflect dynamic, nonequilibrium behavior. As detailed later, films of endogenous lung surfactant compressed at rapid physiologic rates of 10–50 cycles/min reach very high dynamic surface pressures equivalent to minimum surface tensions <1 mN/m at 37°C. This extraordinary degree of surface tension lowering results primarily from the action of dipalmitoyl phosphatidylcholine (DPPC) and related rigid disaturated phospholipids in forming a tightly packed solid-like monolayer at high degrees of compression. Many interfacial studies in the physical science literature utilize slow, quasi-static rates of 5–30 minutes/

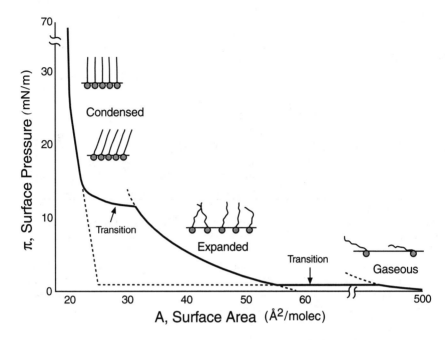

Figure 2-9 Film states and the surface pressure–area (π-A) isotherm. A composite π-A isotherm is shown for compression of a hypothetical surfactant film that exhibits gaseous, expanded, and condensed states and transitions on a water subphase at 37°C. Area decreases as the film is compressed from right to left, causing surfactant concentration to increase and surface pressure to rise. The corresponding surface tension–area (σ-A) isotherm is shown in Figure 2-10. These isotherms are idealized composites for surfactants with a limiting molecular area near 20 Å^2/molecule; individual compounds may not exhibit all the states and transitions shown. (Adapted from Ref. 295.)

cycle to approximate equilibrium behavior to allow thermodynamic analysis. At slow cycling rates, films of endogenous lung surfactant do not reach minimum surface tensions < 1 mN/m at body temperature. The physical factors and chemical components contributing to the surface tension lowering ability of lung surfactant films under rapid dynamic compression at physiologic rates are discussed in later chapters (Chapters 3, 7, and 8).

E. Collapse of Compressed Surface Films

If a surfactant monolayer is compressed at the air-water interface, it eventually reaches the point where not enough room is available to accommodate all of the molecules in a single layer. If compression continues beyond this point, the monolayer must *collapse*, ejecting surfactant molecules into one or more surface

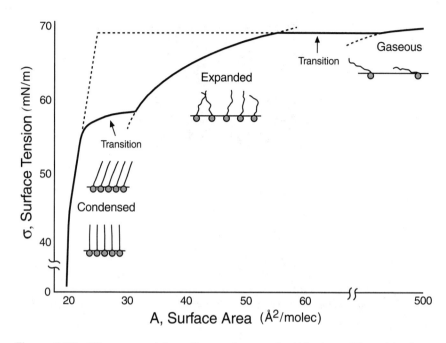

Figure 2-10   Film states and the surface tension–area (σ-A) isotherm. The σ-A isotherm corresponding to the π-A isotherm in Figure 2-9 is shown. The σ-A and π-A isotherms are rotated mirror images. As the film is compressed from right to left, surface tension falls as surfactant concentration in the film increases. See text for details.

or subsurface collapse structures or phases. Film collapse with the formation of a multilayered collapse phase on top of an underlying monolayer is shown schematically in Figure 2-11, along with an electron micrograph demonstrating collapse ridges in such a film. Surfactant films generate their maximum amount of surface tension lowering when compressed into the collapse regime. The presence of monolayer collapse is accompanied by a change in slope or by a horizontal plateau on the π-A and σ-A isotherm. Three ways of quantitating film collapse areas from such changes are illustrated in Figure 2-12. The lower limit collapse area for a monolayer containing surfactant molecules with a single saturated fatty acyl chain is the 18–20 $Å^2$ cross section occupied by such a chain in the solid crystalline state [10, 108, 295, 301, 985, 1007]. Phospholipid molecules have two acyl chains, giving a theoretical minimum collapse area of 36–40 $Å^2$/molecule [108, 301, 985, 1007]. These values are lower limits, and ejection of surfactant molecules from the monolayer can take place earlier during compression. This occurs, for example, in mixed films whose components have different levels of affinity for the interface. Components with less prefer-

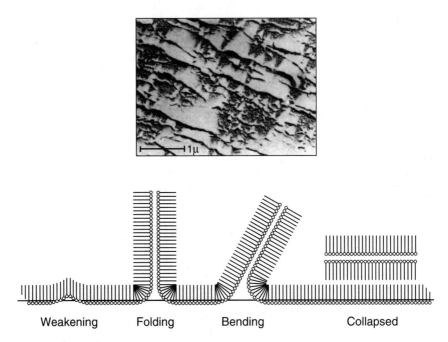

Weakening   Folding   Bending   Collapsed

Figure 2-11 Monolayer collapse with surface multilayer formation. Illustrated schematically is the progression of monolayer collapse showing weakening, folding, bending and multilayer formation on top of the underlying monolayer. Collapse occurs when a surfactant film is compressed so that there is no longer room for all of the molecules in a single layer. Inset at top is an electron micrograph showing collapse ridges in a compressed monolayer. Sophisticated imaging methods have also shown a variety of other surface aggregate forms and phases during compression and collapse of surfactant films. See text for details. (From Ref. 10 as adapted from Ref. 902.)

ence for the interface based on their molecular free energy can be selectively ejected or *squeezed out* of the compressed mixed film. Squeeze-out during compression is thought to be functionally important in lung surfactant films (see later section).

Formation of multilayered interfacial-region aggregates or phases during the compression and collapse of surfactant films is well documented. In addition to collapse ridges such as those in Figure 2-11, a variety of suprasurface structures in compressed monolayers have been demonstrated by electron microscopy, scanning force microscopy, Brewster-angle microscopy, fluorescence microscopy, and other imaging techniques.[3] Interfacial region aggregates extending beneath the

---

3. For example, see Refs. 206, 487, 602, 653–656, 758, 759, 761, 902–905, 1123.

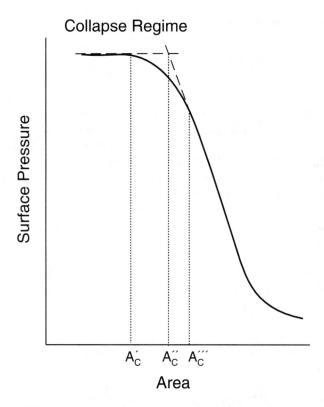

Figure 2-12 Film collapse regime and collapse areas. Compression of a surfactant film into the collapse regime is apparent on the π-A isotherm by a decrease in slope or a horizontal region where surface pressure reaches a maximum (surface tension reaches a minimum). Collapse areas shown are A'_c, the area where maximum pressure is first reached; A''_c, the area where the collapse plateau intersects the line of steepest slope; and A'''_c, the area where the isotherm departs from its steepest slope. See text for details.

surface have also been observed for lung surfactant films [978], correlating with the ability of phospholipid aggregates to organize and self-assemble in the aqueous phase (see Chapter 3 for phospholipid aggregate behavior). Interfacial collapse phases formed during dynamic compression are particularly important for the surface activity of lung surfactants. Collapse behavior, for example, affects film respreading that helps to maintain sufficient interfacial surfactant for effective surface tension lowering during continuing cycles of inspiration and expiration. The collapse behavior of highly compressed lung surfactant films is also important because it is associated with the generation of the lowest surface tensions.

F. Compressibility of Surface Films

The compressibility of a surface film is a measure of how it changes surface tension or surface pressure as area changes:

$$C_s = 1/A \ (\partial A/\partial\sigma)_{T,p} = -1/A \ (\partial A/\partial\pi)_{T,p} \tag{2-3}$$

where C_s is surface compressibility, A is surface area, σ is surface tension, and π is surface pressure. The reciprocal of C_s, defined as the surface compressional modulus, is also sometimes used in describing monolayer properties. The surface compressibility C_s is inversely proportional to the slope of the π-A or σ-A isotherm: if this slope is large, compressibility is small. Films with low compressibility generate a significant reduction in surface tension as area decreases. Films with higher compressibility generate a smaller change in surface pressure or tension as area changes. The lowest compressibilities (or highest *in*compressibilities) are typically found in condensed or solid surfactant films at high surface concentrations just prior to collapse. The close molecular packing in such films generates a large reduction in surface tension over a small area decrease in this region of the isotherm. Interpretations relating compressibility to surface tension lowering in a surfactant film must take into account the isotherm region. The compressibility of many surfactant films varies significantly during compression and is high in some regions of the isotherm and low in others. A surfactant film compressed past collapse, for example, may be at a low surface tension but have a high apparent compressibility because surface tension does not change significantly with area in the collapse regime. Surfactant films may also exhibit high compressibility when surface area is large (dilute films) but have much lower compressibility and reduce surface tension substantially at low surface areas. Lung surfactant films need to have sufficiently low compressibility so that surface tension is reduced effectively over area compressions reached in the alveoli *in vivo*. Limits on the magnitude of film compressibilities that might be associated with active lung surfactants have been suggested in the literature (e.g., [569, 571, 572]), but these are approximate and precise limiting values are not defined by rigorous theory.

G. Miscibility of Components in Surface Films

In any surface film containing two or more components, the concept of miscibility arises. If film components are distributed homogeneously among each other throughout the film, they are said to be fully miscible. In contrast, if the different components exist only as distinct separate islands in the film, they are immiscible. Intermediate behavior where the film contains regions of mixed components and regions of separate pure components denotes partial miscibility. One reason that miscibility is important is that film components must be at least partially miscible if they are to interact at the molecular level in lowering surface tension.

The components might form an ideal mixed film where no functional molecular interactions occur, but the existence of interactions between components requires some degree of miscibility. Endogenous lung surfactant, for example, is a complex mixture whose adsorption and film behavior depends on molecular interactions between its various components.

In binary (two-component) films, σ-A or π-A isotherm data can be analyzed for additivity to indicate if film components interact and are thus at least partially miscible. For an ideal mixed film or a film of immiscible components, any total property for the mixed film is given by a sum of the same property for each pure component times its mole fraction in the mixture. For example, the area A at fixed surface pressure in an ideal or immiscible binary film is given by:

$$A = x_1 A_1 + x_2 A_2 \tag{2-4}$$

In Equation (2-4), x_1 and x_2 are the mole fractions of components 1 and 2, and A_1 and A_2 are the areas in each pure component film at the same surface pressure at equilibrium. In a binary mixture, $x_2 = 1 - x_1$ and Equation (2-4) defines a so-called *additivity line* (e.g., [108, 985]). Experimentally determined values of A are typically plotted, at fixed surface pressure, as a function of composition in the mixed film (e.g., as a function of x_1). If the data indicate that additivity is followed as in Equation (2-4), then the film is either ideal or composed of immiscible components. In either case, molecular interactions between the different film components are minimal. Deviations from additivity indicate that the two components do interact at the molecular level and are at least partially miscible. Additivity analyses can also be done for film properties other than area, such as surface potential and surface viscosity, to obtain more complete information about molecular behavior in the mixed film [108, 985]. Interactions in mixed films with more than two components can in principle be elucidated by additivity analyses of the possible binary combinations that could be present. However, this approach rapidly becomes cumbersome as the number of components increases and is difficult to apply even for ternary (three-component) films.

H. Compression-Expansion Hysteresis in Cycled Surfactant Films

Hysteresis in the surface tension lowering behavior of a surfactant film refers to a difference in the π-A or σ-A isotherm found during compression and expansion (Figure 2-13). Hysteresis arises from film processes that differ in rate, extent, or character between compression and expansion (path-dependent processes). Molecules ejected or squeezed out of the interface during compression may not return or be integrated back into the film at an identical rate during expansion. Components returning to a mixed film during expansion may also not be in the same compositional ratio as in the original film. Hysteresis in the π-A isotherm can also be generated by the relaxation or time decay of dynamic surface

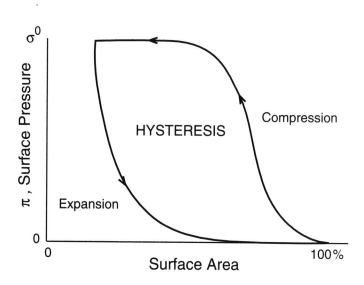

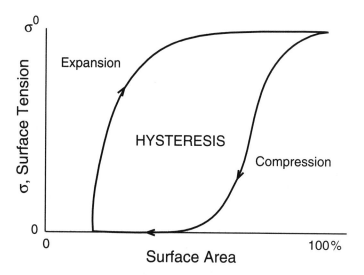

Figure 2-13 Compression-expansion hysteresis in surface films. Hysteresis, or path-dependent behavior, is present if the π-A or σ-A isotherm for a surfactant film differs on compression and expansion. Films of lung surfactant and many phospholipids exhibit substantial physiologically relevant compression-expansion hysteresis.

pressures if cycling is halted even for relatively brief periods of time [321, 795, 1059]. Complex mixed films of endogenous and exogenous lung surfactants typically exhibit substantial π-A hysteresis during cycling. The π-A hysteresis of alveolar surfactant films contributes the major portion of the difference between quasi-static pressure-volume curves measured during inflation and deflation of the lungs with air [41, 136, 137, 304, 714] (Chapter 7). Specific aspects of lung surfactant π-A hysteresis may also have direct benefits in improving the uniformity of alveolar recruitment during inspiration [797].

I. Selective Squeeze-out of Components from Multicomponent Surface Films

In a multicomponent film (often called a mixed film), not all components have an identical free energy at the interface or equal mobility to leave the surface during cycling. Molecules with a lower affinity for the interface will tend to be ejected preferentially from the film during compression. This is called *selective squeeze-out* from the film (Figure 2-14). Selective squeeze-out causes the surface film to become enriched in some components and depleted in others, altering its character and surface tension lowering ability. Squeeze-out into surface phases during monolayer compression has been directly demonstrated in surfactant films using spectroscopic and microscopic methods (e.g., [130, 206, 487, 602, 655, 759, 761, 762, 828, 884]) and is thought to be important in lung surfactant surface behavior. As detailed in Chapter 8, lung surfactant films become enriched in rigid, saturated phospholipids like DPPC at the expense of more fluid film components upon compression to low film areas. This change contributes to several important surface behaviors, giving a lower minimum surface tension, more variation in surface tension with area during cycling, and increased π-A hysteresis.

J. Dynamic Respreading in Cycled Interfacial Films

The process by which surfactant molecules ejected from a monolayer during compression are able to reenter and spread back into the film during expansion is called *dynamic respreading* [790, 796, 1093] (Figure 2-15). The term *re*spreading rather than spreading is used to emphasize that the surfactant molecules involved were originally in the film and had already spread initially at the surface. The further description of this process as dynamic not only describes its intrinsic nature but also emphasizes that the film collapse structures involved are formed dynamically and may be nonequilibrium in nature. The dynamic, nonequilibrium nature of film respreading, and the involvement of interfacial region collapse structures, differentiates this surface property from adsorption as discussed later. Effective dynamic respreading is one of the functional surface behaviors required of lung surfactant films (Chapter 6).

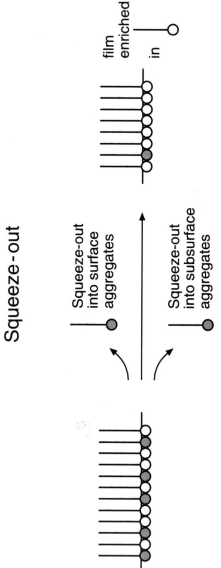

Figure 2-14 Refining of a multicomponent surfactant film by squeeze-out during compression. When a multicomponent surfactant film is compressed, some constituents can be selectively ejected or "squeezed out" of the interface. This enriches the film in other components, altering its surface tension lowering behavior. Selective squeeze-out is thought to occur in lung surfactant films, allowing low minimum surface tensions to be reached while maintaining a surface reservoir that respreads effectively on successive cycles of the surface. Squeeze-out can also be an important contributor to compression-expansion hysteresis.

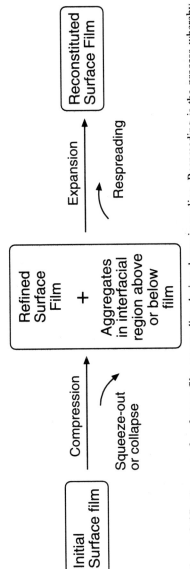

Figure 2-15 Concept of surfactant film respreading during dynamic cycling. Respreading is the process whereby surfactant molecules ejected into interfacial region aggregates or collapse phases during dynamic compression reenter and integrate into the film during expansion. Dynamic respreading, along with adsorption, maintains adequate film material to lower surface tension effectively during repetitive cycling.

Respreading Calculations. Respreading in a surfactant film can be determined from collapse plateau ratios [796, 1093] or areas [1145] calculated from σ or π-A isotherms on successive cycles (Figure 2-16). Both methods of quantitating respreading are based on the understanding that the displacement between compression curves on successive cycles reflects film material that has been lost from the interface and has failed to respread. As described above, some surfactant molecules are selectively squeezed out from a surfactant monolayer during compression, and additional film material is ejected within the collapse regime. If all the surfactant molecules lost from the surface during compression reentered the film and respread during the immediately following expansion, the π-A isotherm of the next compression would be identical to that on the first (hysteresis for each compression-expansion cycle might or might not be present depending on the rapidity of the respreading). Conversely, if none of the molecules lost from the film during the first compression were to respread, the second

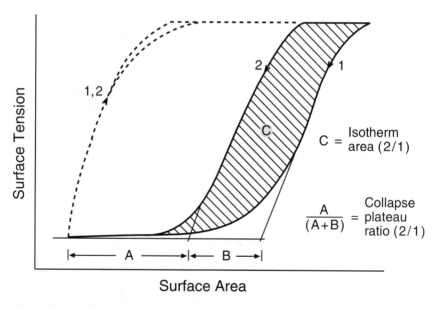

Figure 2-16 Quantitation of dynamic respreading from the σ-A isotherm. Respreading is reflected in the displacement between successive compression curves based on measured collapse plateaus (lengths A and B) or isotherm area (area C). If all surfactant molecules lost from the film during compression respread during expansion, the second compression curve would be identical to the first (no displacement). Complete respreading is thus a collapse plateau ratio of 1 (B = 0) or an isotherm area difference of zero (C = 0). If none of the molecules lost from the film respread, the second compression will be fully displaced to the left and the film will just reach collapse at end-compression. Zero respreading is thus equivalent to a collapse plateau ratio of 0 (A = 0) or to a maximal isotherm area C. See text for details.

compression would be displaced maximally from the first. Small displacements between successive π-A compressions thus equate to better respreading, indicated by collapse plateau ratios close to 1 or by isotherm area differences close to zero (Figure 2-16). The length scale used in calculations of respreading from either collapse plateaus or isotherm areas is arbitrary but must be the same for all cycles [796, 1093, 1145].

K. Surface Potential and Surface Viscosity in Surface Films

A number of properties other than those associated with surface tension and the π-A isotherm can be defined for surfactant films as detailed in standard texts [5, 10, 183, 295, 991]. Two of the more widely studied of these are surface potential and surface viscosity. *Surface potential* is a measure of the change in the electrical potential at the interface due to the presence of a surfactant film. This film property is of most interest for charged surface films, although measurable values of surface potential are also found for uncharged surfactant films. Magnitudes of surface potential of the order of several hundred millivolts are not uncommon experimentally. Surface potential measurements are helpful in elucidating molecular orientations and in analyzing miscibility and interactions in multicomponent surface films including those containing phospholipids [10, 108, 790, 985]. However, fundamental interpretations of surface potential data become more difficult and less precise for complex mixed films such as those relevant for endogenous and exogenous lung surfactants.

Surface viscosity is a two-dimensional analog of viscosity in bulk phases that reflects the resistance of a surfactant film to flow and deformation. Both shear and dilational (also called dilatational) surface viscosities can be defined [5, 10, 183, 295, 965, 991]. The surface shear viscosity reflects the resistance of the film to flow under an applied shear stress, while the surface dilational viscosity reflects the resistance of the film to dilation (isotropic expansion). Surface viscosity is measured in units of surface poise (gm/sec), in analogy to the units of poise (gm/cm-sec) used for viscosity in bulk phases. Surfactant films can have significant shear and/or dilational surface viscosities. These viscosities are sometimes related quantitatively to bulk phase viscosities by dividing by film thickness. Although this procedure is inexact, resultant values for many surfactant films are large and equivalent to those for viscous bulk materials such as butter, heavy grease, or waxes. Because viscous effects are dynamic, surface viscosity contributes in principle to measurements of π-A isotherm hysteresis and interfacial pressure drop during dynamic cycling of surfactant films [295, 965]. Although research has examined the surface viscosity of phospholipid and lung surfactant films (e.g., [378, 539, 598, 715, 1013, 1120]), much more attention has been devoted to the physiologically relevant surface tension–area behavior of these materials.

The surface shear viscosity (η_s) can be defined mathematically as follows [295]:

η_s = tangential force per unit area of surface/rate of strain   (2-5)

Surface shear viscosity is typically measured with a canal-type viscometer that quantitates surface flow through a slit as a function of applied tangential force in analogy with the Ostwald-Poiseuille method for bulk systems (e.g., [10, 295, 407, 408]). Measured values of η_s between 0.001 and 0.1 surface poise are not atypical for surfactant films. If these values are divided by film thickness, which is of the order of 25 Å for many substances, "equivalent" bulk shear viscosities are of the order of 4,000–400,000 poise. By comparison, the viscosity of water is of order 1 centipoise. The surface dilational viscosity (η_d) is defined as:

$$\eta_d = \Delta\sigma \, A \, (dA/dt)^{-1}$$ (2-6)

where A is surface area, t is time, and $\Delta\sigma$ is the deviation in surface tension from equilibrium associated with the change in surface area with time. Since $\Delta\sigma$ is equivalent to a surface pressure, $1/\eta_d$ can be interpreted as the fractional change in area with time per unit of applied surface pressure.

VII. Langmuir-Blodgett Multilayers

In a number of research applications, surface films are studied after being removed from the air-water interface in the form of layers deposited on a slide repeatedly dipped through the surface. Deposited multilayers of this kind are called *Langmuir-Blodgett multilayers*. Moving surface barriers can be used to hold surface tension constant during deposition to improve the uniformity of the resultant layers. The orientation of the initial layer in terms of whether chains or headgroups first contact the support can be altered by changing the direction of dipping or altering the hydrophobicity/hydrophilicity of the slide support surface. Langmuir-Blodgett multilayers have been widely studied with methods ranging from electron microscopy to X-ray diffraction and various kinds of spectroscopy to help understand molecular structure and interactions in surface films and bilayers.

VIII. Wetting, Contact Angles, and Spreading

Wetting and spreading are two additional interfacial phenomena relevant for the behavior of surfactant substances. *Wetting* refers to how a liquid spreads out on a solid and is reflected in the *contact angle* θ (Figure 2-17). Complete wetting is defined as the absence of a finite contact angle. Wetting with a zero contact angle is important to achieve, for example, in surface tension measurements using a Wilhelmy slide and surface balance (Chapter 4). As wetting decreases, the contact angle increases. Hydrophobic solids that resist wetting exhibit large contact

Wetting

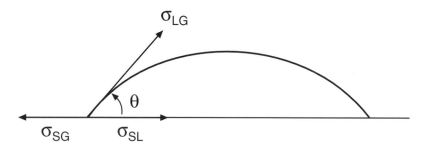

Figure 2-17 Wetting and the contact angle between a liquid and solid surface. A liquid droplet makes a contact angle θ with a solid surface. The smaller the contact angle, the better the wetting. Surfactants act to promote wetting by lowering surface tension. The interfacial tensions contributing to the contact angle are those between the solid and liquid (σ_{SL}), the solid and gas (σ_{SG}), and the liquid and gas (σ_{SL}). See text for details.

angles with water droplets. The magnitude of the contact angle θ can be determined from the Young equation, which reflects the balance of the interfacial forces on the liquid droplet at equilibrium:

$$\text{Cos } \theta = (\sigma_{SG} - \sigma_{SL})/\sigma_{LG} \qquad (2\text{-}7)$$

In Equation (2-7), σ_{SG} is the interfacial tension between solid and gas, σ_{SL} is the interfacial tension between the solid and the liquid, and σ_{LG} is the interfacial tension between the liquid and the gas (i.e., the surface tension) (Figure 2-17). Surface active agents act to reduce surface tension and hence to improve wetting based on Equation (2-7), although effects on σ_{SL} as well as on σ_{LG} must be taken into account. If surface tension is reduced so that cos $\theta > 1$, then there is no finite contact angle satisfying the Young equation and complete wetting occurs. When incomplete wetting with a nonzero contact angle is observed experimentally, the contact angle measured when the liquid surface is advancing over the clean solid (advancing contact angle) can differ from that found when the liquid surface is withdrawing across previously wetted solid (receding contact angle). When contact angle hysteresis of this kind occurs, receding contact angles are typically smaller than advancing contact angles.

Spreading of one immiscible liquid on another is also relevant for surface film studies. For example, this phenomenon is exploited when a solvent such as 9:1 (v:v) hexane:ethanol or chloroform is used to spread an interfacial surfactant film at the air-water interface in a surface balance prior to measurement of π-A or σ-A behavior (Chapter 4). In determining whether liquid B will spread spontaneously on liquid A, a spreading coefficient $S_{B/A}$ can be defined as [10, 183, 295, 991]:

$$S_{B/A} = \sigma_A - (\sigma_B + \sigma_{AB}) \tag{2-8}$$

where σ_A is the surface tension of pure A, σ_B is the surface tension of pure B, and σ_{AB} is the interfacial tension between liquids A and B. If $S_{B/A} > 0$, liquid B will spread spontaneously on liquid A, while if $S_{B/A} < 0$, liquid B will form a nonspreading lens. Equation (2-8) is actually valid only for initial spreading, because the two contacting liquids can become altered over time by mutual solubility or other factors. For example, as discussed by Gaines [295], benzene has an initial spreading coefficient on water of +8.9 mN/m and is found to spread spontaneously. However, benzene decreases the surface tension of water and subsequently added benzene droplets eventually form lenses on water [295].

IX. Adsorption and Behavior of Surfactants Suspended in the Liquid Phase

The study of many surfactants requires that aqueous phase behavior be considered in conjunction with film behavior. For example, surfactant compounds in many physical and biological systems are present in a bulk liquid phase and adsorb to the air-water interface to form a film. Adsorption is a particularly important property for endogenous and exogenous pulmonary surfactants (Chapters 6, 8). This section summarizes several concepts and properties relevant for the adsorption and suspension behavior of surfactants (see Refs. 5, 10, 86, 183, 198, 203, 991 for more detailed coverage of the behavior and thermodynamics of surfactants in solution).

A. Colloids

A colloid is defined as a suspension in a continuous medium of particles sufficiently small in size that gravity does not significantly influence their motion and sedimentation is very slow. Particle sizes in colloids typically range from about 5 Å to 10,000 Å (0.0005 to 1 micron). Paints, either water based or oil based, are an example of colloidal suspensions. Lung surfactants are another. The colloidal nature of most lung surfactants is primarily due to the propensity of phospholipid molecules, with or without added apoproteins, to form aggregates when dispersed in water. A significant percentage of endogenous surfactant exists as a colloidal suspension in the alveolar hypophase, and exogenous lung surfactants are also commonly administered as colloidal dispersions.

B. Critical Micelle Concentration (CMC)

If surface active molecules are added individually to a subphase, the concentration at which they begin to aggregate is termed the critical micelle concentration or CMC. Although some authors use the term micelle specifically to refer to spherical aggregates such as those formed by simple soaps, the concept of the CMC is generally used broadly as a reflection of surfactant molecule aggregation. A number of properties change in solutions or dispersions at the CMC, making it important in many applications. The ability of soaps and detergents to aggregate in the form of spherical micelles in water is responsible for their ability to segregate grease and oils and hence to act as cleaning agents. Many phospholipids have extremely low CMCs, consistent with a propensity to form a variety of aggregate forms in water [301, 1007]. The CMC for DPPC, the most prevalent phospholipid in lung surfactant, has been measured at $\sim 10^{-10}$ M [1010]. The complex aggregate and phase behavior of phosphollipids is directly relevant for the surface properties of lung surfactants, particularly adsorption. The biophysical behavior and molecular aggregation of phospholipids are discussed in more detail in Chapter 3.

C. Adsorption of Surfactants

Although all surface active molecules prefer to be at the interface, they may start out in the liquid subphase as individual molecules or groups of molecules as noted above. *Adsorption* is the process by which surfactant molecules in the subphase enter the air-water interface and form a surface film (Figure 2-18). *Ad*sorption is analogous to *ab*sorption, except that it refers to movement into an interface as opposed to movement into a bulk phase. Surface tension lowering during adsorption is limited by the equilibrium spreading pressure of the adsorbing substance. Adsorption can involve the movement of individual molecules and/or groups of molecules into the interface. In the case of lung surfactants, the adsorption of groups of phospholipid molecules is thought to be an important contributing mechanism (e.g., [118, 527, 561]) (Chapter 7). At equilibrium, there is a balance of molecular adsorption into the interface and molecular desorption from it. Net adsorption, the difference between these two processes, is always positive for surface active substances (see discussion of Gibbs adsorption equation in Appendix to this chapter). In the lung surfactant literature, the terms adsorption and net adsorption are typically used interchangeably. Adsorption is most precisely considered as distinct from transport by diffusion and convection in the bulk liquid. Diffusion and convection act to transport molecules within the bulk phase and into the region of the interface (subsurface region). Adsorption then involves the entry of molecules into the interface to form a surface film. Since the interfacial region in real systems is finite in thickness rather than being a

Adsorption

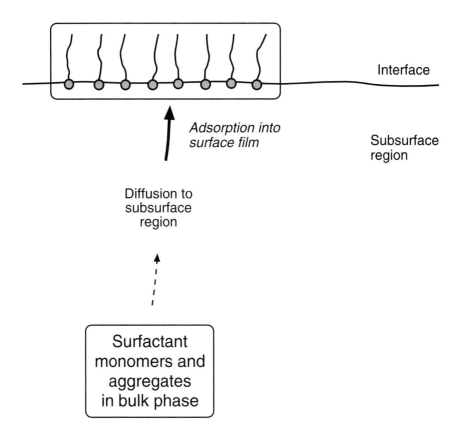

Figure 2-18   Adsorption into the air-water interface. Adsorption is the process by which surfactants in the subphase enter the interface and form a film. Adsorption can involve the movement of single molecules or molecular aggegates from the subsurface region into the film. Various kinds of subphase aggregates are known, all with surfactant molecules organized to decrease free energy. Soaps and detergents, for example, form simple spherical micelles in water while phospholipids organize as lamellae, vesicular liposomes, hex phases, and other forms (see Chapter 3).

Adsorption of Surfactants to the Air-Waste Interface

Surfactant molecules in the liquid subphase must adsorb to the air-water interface to form a film. Surface tension lowering during adsorption is limited by the equilibrium spreading pressure of the adsorbing substance.

Net adsorption is a balance between ongoing adsorption and desorption. All substances which lower surface tension exhibit positive net adsorption.

Adsorption is distinct from diffusion and involves the movement and insertion of individual molecules and/or groups of molecules into the interface.

Alveolar surfactant, for example, exists as phospholipid-rich aggregates in the aqueous phase that adsorb at least partly as groups of molecules, a process facilitated by interactions between surfactant lipids and proteins.

two-dimensional plane (Figure 2-2), the point where diffusive transport leaves off and adsorption into the interface begins is necessarily nebulous.

Although all surface active materials exhibit positive net adsorption, they vary widely in the rates at which they adsorb and in the equilibrium surface tensions they generate. Conceptually, adsorption is influenced by the relative free energies of surfactant molecules in the bulk liquid and in the surface film, together with the energy barrier associated with leaving the bulk and entering into the interface (Figure 2-19). Surfactant molecules with less stability in the bulk phase (higher bulk free energy), less difficulty in entering the surface, and greater affinity for the surface (lower surface free energy) have the best overall adsorption. A functional requirement for all pulmonary surfactants is that they adsorb rapidly to a high equilibrium surface pressure (low equilibrium surface tension) at the air-water interface. Rapid adsorption requires molecular characteristics different from those associated with the reduction of surface tension to very low values in dynamically compressed surface films. In order to achieve both these behaviors, endogenous lung surfactant requires a mixture of rigid and fluid lipids as well as three biophysically active proteins (Chapters 6, 8). By interacting with phospholipid bilayers and aggregates, surfactant proteins reduce bulk phase stability and also lower the effective energy barrier for interfacial entry. The importance of both adsorption and dynamic film behavior in the activity of lung

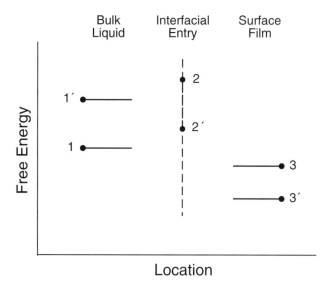

Location

Figure 2-19   Energy barriers in adsorption. From a simplified conceptual perspective, adsorption reflects the relative free energies of surfactant molecules in the bulk liquid and surface film, plus an energy barrier for entry into the interface. A low free energy in the bulk liquid, such as associated with highly stable aggregate formation, opposes adsorption relative to a higher bulk free energy (level 1 vs 1′). A large energy requirement for interfacial entry similarly lowers adsorption (level 2 vs 2′). All surfactants have reduced free energy when actually in the surface film, but those with lower surface free energy will adsorb more readily and desorb less readily so that net adsorption is increased (level 3′ vs 3).

surfactants is one reason that multiple experimental methods are used in measuring their surface active properties (Chapter 4).

D.   Comparing Adsorption and Film Respreading

Adsorption and dynamic film respreading both involve the transport of surfactant molecules into an interfacial film, but they have clear and distinct differences. Adsorption involves the movement of molecules from the bulk liquid into the interface. It incorporates an equilibrium between interfacial molecules and those in the subphase and is not dependent on dynamically generated phases or structures. In contrast, film respreading reflects dynamic phenomena that are operative only in the immediate region of the interface and may not be in equilibrium with the subphase. Respreading involves surfactant molecules in collapse structures formed either above or below the surface film during dynamic, nonequilibrium compression.[4] Respreading also necessitates that surfactant molecules reintegrate at the interface with a dynamically refined film. The differentiation of spreading and adsorption is shown, for example, by the fact that many fluid phospholipids respread extremely well in surface films but do not exhibit adsorption approaching that of active endogenous lung surfactant. Dynamic, nonequilibrium influences on respreading are prominent in the behavior of lung surfactant films, which are cycled physiologically at rapid rates. These rates vary from 12–15 breaths/min in adults to 40–50 breaths/min in newborns and are much higher than compressions used in many physical studies with nonbiological surfactants. Both adsorption and respreading are important in lung surfactant behavior, along with the variation of surface tension and its reduction to low values during dynamic cycling (Chapters 6–8).

X. Chapter Summary

Surface tension arises at any liquid-gas interface due to the differing balance of forces on molecules in the interfacial region relative to the bulk liquid. Surface tension or interfacial free energy per area is defined from thermodynamics and is related to the work necessary to expand the surface. Surface active agents or "*surfactants*" are substances with an energetic preference for the interface. There are many such substances, including soaps, detergents, fatty acids, and alcohols, and a variety of biological molecules like phospholipids and proteins. All surfactants active at the air-water interface have polar (hydrophilic) and nonpolar (hydrophobic) regions of molecular structure. They prefer an interfacial location because the polar regions can contact the water phase while the nonpolar regions extend primarily into the air. Surfactant molecules at an interface form a monomolecular film or monolayer that lowers surface tension as a function of concentration. Although surfactants share common characteristics and are studied by similar techniques, they vary widely in their specific properties and surface tension lowering power. Surfactants with high activity in the lungs require particular combinations of surface behaviors as detailed in later chapters.

There is a wealth of basic science information about surfactant films and their properties. Molecular states in surfactant films have classically been related to two-dimensional analogs of three-dimensional gases, liquids, and solids. These views still have conceptual utility although they are being refined by modern technology and theory. The quantitative surface tension lowering and respreading behavior of cycled surfactant films is defined by a surface pressure–area (π-A) or

---

4. For example, see Refs. 206, 653, 654, 656, 758, 759, 902–905, 978, 1123.

surface tension–area (σ-A) isotherm. These isotherms, measured with instruments called surface balances, also give information on monolayer collapse, film refining during compression, and surface tension-area hysteresis on successive cycles. Isotherms can also be analyzed to infer component miscibility and molecular interactions in mixed films. Also important for many surfactants is their behavior in solution and their adsorption to the air-water interface. Adsorption is the process whereby surfactant molecules move from the subphase into the interface to form a surface film. All surfactants have a positive tendency to adsorb, but some do so much more rapidly and extensively than others. Lung surfactants, for example, must be able to adsorb rapidly to high equilibrium surface pressures at the alveolar interface (Chapters 6–8). Also described more briefly in the current chapter have been a number of other properties of surfactants and surfactant films such as surface potential, surface viscosity, wetting, and spreading. The following appendix also introduces some basic principles of surface thermodynamics.

Appendix: Introduction to Surface Thermodynamics and Interfacial Tension

Thermodynamic Definition of Surface Tension. There is an extensive base of thermodynamic theory underlying the concept, definition, and behavior of interfacial tension (e.g., [5, 10, 86, 183, 198, 203, 295, 991]). Some of the thermodynamic functions relevant for interfacial phenomena are:

U = internal energy
H = enthalpy ($= U + PV$)
S = entropy
G = Gibbs free energy ($= H - TS$)
F = Helmholtz free energy ($= U - TS$)
μ_i = Chemical potential of component i

where T is temperature, P is pressure, and V is volume. Conceptually, the internal energy U is a measure of the energy intrinsic to a molecule (from internal vibration, rotation, etc.) independent of external potential energy or translational kinetic energy. Enthalpy H, a related energy function, adds pressure-volume work to internal energy. H and U can be related mathematically to heat and work and are used in formulating the basic energy balance of the first law of thermodynamics in bulk systems. Entropy S has several conceptual interpretations. It is defined as the change in heat in a reversible process divided by the temperature and from a statistical point of view can be associated with the disorder in a system: entropy increases as disorder increases. Entropy is basic to the formulation of the second law of thermodynamics, which states that the entropy change of a closed system is positive for any irreversible process and reaches a limiting value of zero only for a reversible process.

The Gibbs free energy G and the Helmholtz free energy F are work-related energy functions that are particularly important in problems involving phase equilibria and interfacial phenomena. For an open, multiphase system such as that in Figure 2-2, the equations describing changes in Gibbs and Helmholtz free energies can be written as:

$$dG = -SdT + VdP + \sum \mu_i dn_i + \sigma dA \qquad (A2\text{-}1)$$

$$dF = -SdT - PdV + \sum \mu_i dn_i + \sigma dA \qquad (A2\text{-}2)$$

where n_i is the number of moles of component i, A is the interfacial area separating the bulk phases, and σ is interfacial tension. Using the properties of partial derivitives as described in so-called Maxwell relations, Equations (A2-1) and (A2-2) contain the thermodynamic definition of interfacial tension as:

$$\sigma = (\partial G/\partial A)_{T,P,n} = (\partial F/\partial A)_{T,V,n} \qquad (A2\text{-}3)$$

Equation (A2-3), expressing interfacial tension as a partial derivitive of the Gibbs or Helmholtz free energy, is equivalent to Equation (2-1) in the body of the chapter.

Surface Phase Rule for Interfacial Films. Because surface tension is a thermodynamic quantity, a surface phase rule can be applied to surfactant film behavior in analogy with the Gibbs phase rule in three-dimensional (bulk) phases. Phase rules specify the number of free intensive variables (variables independent of the mass of the system) necessary to fix the state of a system at equilibrium. By analogy with the Gibbs phase rule, the surface phase rule is given by [10, 86, 183, 295]:

$$F = C^B + C^S - P^B - P^S + 3 \qquad (A2\text{-}4)$$

In Equation (A2-4), F is the number of free intensive variables, C^B is the number of bulk phase components equilibrated throughout the system, C^S is the number of components confined to the surface, P^B is the number of bulk phases, and P^S is the number of surface phases. Among other applications, the surface phase rule can be used to analyze and help understand the π-A isotherm characteristics of surfactant films.

As an example of the use of the surface phase rule, consider a system with air and water as bulk phases ($P^B = 2$, $C^B = 2$). If an insoluble surfactant film containing a single substance occupies the interface ($C^S = 1$), the surface phase rule simplifies to $F = 4 - P^S$. If the surfactant film has only a single surface phase ($P^S = 1$), this implies that the system is fixed by any three intensive variables ($F = 3$). Three intensive variables relevant for the π-A isotherm are surface area (or surface concentration), temperature, and pressure. The surface phase rule thus indicates that at equilibrium, surface pressure in a single phase surfactant film must vary as a unique function of surface concentration at fixed temperature and pressure. This behavior is observed in the π-A isotherm of pure surfactant films (Figure 2-8). If the film is undergoing a transition between two surface phases ($P^S = 2$), the phase rule in this case simplifies to $F = 2$. If temperature and pressure are fixed, the system is completely defined and surface pressure must be constant as surface concentration is changed. This is reflected by a plateau in the π-A isotherm for pure surfactant films undergoing transitions between the gaseous-expanded and expanded-condensed film phases (Figure 2-8). A similar plateau is also found during the collapse of a surfactant monolayer into a collapse phase at the interface ($P^S = 2$). Although the surface phase rule can be very useful in understanding the properties of simple surfactant films, it is much more difficult to apply to complex multicomponent films under dynamic conditions that apply for most lung surfactants.

Surface Excess Quantities and the Gibbs Adsorption Equation. Another group of useful thermodynamic relations can be derived in terms of so-called surface excess quantities defined by reference to the real and idealized interfacial systems in Figure 2-2. The interfacial region in the real system has a finite extent over which properties vary between those in the two surrounding bulk phases. In the idealized system, the interface is arbitrarily located at the two-dimensional Gibbs dividing surface. All of the extensive properties (properties dependent on the mass or extent of the system) of the two phases are constant right up to the dividing surface, and there is a step change in these properties at this surface. If the real and ideal systems are compared, there will be an *excess* (positive or negative) in these extensive properties compared to the idealized system. Extensive properties (e.g., U, S, H, G, F, V) for the real system can thus be defined as:

$$U = U^1 + U^2 + U^{ex}$$
$$S = S^1 + S^2 + S^{ex}$$
$$H = H^1 + H^2 + H^{ex}$$
$$G = G^1 + G^2 + G^{ex}$$
$$F = F^1 + F^2 + F^{ex}$$
$$V = V^1 + V^2$$

where the superscripts 1 and 2 refer to phases 1 and 2 in the idealized system, and the superscript "ex" refers to the excess in the real system relative to the idealized system. For a system of negligible interfacial curvature, relations for the surface excess quantities in the above definitions can be derived that are independent of the position of the dividing surface. In particular, a surface analog of the Gibbs-Duhem equation that gives the dependence of surface tension on the chemical potentials of the different components in the system can be derived as [10, 86, 183, 295]:

$$S^{ex}dT + \sum n_i^{ex}d\mu_i + Ad\sigma = 0 \tag{A2-5}$$

where n_i^{ex} is the surface excess number of moles of component i ($n_i = n_i^1 + n_i^2 + n_i^{ex}$). Equation (A2-5) can be used to derive the Gibbs adsorption equation that relates changes in surface tension to changes in concentration during adsorption. At constant temperature, Equation (A2-5) simplifies to:

$$\sum n_i^{ex}d\mu_i + Ad\sigma = 0 \tag{A2-6}$$

For a binary system with only two components (solute a and solvent b), Equation (A2-6) becomes:

$$n_a^{ex}d\mu_a + n_b^{ex}d\mu_b + Ad\sigma = 0 \tag{A2-7}$$

Since Equation (A2-7) holds for a plane interface independent of the position of the dividing surface, this position can be chosen so as to cause the surface excess of solvent b to vanish ($n_b^{ex} = 0$). Using this convention, and defining surface excess concentration as $\Gamma_i = n_i^{ex}/A$, allows simplification of Equation (A2-7) to:

$$d\sigma = -\Gamma_a^{(b)} d\mu_a \tag{A2-8}$$

The superscript (b) on Γ in Equation (A2-8) indicates that the dividing surface has been chosen so that the surface excess concentration of solvent b is zero. Finally, from basic thermodynamics, the chemical potential of solute (a) can be written as:

$$\mu_a = \mu_a^\circ + RT \ln \gamma_a x_a \tag{A2-9}$$

where μ_a° is a standard state chemical potential that depends only on temperature and pressure, R is the gas constant, γ_a is the activity coefficient of a, and x_a is the mole fraction of a in the liquid phase. Since the change in standard state chemical potential is zero at constant temperature and pressure, Equations (A2-8) and (A2-9) can be combined to give:

$$\Gamma_a^{(b)} = -(\gamma_a x_a/RT)\partial\sigma/\partial\gamma_a x_a \tag{A2-10}$$

Equation (A2-10) is the *Gibbs adsorption isotherm*, which allows the surface excess concentration of a solute to be calculated from the variation in surface tension. For an ideal solution, the activity coefficient $\gamma_a = 1$. Furthermore, for a dilute system, total concentration can be assumed constant. Under these conditions, Equation (A2-10) simplifies to:

$$\Gamma_a^{(b)} = -(C_a/RT)\partial\sigma/\partial C_a \tag{A2-11}$$

where C_a is the concentration of component a (mass or moles a/volume). Note that the Gibbs adsorption equation in the form of either (A2-10) or (A2-11) shows that if the surface tension of a system decreases as the concentration of solute increases, then $\Gamma_a^{(b)} > 0$. This means that a positive surface excess concentration is present for such a solute, i.e., it is undergoing net adsorption into the interfacial region. This is the case with all solutes that have surface activity. The Gibbs adsorption equation can also be used to infer information about area per molecule in adsorbed films if monomolecular surface coverage is assumed (e.g., [5, 10, 198, 295]).

3

Phospholipids: Introduction to Structure and Biophysics

I. Overview

This chapter summarizes the molecular structure and biophysical behavior of phospholipids. Glycerophospholipids are major components of endogenous lung surfactant and most exogenous lung surfactants. All phospholipids are surface active due to their amphipathic (polar and nonpolar) molecular structure, but individual phospholipids vary significantly in the specifics of their film behavior and surface properties. Some aspects of phospholipid surface behavior can be correlated with the properties of their hydrated bilayers and whether they are in the rigid gel phase or the more fluid liquid crystal phase at the temperature of interest. Another factor relevant for the surface activity of phospholipids is their aggregation in water to form lamellae, liposomes, hex phases, and less regular structures. Phospholipid aggregation affects adsorption and is largely responsible for the complex aqueous phase microstructure of endogenous lung surfactant and many exogenous lung surfactants. Phospholipid aggregation can be further modified by interactions with surfactant apoproteins leading, for example, to tubular myelin formation in endogenous surfactant. Coverage in this chapter provides an overview of phospholipid film, phase, aggregation, and adsorption behavior.

II. Phospholipids as Lung Surfactant Constituents

Phospholipids are ubiquitous biological molecules. Their most obvious importance in living systems is forming the structural basis for all cell and organelle membranes. Additional biological roles of phospholipids range from participating in cell membane signaling to serving as substrates for metabolism. Less widely appreciated, but certainly as crucial, is the activity of phospholipids as constitu-

---

Phospholipids and Their Biophysical Behavior

Phospholipids have multiple roles in living systems, ranging from being essential components of biological membranes to serving as constituents of endogenous and exogenous lung surfactants.

The biophysical behavior of phospholipids is integral to all of their biological functions.

Phospholipids are intrinsically surface active due to their amphipathic (polar/nonpolar) molecular structure.

Areas of phospholipid biophysics discussed in following sections as relevant for lung surfactant activity include:
Surface film behavior
Gel to liquid crystal transition temperature
Aggregation behavior in water
Adsorption behavior

---

ents of lung surfactant. Phospholipids make up approximately 85% of endogenous lung surfactant by weight [365, 568, 569, 936, 1105], and they are also major constituents of most clinical exogenous surfactants. The molecular structure and biophysical behavior of phospholipids are directly linked to their surface active function in lung surfactants (Chapter 8).

III. Phospholipid Classes and Their Molecular Structure

With the exception of lyso-derivatives, glycerophospholipids have a molecular structure containing a polar headgroup and two nonpolar fatty acyl chains (Figure 3-1) [22, 301, 1007]. Glycerophospholipid classes are defined by the structure of the headgroup attached via a phosphate moiety (PO_4) to the three-carbon glycerol backbone. The carbon atom to which the headgroup is attached is designated as number 3 (*sn*-3 in stereospecific numbering). The acyl chains of phospholipids are attached by ester linkages to carbons 1 and 2 (*sn*-1 and *sn*-2) in the glycerol backbone. These fatty chains vary in length and saturation (number of double bonds) and strongly affect molecular behavior. Glycerophospholipid classes shown in Figure 3-1 are *phosphatidylcholine* (PC), *phosphatidylethanolamine* (PE), *phosphatidylglycerol* (PG), *phosphatidylinositol* (PI), and *phosphatidylserine* (PS). Phosphatidylcholines are also known as lecithins. The headgroups of PC and PE compounds are zwitterionic (contain both a

PHOSPHOLIPIDS

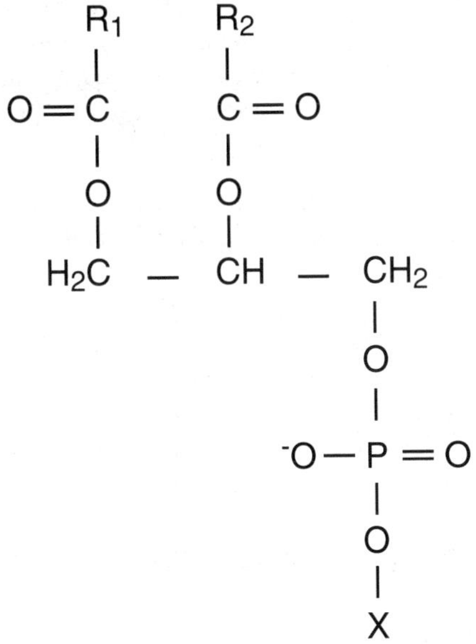

$$
\begin{array}{cc}
R_1 & R_2 \\
| & | \\
O = C & C = O \\
| & | \\
O & O \\
| & | \\
H_2C - CH & - CH_2 \\
\end{array}
$$

$$
\begin{array}{c}
CH_2 \\
| \\
O \\
| \\
{}^-O - P = O \\
| \\
O \\
| \\
X
\end{array}
$$

PHOSPHOLIPID CLASSES

Phosphatidylcholine:	$X = CH_2 - CH_2 - {}^+N(CH_3)_3$
Phosphatidylethanolamine:	$X = CH_2 - CH_2 - {}^+NH_3$
Phosphatidylglycerol:	$X = CH_2 - CHOH - CH_2OH$
Phosphatidylinositol:	$X = C_6H_6(OH)_5$
Phosphatidylserine:	$X = CH_2 - CH(COO^-) - {}^+NH_3$

Figure 3-1 Molecular structure of glycerophospholipids. Glycerophospholipids have two fatty acyl chains (R_1 and R_2) attached by ester linkages to a three-carbon glycerol backbone. A polar headgroup, which determines the phospholipid class, is also attached to the glycerol backbone. The classes shown are PC, phosphatidylcholine; PE, phosphatidylethanolamine; PG, phosphatidylglycerol; PI, phosphatidylinositol; and PS, phosphatidylserine. The charges shown on the headgroups are those near neutral pH.

positive and a negative charge) near neutral pH, while those of PG, PI, and PS compounds are anionic (negatively charged). *Lysophospholipids* have the same general structure as in Figure 3-1 but have only one fatty chain. An additional class of phospholipids is *sphingomyelin* (SPH). Sphingomyelin is a phos-phosphingolipid with a PC headgroup, but its hydrophobic group is a ceramide and it differs from glycerophospholipids in the linkage regions of the fatty chains (Figure 3-2). Lung surfactant also contains some neutral lipids as well as phospholipids. The molecular structures of three types of neutral lipids (*cholesterol*, *diglycerides*, and *triglycerides*) are also shown in Figure 3-2.

IV. Distribution of Phospholipid Classes in Cell Membranes and Lung Surfactant

The composition and activity of the various phospholipid compounds and classes in lung surfactant are described in detail later (Chapters 6, 8). Although many of the same phospholipid classes are found in lung surfactant and cell membranes, their quantitative distributions are very different (Table 3-1). PC is by far the largest phospholipid class in endogenous lung surfactant, accounting for about 80% of total lavageable phospholipid in most animal species [365, 568, 788, 936]. Cell membranes also have a substantial PC content, but it is much lower than in lung surfactant. Moreover, many of the specific PC compounds in lung surfactant,

Table 3-1 Approximate Lipid Compositions of Endogenous Lung Surfactant Compared to Cell and Bacterial Membranes

Lipid	Percentage of total lipid by weight					
	Liver cell membrane	Red cell membrane	Myelin	Mitochondrial membranes	*E. coli*	Lavaged lung surfactant
Phosphatidylcholine	24	17	10	39	0	78
Phosphatidyl-ethanolamine	7	18	15	35	70	3
Phosphatidylserine	4	7	9	2	Trace	5
Phosphatidylglycerol	—	—	—	—	20	7
Sphingomyelin	19	18	8	0	0	2
Glycolipids	7	3	28	Trace	0	—
Cholesterol	17	23	22	3	0	5
Others	22	13	8	21	10	—

Tabulated weight percents are approximate only. Value for phosphatidylserine in endogenous lung surfactant also includes anionic phosphatidylinositol.
Source: Adapted from Refs. 15, 301, 788.

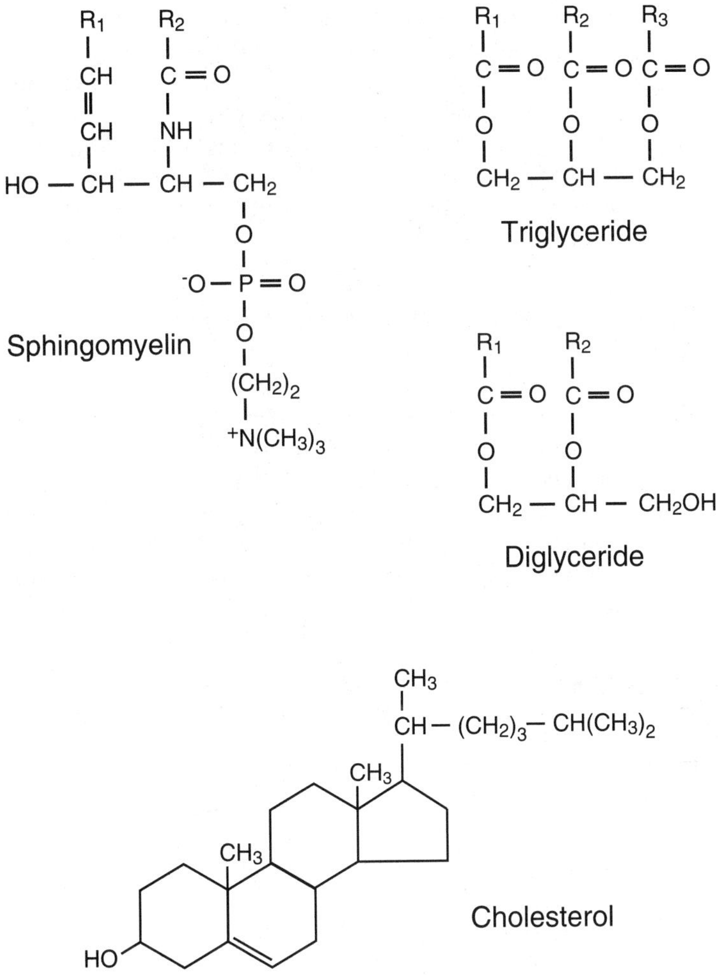

Figure 3-2 Molecular structure of additional lipids. Sphingomyelin has a PC headgroup but differs from the glycerophospholipids in Figure 3-1 in the linkage group region of the fatty chains. Also shown are the structures of several neutral lipids: cholesterol, diglycerides, and triglycerides. R_1, R_2, and R_3 are hydrocarbon chains that can vary in length and saturation (number of double bonds). R_1 in sphingomyelin is typically $(CH_2)_{12}CH_3$.

particularly those containing two saturated fatty chains (see following section), are not common cell membrane constituents. Cell membranes generally have a much higher percentage of other zwitterionic phospholipids such as PE and SPH compared to lung surfactant. Cell membranes and lung surfactant also contain differing ratios of anionic phospholipids (Table 3-1). The presence of anionic PG compounds in lung surfactant has attracted attention because this class of phospholipids is not common in mammalian cells and is more typically a constituent of bacterial membranes. PG has proved to be a very useful biochemical marker for the maturity of the surfactant system in the developing fetus [517, 1061, 1075]. The significance of PG for biophysical function in lung surfactant is, however, still an open question (Chapter 8).

V. Fatty Chains in Individual Compounds Within a Phospholipid Class

Although the polar headgroup determines the phospholipid class, the fatty chains define individual compounds within each class. Common names for several fatty acids found in phospholipids are listed in Table 3-2. These names are used in conjunction with the headgroup-specific class to designate individual phospholipid compounds. For example, myristic acid contains a 14 carbon atom chain with no double bonds (C14:0) and oleic acid contains an 18 carbon atom chain

Table 3-2 Common Names for Some Common Fatty Acids Found in Phospholipids

Common name	Chain length:double bonds
Lauric	12:0
Myristic	14:0
Palmitic	16:0
Palmitoleic	16:1 (9-*cis*)
Stearic	18:0
Oleic	18:1 (9-*cis*)
Vaccenic	18:1 (11-*cis*)
Linoleic	18:2 (9-*cis*, 12-*cis*)
γ-Linolenic	18:3 (6-, 9-, 12-, all *cis*)
α-Linolenic	18:3 (9-, 12-, 15-, all *cis*)
Arachidic	20:0
Behenic	22:0
Arachidonic	20:4 (5-, 8-, 11-, 14-, all *cis*)

In unsaturated biological phospholipids, *cis* double bonds are much more common than trans double bonds. See text for discussion of the influence of fatty chains on phospholipid behavior.

with a *cis* double bond at carbon 9 (C18:1, *cis* 9). The PC molecule with C14:0 and C18:1 fatty acids at positions *sn*-1 and *sn*-2, respectively, is therefore designated 1-myristoyl-2-oleoyl-*sn*-3-phosphatidylcholine (MOPC). MOPC is an example of an *unsaturated phospholipid* because it contains a fatty chain with at least one double bond. More specifically, since only one double bond on one chain is present in MOPC, it is also a *monoenoic phospholipid*. In such compounds in animals, the unsaturated fatty acid is usually found attached to the glycerol backbone at the *sn*-2 position. As another example, the PC molecule with two fully saturated C16:0 palmitic chains at positions *sn*-1,2 is designated 1,2-dipalmitoyl-*sn*-3-phosphatidylcholine (DPPC). Because the fatty chains of DPPC contain no double bonds, it is a *saturated phospholipid*. The term *disaturated phospholipid* is also often used to emphasize that both chains are saturated. Although not strictly necessary, this latter terminology is common in the lung surfactant literature and is used interchangeably with saturated phospholipid in this book.

VI. Biophysical Importance of Phospholipid Fatty Chains

The hydrophobic chain region of phospholipids is particularly important in biophysical behavior. Fatty chain length and saturation strongly influence the fluidity, film properties, and phase behavior of phospholipids. Disaturated phospholipids have significantly higher melting and phase transition temperatures than otherwise equivalent monoenoic phospholipids with one double bond. The type of double bond (*cis* vs *trans*) also affects physical properties and phase behavior. Lung surfactant contains a spectrum of saturated and unsaturated phospholipids that contribute to surface activity. The fatty chains of these phospholipids, along with their headgroups, determine their ability to interact in surface films and subphase aggregates. In order to interact in a surface film or bilayer, molecules of different phospholipids must be at least partially miscible (Chapter 2). The degree of miscibility of different phospholipids is highly dependent on the similarity between their fatty chains, with miscibility greatest for identical chains.

Figure 3-3 shows a molecular model of the structure of a crystalline dihydrate of 1,2-dimyristoyl-3-phosphatidylcholine (DMPC) [301]. The figure illustrates several molecular regions, cross sections, and characteristics important in physicochemical behavior. With the exception of the initial part of the *sn*-2 fatty acid, the saturated chains are shown in the fully extended *trans*[1] configuration that is the most stable [301]. An energy barrier of ~3.5 kcal/mol is required in order for the saturated chain to rotate to

1. The fully extended all-*trans* configuration of saturated fatty chains in phospholipids is not to be confused with *trans* double bonds.

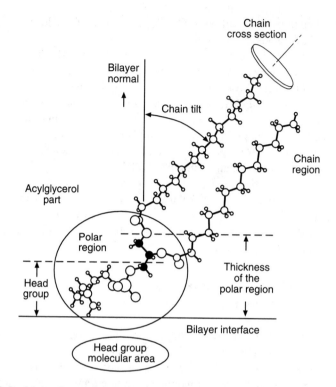

Figure 3-3 Molecular model showing structural regions in a phospholipid. The specific compound shown is 1,2-diimyristoyl phosphatidylcholine (DMPC), but many of the molecular features hold for other glycerophospholipids. The saturated chains are shown in the fully extended *trans* configuration, with a tilt relative to the bilayer normal. The cross-sectional areas occupied by the headgroup and by the larger polar region that includes the glycerol backbone and chain junctional regions are also indicated. The cross-sectional area of each hydrophobic fatty chain is determined perpendicular to the chain axis. (Adapted from Ref. 414 as reported in Ref. 301.)

the *gauche* form, which alters chain direction and results in a larger effective cross-sectional area and less ordered packing [301]. The cross-sectional area occupied by fatty chains is increased even more by the introduction of double bonds. The chain kink introduced by the predominant *cis* double bonds found in mammalian phospholipids gives a change in direction that can significantly disorder chain packing in bilayers and films. The presence of even one double bond substantially enlarges the cross-sectional area swept out by a fatty acyl chain. Because of this, unsaturated phospholipids occupy much larger areas per molecule than saturated phospholipids at fixed surface tension or pressure in surface films [108, 120, 301, 985, 1007]. Films of unsaturated phospholipids are expanded rather than condensed and solid-like as is the case for saturated phospholipid films as described below.

VII. Phospholipid Surface Films and Their Characteristics

Although all phospholipids are surface active, individual compounds vary significantly in their specific film properties. Phospholipid films can differ not only in their ability to lower surface tension during dynamic compression but also in their compressibility, compression-expansion hysteresis, and dynamic respreading characteristics. Films of long-chain saturated phosphatidylcholines such as DPPC typically reduce surface tension to low values during dynamic compression and exhibit a large π-A isotherm hysteresis and poor respreading on successive cycles. In contrast, films of unsaturated phospholipids do not reduce surface tension as effectively during compression but have better respreading and less hysteresis during cycling. The headgroup region of phospholipid molecules also affects film behavior, leading to further differences in isotherm characteristics between phospholipid classes. Films of anionic phospholipids, for example, are strongly affected by subphase ions such as calcium, while zwitterionic phospholipids are relatively unaffected by divalent metal ions. Films of phosphatidyethanolamines also can have very different properties from films of phosphatidylcholines due to the smaller molecular cross section of the ethanolamine moiety and its ability to participate in intermolecular hydrogen bonding. Finally, the behavior of all phospholipid surface films is a function of physical variables such as temperature, humidity, cycling rate, and the extent of area compression.

A. Behavior and Properties of DPPC Surface Films

Perhaps the most impressive characteristic of DPPC surface films is their ability to reduce surface tension to extremely low values <1 mN/m during dynamic compression. Figure 3-4 shows π-A isotherms for monolayers of DPPC at 23°C and 37°C studied in a Wilhelmy surface balance [796, 1058] (see Chapter 4 for a description of the Wilhelmy balance and its use). Although these DPPC films reach very low surface tensions during dynamic compression, they also exhibit very poor respreading on successive cycles of compression and expansion. At both room and body temperature, the second compression curve for DPPC is displaced almost fully to the left, indicating very little respreading of film material ejected from the surface during the first compression (Chapter 2). In contrast, unsaturated phospholipids tend to have π-A isotherms with lower maximum surface pressures ($\pi_{max} < 50$ mN/m), smaller compression-expansion hysteresis, and significantly better respreading. Fluid lung surfactant phospholipids greatly improve respreading when present together with DPPC in mixed films (Chapter 8).

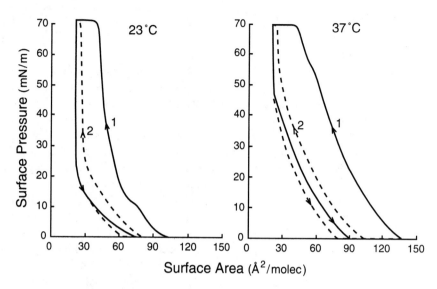

Figure 3-4 Surface pressure-area isotherms for DPPC monolayers at room and body temperature. *Left,* 23°C; *right,* 37°C. DPPC was spread from an organic solvent at the air-water interface in a Wilhemy balance and compressed at a rate of 5 min/cycle. The maximum surface pressures reached by these DPPC films are equivalent to surface tensions <1 mN/m. The displacement of the first and second compression curves indicates the poor respreading of DPPC. See text for details. (Data adapted from Refs. 796, 1058.)

Molecular Packing in Films of Saturated and Unsaturated Phosphatidylcholines. The differing film characteristics of saturated and unsaturated phosphatidylcholines stem from their molecular structural differences. Long chain saturated phosphatidylcholines like DPPC are able to pack tightly together in an extremely stable manner at the interface, leading to condensed, solid-like films that lower surface tension very effectively during dynamic compression (Figure 3-5). In contrast, unsaturated phospholipids have an increased chain cross section that disrupts packing and leads to more expanded films with greater molecular mobility. The condensed nature of films of saturated vs unsaturated phosphatidylcholines is shown, for example, by comparing area data for monolayers of DPPC and dioleoyl phosphatidylcholine (DOPC) [796]. At a fixed temperature of 23°C and a surface pressure of 10 mN/m, DPPC occupies a molecular area of about 77 $Å^2$ while DOPC occupies over 90 $Å^2$ [796]. At a surface pressure of 20 mN/m, molecular areas for DPPC and DOPC are approximately 59 $Å^2$ and 75 $Å^2$, respectively, and at 40 mN/m they are 45 $Å^2$ and 53 $Å^2$ [796]. Corresponding with their more condensed, solid-like nature, DPPC films are able to reach minimum surface tensions <1 mN/m compared to minimum values near 20 mN/m for DOPC films. The ability of DPPC and other long chain disaturated phosphatidylcholines to form condensed, solid films with excellent surface tension lowering

A Condensed Disaturated PC Monolayers

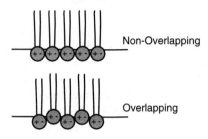

Non-Overlapping

Overlapping

B Expanded Unsaturated PC Monolayers

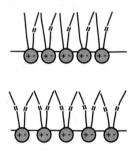

Figure 3-5 Molecular packing in films of saturated vs unsaturated phosphatidyl-cholines. (A) Long chain disaturated PCs form condensed, solid-like films. (B) Unsaturated PCs form fluid, expanded films. The rigid, gel-like properties of long chain disaturated PCs allow them to form tightly packed condensed films with great surface tension lowering power. Molecular accommodations such as headgroup overlapping may occur at high degrees of compression because the bulky PC headgroup occupies a larger cross section than the chains. The double bonds in unsaturated PCs greatly increase the area swept out by the fatty chains, leading to expanded films that lower surface tension less effectively but exhibit better respreading. (Adapted from Ref. 301.)

ability correlates in part with the biophysical behavior of hydrated bilayers of these compounds and their existence in the rigid gel phase as described later. It is important to note, however, that because of the important influence of the headgroup region on film and phase behavior, some disaturated phospholipids have quite different properties from those exhibited by DPPC. Films of the disaturated phosphatidylethanolamine DPPE, for example, have very different surface pressure–area isotherm behavior from DPPC despite the fact that both compounds have identical fatty chains and are in the gel phase at room and body temperature [657, 658, 1057].

VIII. Dynamic versus Equilibrium Surface Tension Lowering in Compressed Phospholipid Films

Minimum surface tension values <1 mN/m as generated by DPPC films under rapid, dynamic compression are well below the minimum values that can be generated under equilibrium conditions. Saturated phosphatidylcholine films compressed at slow quasi-static rates reach minimum surface tensions of only 10–20 mN/m, nearer the upper limit of equilibrium spreading pressures for phospholipids from solid crystals [852]. Nonequilibrium surface pressures achieved during dynamic compression reflect the formation of metastable surface states within phospholipid films.[2] When compression is halted, these dynamic surface pressures decay (relax) toward equilibrium values. The time scale over which this occurs is a function of several variables including surface concentration and temperature. Relaxation behavior in compressed surface films has been studied for a variety of surfactant compounds and mixtures.[3] The dynamic nature of surface tensions in compressed films of lung surfactant, and their relaxation with time, was recognized in the 1960's [724, 1157]. Particular attention has been paid to dynamic surface pressure relaxation in films of DPPC [321, 755, 795, 1059, 1094, 1157]. Figure 3-6 shows the significant surface pressure relaxation found during 5 minute pauses at multiple points along the π-A isotherm of a dynamically compressed DPPC surface film at 25°C [1059].

The detailed time dependence of dynamic surface pressure relaxation for the compressed DPPC film at different surface concentrations (area per molecule) is shown in Figure 3-7. Major characteristics of the relaxation behavior are as follows. At DPPC film areas above 40 Å^2/molecule, prior to monolayer collapse based on fatty chain limiting areas, relaxation is apparent as an exponential-like decay of surface pressure toward an asymptotic equilibrium value.[4] The "relaxation time" required for dynamic surface pressure to approach to within 10% of the asymptotic value increases as molecular area decreases. Relaxation times at selected areas in Figure 3-7 are, for example, approximately 50 seconds at 68 Å^2/molecule, 100 seconds at 49 Å^2/molecule, 300 seconds at 44 Å^2/molecule, and

---

2. The formation of metastable dynamic states in phospholipid films, and the ramifications of this behavior for the generation of very low minimum surface tensions in lung surfactant films during cycling, is discussed further in Chapter 7.

3. For example, Refs. 321, 432, 724, 754, 755, 795, 879, 1003, 1004, 1059, 1094, 1157.

4. Although this precollapse surface pressure relaxation is readily apparent in Figure 3-7, the time scale of dynamic surface pressure decrease (surface tension increase) is still long relative to film cycling rates occurring at the alveolar interface during breathing. In this sense, the dynamic surface pressures achieved are metastable and meaningful for surface active function.

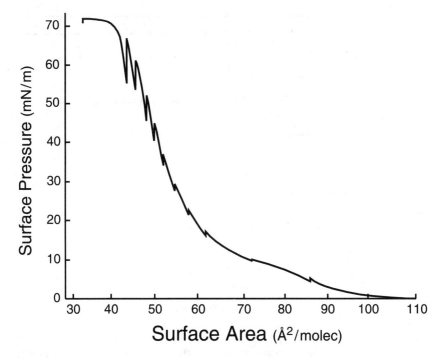

Figure 3-6 Dynamic surface pressure relaxation in a dynamically compressed DPPC surface film. Compressed DPPC films at the air-water interface achieve dynamic surface pressures in excess of equilibrium values. When compression is halted at fixed area, dynamic surface pressures relax toward equilibrium over a time scale that varies with the extent of film compression. See text for details. (Redrawn from Ref. 1059.)

850 seconds at 40 A^2/molecule [1059]. The asymptotic values for precollapse dynamic surface pressure relaxation are essentially the same as those found at the same surface concentration by direct interfacial deposition of DPPC in a spreading solvent (Figure 3-8). After compression beyond monolayer collapse, however, there is a substantial further increase in the stability of dynamic surface pressure. Surface pressure remains within a few mN/m of the dynamic collapse value over an extended time scale (Figure 3-7).

The extreme stability of dynamic collapse pressures in DPPC films occurs despite the fact that these surface pressures are 20 mN/m or more above the maximum values achieved by static spreading methods (Figure 3-8). This indicates that the dynamic collapse of DPPC films generates surface conditions where the ability of molecules to reorient and reorganize in the film is highly constrained. The mechanisms of DPPC film collapse are not fully understood, although it has been suggested that

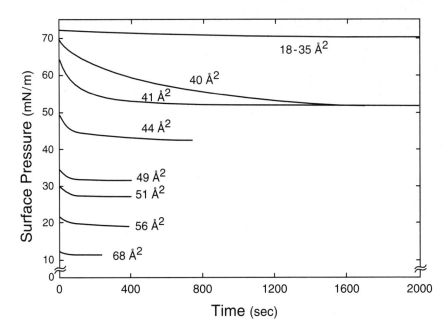

Figure 3-7 Time course of dynamic surface pressure relaxation for DPPC as a function
of film molecular area. Each relaxation curve was initiated following dynamic compression
of a DPPC monolayer to the indicated molecular area at a rate of 26 $Å^2$/molecule/min.
Surface pressure was then measured as a function of time. Data at 25°C. (Redrawn from
Ref. 1059.)

some processes distinct from other solid films may be involved [1011]. In any case,
the constrained postcollapse relaxation observed in dynamically compressed DPPC
films is consistent with the rigid molecular characteristics of disaturated phosphatidylcholines below their gel to liquid crystal transition temperature Tc (see following
section). Decreased postcollapse surface pressure relaxation in DPPC films is also
consistent with their poor respreading when cycled into the collapse regime (cf. Figure
3-4). The poor respreading in DPPC films also implies that molecular movement in the
interfacial region is limited. Both respreading and relaxation behavior can be changed by
the presence of compounds that disrupt packing and increase fluidity. For example,
addition of cholesterol or unsaturated phospholipids like DOPC in mixed films with
DPPC greatly increases dynamic respreading and decreases the stability of postcollapse
dynamic surface pressures [795, 796]. If large enough percentages of DOPC or
cholesterol are added to DPPC, the mixed films become fluidized to the extent that it is
not possible to generate dynamic collapse pressures appreciably above static values.
Similarly, if phospholipid monolayers are studied at temperatures above the gel to liquid
crystal T_c of film components, dynamic collapse pressures and monolayer stability are also
decreased [321]. Lung surfactant requires a proper balance of fluid and rigid phospholipids,

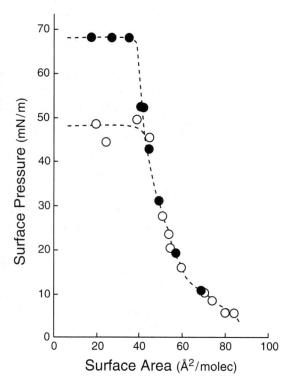

Figure 3-8 Comparison of relaxation asymptotes for DPPC films to surface pressures found by equilibrium spreading. *Filled circles*, long time asymptotes from Figure 3-7. *Open circles*, surface pressure following interfacial deposition in a spreading solvent at the indicated surface concentration. Dynamic surface pressures in DPPC films prior to collapse relax over a time scale of minutes to final asymptotic values equal to those found by equilibrium spreading to the same area. Maximum dynamic pressures (minimum surface tensions) after compression of DPPC into the collapse regime are stable over longer times. (Redrawn from Ref. 1059.)

along with contributions from surfactant apoproteins, to generate the overall surface tension lowering, respreading, and adsorption behavior required for functional activity (Chapter 8).

IX. Phospholipid Bilayers and Their Properties

The ability of phospholipids to organize in a thermodynamically stable bilayer in the presence of water is fundamental to their biological activity. In many phospholipid bilayers as in typical cell membranes, the fatty chains in one layer

---

Summary of Phospholipid Film Behavior

DPPC and related rigid disaturated phospholipids can form solid, tightly packed surface films capable of lowering surface tension to <1 mN/m under dynamic compression. Molecular interactions involving both the headgroup and the fatty acyl chains participate in generating these behaviors.

Dynamic surface pressures above the equilibrium spreading limit in saturated phospholipid films are metastable and relax over time toward equilibrium values. However, the time scale of relaxation is long compared to cycling rates in the lungs.

Unsaturated phospholipids tend to form more fluid films that have lower maximum surface pressures (higher minimum surface tensions) but better respreading relative to disaturated phosphatidylcholines.

The behavior of phospholipids in surface films can be correlated at least in part with the phase behavior of the hydrated bilayers (gel and liquid crystal phases) as described in the following sections.

---

of molecules do not overlap those of the next. These bilayers are described as "opposed." If the fatty chains of the two layers of phospholipids do overlap to some degree, a bilayer is said to be "interdigitated." In both opposed and interdigitated phospholipid bilayers, the tilt and degree of interaction of fatty chains vary with temperature and hydration, and the polar headgroups can assume different orientations and in-plane layering depending on their polarities and fixed charges. In lung surfactant, phospholipid bilayer and aggregation behavior is additionally modified by interactions with surfactant apoproteins. Structure and molecular interactions in phospholipid bilayers and aggregates are studied using multiple physicochemial techniques, including electron microscopy, fluorescence microscopy, light scatter, X-ray diffraction, differential scanning calorimetry, and spectroscopic methods such as absorption, electron spin resonance, nuclear magnetic resonance (NMR), and infrared (IR) and Fourier transform infrared (FTIR) spectroscopy. Only a brief summary of the bilayer and phase behavior of hydrated phospholipids is given here, with further details provided in research texts and review articles (e.g., [22, 108, 301, 985, 1007]).

A. Bilayer Phase Behavior and the Gel to Liquid Crystal Transition Temperature (T_c)

The thermal phase behavior of hydrated phospholipid layers can be very helpful in understanding the film properties of these compounds. At sufficiently low temperature, phospholipids exist as a highly ordered solid crystal. As they are heated, hydrated phospholipids pass through one or more intermediate states called "mesophases" before melting to an isotropic liquid. These mesophases are characterized by different degrees of regional and directional order. Two important mesophases formed by hydrated phospholipid bilayers are the rigid *lamellar gel phase* and the more fluid lamellar liquid crystalline phase (often called simply the "gel phase" and the "liquid crystal phase") (Figure 3-9). The temperature of the transition between these phases is called the gel to liquid crystal transition temperature (T_c). The gel to liquid crystal transition temperature for a phospholipid is typically much lower than its solid-liquid melting point. For example, the temperature of the main gel to liquid crystal transition for DPPC is just above 41°C, far below its solid-liquid melting point of 220°C. Chain length and saturation both affect T_c, with shorter chains or the presence of double bonds giving rise to lower transition temperatures (Table 3-3). For example, DMPC with two C14:0 chains has a T_c of 23–24°C, significantly lower than that of DPPC with

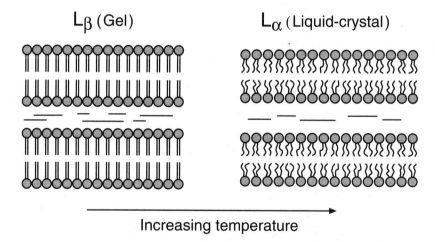

$$L_\beta \text{ (Gel)} \qquad L_\alpha \text{ (Liquid-crystal)}$$

Increasing temperature

Figure 3-9 Schematic of the lamellar gel phase (L_β) and the liquid crystal phase (L_α) of hydrated phospholipid bilayers. The rigid gel phase exists at lower temperatures and the fluid liquid crystal phase at higher temperatures. The gel phase has a greater bilayer thickness because the phospholipid chains are fully extended rather than "melted." Phospholipid thermal phase behavior can be correlated to some degree with film properties as described in the text. (Adapted from Ref. 301.)

Table 3-3 Transition temperatures between the gel and liquid crystal phases for hydrated bilayers of selected phospholipids

Compound	Fatty chains (sn-1, sn-2)	Gel to liquid crystal T_c (°C)*	Transition enthalpy ΔH (kcal/mol)
	Phosphatidylcholines		
Saturated PCs			
DMPC	(C14:0, C14:0)	23.0–24.0	5.0–6.8
DPPC	(C16:0, C16:0)	41.1–41.5	8.5–8.7
DSPC	(C18:0, C18:0)	53.5–54.9	10.6–10.8
PMPC	(C16:0, C14:0)	27.1–27.5	5.2–7.8
MPPC	(C14:0, C16:0)	34.8–35.1	6.9–8.1
MSPC	(C14:0, C18:0)	38.6–42.0	6.9–8.2
SMPC	(C18:0, C14:0)	29.4–34.0	5.2–6.0
SPPC	(C18:0, C16:0)	43.7–44.1	7.2–7.3
PSPC	(C16:0, C18:0)	48.8–49.0	8.3–9.8
Unsaturated PCs			
DPoPC	(C16:1, C16:1) *cis* 9	−35.5 to −36	9.1
DOPC	(C18:1, C18:1) *cis* 9	−18 to −22	7.6–8.0
POPC	(C16:0, C18:1) *cis* 9	−5–4	5.4–8.1
OPPC	(C18:1, C16:0) *cis* 9	−9.3 to −10.3	4.6
SOPC	(C18:0, C18:1) *cis* 9	6.3–13	5.4
	Other glycerophospholipid classes		
DMPE	(C14:0, C14:0)	47.5–49.5	5.8–6.4
DPPE	(C16:0, C16:0)	60–63.5	8.5–8.9
DPPS	(C16:0, C16:0)	50.8–55.0 (Na^+)	9.0
		69–< 85 (Ca^{++})	—
DMPG	(C14:0, C14:0)	23–23.7 (Na^+)	6.8–6.9
DPPG	(C16:0, C16:0)	39.5–41.5 (Na^+)	7.9–9.1
		80–90 (Ca^{++})	19
	Diether phospholipids		
DHPC	(C16:0, C16:0)	43.4–43.8	9.2
DHPE	(C16:0, C16:0)	68.0	8.5

*Tabulated transition temperatures and enthalpies are representative values or ranges from the literature. The main gel to liquid crystal transition near neutral pH is tabulated; most PCs also have a pretransition (not shown). Transition entropy can be approximated by dividing enthalpy by T_c in °K.

Abbreviations: DHPC, dihexadecyl PC; DHPE, dihexadecyl PE; DMPC, dimyristoyl PC; DMPE, dimyristoyl phosphatidylethanolamine; DMPG, dimyristoyl phosphatidylglycerol; DOPC, dioleoyl PC; DPoPC, dipalmitoleoyl PC; DPPC, dipalmitoyl PC; DPPE, dipalmitoyl PE; DPPG, dipalmitoyl PG; DPPS, dipalmitoyl phosphatidylserine; DSPC, distearoyl PC; MPPC, myristoyl-palmitoyl PC; MSPC, myristoyl-stearoyl PC; OPPC, oleoyl-palmitoyl PC; OSPC, oleoyl-stearoyl PC; PMPC, palmitoyl-myristoyl PC; POPC, palmitoyl-oleoyl PC; PSPC, palmitoyl-stearoyl PC; SMPC, stearoyl-myristoyl PC; SPPC, stearoyl-palmitoyl PC; SOPC, stearoyl-oleoyl PC; SPPC, stearoyl-palmitoyl PC.

Source: Compiled from Refs. 73, 123, 156, 177, 255, 301, 541, 558, 659, 680, 1007, 1019.

its C16:0 chains. The influence of double bonds on T_c is even more pronounced. POPC, containing one C16:0 chain and one C18:1 chain, has a T_c value about 40°C below that of DPPC. DOPC, containing two C18:1 chains, has an even lower gel to liquid crystal transition near –20°C. The T_c of phospholipids also depends on headgroup structure and on chain linkage group. For example, phosphatidylethanolamines have higher phase transition temperatures than comparable phosphatidylcholines due to intermolecular hydrogen bonding. Compounds with ether linkages at the glycerol backbone also have slightly elevated T_c values relative to comparable ester-linked phospholipids (e.g., dihexadecyl-PC or -PE in Table 3-3). T_c can also be significantly increased by ionic interactions, such as in the case of anionic phosphatidylglycerols or phosphatidylserines in the presence of calcium (e.g., DPPG, DPPS in Table 3-3 [73, 255, 1007]).

At room or body temperature, DPPC and similar long chain saturated phosphatidylcholines are in the rigid gel phase, correlating with their ability to form tightly packed, condensed films with excellent surface tension lowering ability. In contrast, unsaturated phosphatidylcholines are in the liquid crystal phase and form films with expanded molecular packing and more fluid behavior relative to DPPC. Such films typically have lower maximum pressures (higher minimum surface tensions) but increased respreading relative to gel phase phosphatidylcholines. It is important to note, however, that correlations between the gel phase and the ability to lower surface tension in compressed films are not absolute and do not hold for all phospholipids. For example, DPPE and other long chain saturated phosphatidylethanolamines have higher phase transition temperatures than DPPC (Table 3-3). Despite this, DPPE forms surface films that generate much lower maximum surface pressures and have better respreading than DPPC films at room and body temperature [657, 658, 1057]. This reflects the significant effect of intermolecular hydrogen bonding on the film and bilayer behavior of PE compounds, as well as the steric influence of the much smaller ethanolamine headgroup that alters molecular cross section and further changes film properties relative to phosphatidylcholines.

Measurements of T_c and Thermodynamic Parameters for Phospholipid Phase Transitions. Transition temperatures between phases in hydrated phospholipid bilayers are measured experimentally by differential scanning calorimetry (DSC) or electron spin resonance (ESR) spectroscopy (Chapter 4). Thermodynamic parameters for phase transitions can also be determined including the enthalpy and entropy of transition (ΔH and ΔS) [22, 120, 301, 985, 1007]. On a DSC scan, differential heat input is measured between a phospholipid sample and a reference sample containing an equal mass of solvent and plotted as a function of temperature. Phase transitions are apparent as peaks in the differential heat input over a small change in temperature. The phase transition temperature is generally defined as the temperature of baseline departure at the onset of the transition, although the midpoint temperature where the transition is 50% complete (sometimes also called the melting temperature) is also utilized. The transition enthalpy ΔH is defined as the heat required for the transition, i.e., the area under the transition peak on the heat input-temperature scan. The transition entropy ΔS is determined as $\Delta H/T_m$, where T_m is the transition temperature in $°K$. The width of the transition peak can also be analyzed to assess "cooperativity." A highly cooperative phase transition is one where the transition peak is tall but narrow, indicating that large units of phospholipid molecules are melting together. A less cooperative phase transition has a broader width and a lower height, indicative of successive melting of smaller groups of molecules. However, phase transition peak widths and thermodynamic parameters need to be interpreted with some caution since they can be sensitive to variables such as differences in sample mass or the presence of small amounts of impurities.

Molecular Characteristics of Mesomorphic Phospholipid Phases. Because phospholipids form mesophases as a function of temperature, they are said to exhibit thermotropic mesomorphism. Since liquid crystals that form through interaction with a liquid (in this case water) are defined as lyotropic liquid crystals, phospholipids are also said to exhibit lyotropic mesomorphism. In the literature on phospholipid biophysics, the rigid gel phase is commonly designated as L_β or $L_{\beta'}$ and the more fluid liquid crystal phase as L_α [22, 301, 1007] (Figure 3-9). The designation $L_{\beta'}$ generally indicates that the phospholipid chains are tilted at an angle to the bilayer normal. In the gel phase (either L_β or $L_{\beta'}$), phospholipid chains are packed in a near-hexagonal array, and there is little lateral diffusion in the plane of the bilayer. In the liquid crystal phase, the fatty chains are "melted" and have greatly increased freedom of motion. Lateral diffusion in the plane of the bilayer is also increased. Because of the increased chain fluidity in the L_α phase, bilayer thickness decreases compared to the rigid L_β or $L_{\beta'}$ phases where the phospholipid chains are fully or nearly fully extended. Sometimes an additional phase exists between the main L_β (or $L_{\beta'}$) and L_α phases. A ripple phase (P_β), for example, is found for DPPC and many other phosphatidylcholines as they move from the gel to liquid crystal phases. In the ripple phase, phospholipid fatty chains and bilayer order are perturbed from the gel phase but are not fully melted into the fluid liquid crystal phase. The gel to liquid crystal phase transition temperature of phospholipids is highly dependent on the fatty acyl chain portion of the molecules because van der Waals forces (interactions between transient induced dipoles in adjacent hydrocarbon chains) make a primary contribution to bilayer character and stability. The tight packing associated with the extended *trans* chains of disaturated phospholipids maximizes van der Waals interactions to stabilize the gel phase and increase the transition temperature. Intermolecular hydrogen bonding as in PE compounds, or increased headgroup association through ionic interactions, also promotes gel phase stability and hence increases phase transition temperatures as noted earlier.

X. Effects of Cholesterol in Phospholipid Films and Bilayers

Cholesterol is one of many molecules that can interact with phospholipids in films and bilayers (for detailed review and discussion see Refs. 22, 301, 985, 1007). Interactions between phospholipids and cholesterol are important in cell membrane stability and behavior, and they are also relevant for lung surfactant (Chapter 8). The primary effects of cholesterol on phospholipids appear to be through interactions affecting the fatty chain region. In bilayers, cholesterol tends to insert normal to the plane of the bilayer with its OH group in proximity to the ester linkages in the phospholipid molecules. In surface films, the compact structure of cholesterol allows it to pack efficiently between phospholipid molecules, particularly those with kinked, unsaturated fatty chains. However, cholesterol disrupts tight packing such as found in films of disaturated phospholipids at high degrees of compression. As a consequence, cholesterol tends to fluidize films or bilayers containing phospholipids in the gel phase, while having the opposite effect in mixtures with phospholipids in the liquid crystal phase. Cholesterol has been shown to increase the fraction of fatty chains in the more fluid *gauche* configuration below T_c, while it decreases *gauche* rotamers above T_c. Cholesterol esters also can affect film and bilayer behavior but are less miscible with phospholipids than cholesterol and thus have a smaller influence on biophysical behavior.

> *Phase Behavior of Cholesterol and Phospholipids.* The degree and character-istics with which cholesterol mixes with phospholipids are of obvious importance for molecular interactions between these compounds. The phase behavior of mixtures of cholesterol and phospholipids is not fully defined, but a good deal of information is available from studies using DSC, NMR, and FTIR spectroscopy; X-ray and neutron scattering; and other methodology ([301, 1007] for review). Cholesterol is at least partially miscible with phospholipids in surface films and bilayers; i.e., it is at least partially distributed within and among phospholipid molecules. However, it is also clear that particularly for gel phase phospholipids, the distribution is not completely random. Spectroscopic data suggest that cholesterol is relatively miscible with phospholipids above the gel to liquid crystal transition. However, some degree of phase separation is known to occur below T_c, and a variety of different phases have been reported in mixtures containing phospholipids and cholesterol. The precise roles and importance of such phases in the effects of cholesterol on phospholipid biophysical behavior are still uncertain.

XI. Phospholipid Aggregates in the Aqueous Phase

In addition to the temperature-dependent mesophases formed by hydrated phospholipid bilayers, phospholipid molecules form aggregates when suspended in the bulk aqueous phase. Aggregate formation by phospholipids in water is facilated by their amphipathic (polar/nonpolar) structure, which allows a variety of configurations organized to reduce free energy [22, 54, 301, 342, 985, 1007, 1056]. Phospholipid aggregates in water are sometimes described as "micelles," but this term is usually reserved for simple spherical aggregates such as those formed in water by soaps and detergents. Examples of phospholipid aggregates in water are bilayers and multilayers (*lamellae*), unilamellar or multilamellar spheroidal vesicles (liposomes), and cylindrical hexagonal phases (Hex I and II) (Figure 3-10). The sizes of these aqueous phase phospholipid aggregates vary from the nanometer scale to the micron scale depending on preparation methods and composition. The aqueous phase microstructure of endogenous lung surfactant and many exogenous surfactants contains a variety of lamellar and vesicular phospholipid-rich aggregates as well as less regular forms (Chapter 6). Phospholipid aggregation is also significantly influenced by interactions with surfactant proteins, with one prominent example being the formation of tubular myelin as described in the next section.

One group of phospholipid aggregates that has attracted significant interest outside lung surfactant research is liposomes ([54, 197, 301, 342, 466, 1056] for review and preparation methodology). Liposomes are generally defined as bilayer-containing lipid vesicles that enclose a volume in the aqueous phase. Liposomes are commonly subclassified by size and by the number of bilayers they contain (unilamellar or multilamellar). Small unilamellar vesicles (SUVs) typically have diameters of 200–500 Å (20–50 nm), while large unilamellar vesicles (LUVs) have diameters of 500–5,000 Å (50 nm–0.5 microns). Very large unilamellar liposomes of cellular size can also be made [753]. Multilamellar vesicles (MLVs) also cover a broad range of sizes from nanometers to microns. Liposomes are widely used in medicine and cell biological research. Medical uses include encapsulating drugs to increase delivery to specific cells or tissues or to reduce degradation and prolong biologic lifetime. Liposomes are also used to deliver transfecting agents or genes to cells in molecular biological applications and as model membranes in basic research. In addition, liposomal-type vesicles are part of the aqueous phase microstructure of all surfactants with a significant phospholipid content.

When sonication, vortexing, or other energetic processes are used to suspend phospholipids in water, the resultant aggregation states (liposomes or other aggregates) may be thermodynamically metastable rather than true equilibrium states. In a practical sense, however, phospholipid aggregates formed by sonication and vortexing can be stable over long time scales. Although some modification of microstructure may occur, aqueous suspensions of endogenous lung surfactant and phospholipid-rich exogenous surfactants can be stored under refrigeration or freezing and retain high surface activity over periods ranging from days to months.

A. Tubular Myelin

The term tubular myelin designates the distinctive cross-hatched bilayer microstructure observed by electron microscopy in aqueous suspensions of endogenous lung surfactant in the presence of calcium.[5] Tubular myelin is formed by interactions between phospholipids and lung surfactant proteins. Under electron

---

5. For example, Refs. 67, 413, 685, 791, 1034, 1036, 1037, 1053, 1182–1184, 1200.

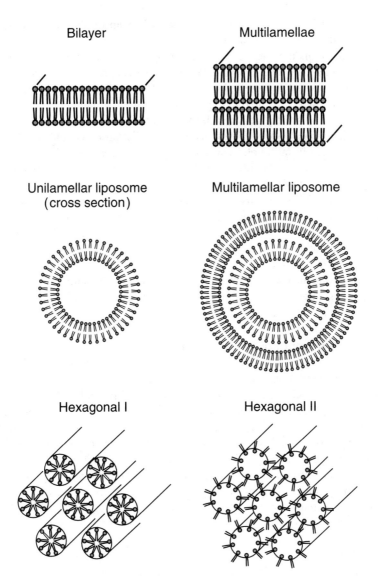

Figure 3-10 Phospholipid aggregates in the aqueous phase. Aggregate forms assumed by phospholipids in water include simple opposed bilayers, multilayered lamellae, unilamellar and multilamellar vesicles (liposomes), and hexagonal phases I and II. A variety of less regular phospholipid aggregates also exist. Sizes of phospholipid aggregates vary from nanometers to microns depending on composition and preparation method. Phospholipid-rich aggregates in lung surfactant are further modified by interactions with apoproteins.

microscopy, it is a three-dimensional lattice made up of stacks of intersecting phospholipid bilayers with a repeat distance of about 50 nm (Figure 3-11). Particles are aligned at regular intervals along parallel phospholipid bilayers in tubular myelin, and when cut in cross section these particles are apparent as rod-like projections. The size and structure of these particles strongly suggest that they are oligomers of surfactant protein (SP)-A. The presence of surfactant protein SP-B is also required in order for tubular myelin to form [1053, 1184], but the location and distribution of this much smaller protein in the lattice are not known. Calcium in millimolar concentrations is also required for the formation of tubular myelin, probably reflecting the known calcium-dependent oligomerization of SP-A (Chapter 8). Surfactant protein SP-C is not necessary for tubular myelin formation, as demonstrated in direct lipid-protein reconstitution studies [1053, 1184]. Tubular myelin is just one of many phospholipid-rich aggregate forms in the heterogeneous microstructure of endogenous lung surfactant in water (e.g., [67, 421, 685, 791, 1053, 1184, 1200]). It has biophysical relevance because of its association with the generation of maximal adsorption of endogenous lung surfactant to the air-water interface [685, 791]. Tubular myelin and lung surfactant aggregate subtypes are discussed further in Chapters 6 and 8.

XII.   Phospholipid Adsorption

Unless phospholipids are spread directly at the air-water interface, they must adsorb from the subphase to form a surface film. Many phospholipids adsorb relatively poorly at the air-water interface in comparison to their ability to lower surface tension in the compressed surface film itself. The adsorption of DPPC is particularly poor, and if dispersed in water this saturated phospholipid is only able to lower surface tension by a few mN/m over times as long as hours [569, 783, 794]. A typical surface pressure-time (π-t) adsorption isotherm for DPPC dispersed in physiological saline is given in Figure 3-12. Shown for comparison is the rapid and extensive adsorption exhibited by either lavaged calf lung surfactant or its chloroform:methanol extract containing surfactant lipids and hydrophobic proteins. The poor adsorption of DPPC in Figure 3-12, along with its ineffective respreading in cycled surface films, makes this phospholipid ineffective biophysically when used alone as an exogenous surfactant to treat RDS in infants [129, 920]. The adsorption of unsaturated phospholipids with fluid chains is better than that of DPPC, but even complex mixtures of synthetic or lung-derived phospholipids do not approach the adsorption facility of whole or extracted endogenous lung surfactant [783, 794]. One of the most important biophysical actions of apoproteins in endogenous lung surfactant is to facilitate the adsorption of phospholipids into the air-water interface (Chapter 8). Exogenous lung surfactants also must exhibit rapid adsorption, and surfactant proteins or substitutes for

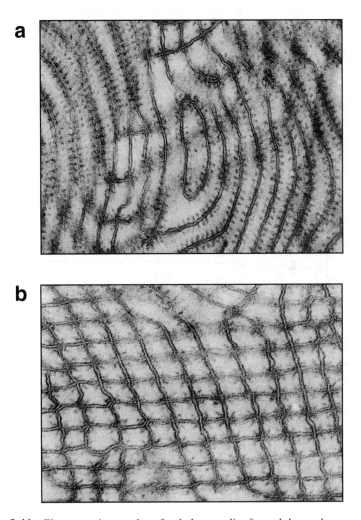

Figure 3-11 Electron micrographs of tubular myelin formed by endogenous lung surfactant. Tubular myelin is shown after fixation with tannic acid and sectioning in the plane of the bilayer (a) or in cross section (b). The spacing of phospholipid bilayers in tubular myelin is about 50 nm. The particles in association with the bilayers are thought to contain oligomeric SP-A. In (a), these particles are aligned in register along parallel bilayers. In (b), they are seen to be located at and to project from bilayer intersections. The location and distribution of SP-B in tubular myelin are uncertain. Magnification approximately × 120,000. (From Ref. 413 as reproduced in Ref. 1034, with permission.)

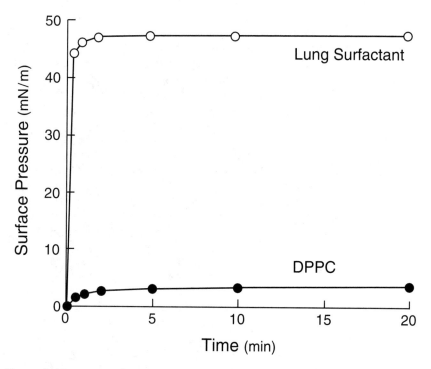

Figure 3-12 Adsorption of DPPC compared to whole or solvent-extracted lung surfac-tant. DPPC adsorbs much more slowly than endogenous lung surfactant or its chloro-form:methanol extract containing surfactant lipids and hydrophobic apoproteins. Fluid phospholipids adsorb better than DPPC but still not nearly as well as lung surfactant (not shown). Lung surfactant adsorption is greatly facilitated by interactions between phos-pholipids and surfactant proteins (Chapter 8). Curves are representative from literature data for DPPC and whole and extracted surfactant obtained by lavage from calf lungs.

them are utilized to improve phospholipid adsorption in many of these prepara-tions as well.

XIII. Chapter Summary

In addition to their many other biological roles, phospholipids are important functional constituents in endogenous and exogenous lung surfactants. All phos-pholipids are surface active due to their amphipathic structure containing a polar headgroup and nonpolar fatty chains. The headgroup determines the phospho-lipid class, while the fatty acyl chains define individual compounds within each class. Glycerophospholipid classes relevant for lung surfactant include phosphatidylcholine (PC), phosphatidylethanolamine (PE), phosphatidylglycerol (PG), phosphatidylinositol (PI), and phosphatidylserine (PS). Sphingomyelin (SPH), a phosphosphingolipid, has a PC headgroup but a ceramide hydrophobic group.

The properties of phospholipids in surface films are integral to their functional actions in lung surfactants. DPPC and related rigid, disaturated phospholipids display several properties important in lung surfactant activity. DPPC forms condensed surface films able to lower the surface tension of water to <1 mN/m under rapid dynamic compression at room or body temperature. DPPC films also vary surface tension significantly during cycling and have a large compression-expansion hysteresis. However, they respread poorly on successive cycles. Respreading in DPPC films can be significantly improved by more fluid unsaturated phospholipids, although these lack the surface tension lowering ability of disaturated phospholipids. The film behavior and surface properties of saturated and unsaturated phospholipids can be correlated in part with the phase behavior of their hydrated bilayers. In particular, DPPC and many other desaturated phospholipids are in the rigid gel phase at body temperature, while unsaturated phospholipids are in the fluid liquid crystal phase.

Another important aspect of phospholipid biophysics is the aggregation of these compounds in the aqueous phase. Phospholipids form a variety of aggregates in water, including bilayers, multilayers (lamellae), spheroidal vesicles (liposomes), cylindrical hexagonal phases, and less regular forms. An additional specialized microstructure, tubular myelin, is formed by phospholipids interacting in the presence of calcium with lung surfactant proteins SP-A and SP-B. Phospholipid aggregation is highly relevant for the aqueous phase microstructure and adsorption of lung surfactants. The adsorption of phospholipids requires molecular transport from subphase aggregates into the air-water interface. DPPC adsorbs very poorly to the air-water interface, lowering surface tension by only a few mN/m over times as long as hours. Unsaturated phospholipids adsorb more rapidly, but even complex phospholipid mixtures do not equal the excellent adsorption of whole or extracted endogenous lung surfactant. One important role of surfactant apoproteins is to facilitate phospholipid adsorption (Chapter 8), and this must also be accomplished by apoproteins or other additives in exogenous lung surfactants (Chapter 14).

4

Physicochemical Methods for Studying Lung Surfactants

I. OVERVIEW

This chapter summarizes physical and chemical methods used in the study of lung surfactants. Many of these methods are standard techniques from interfacial phenomena research modified as necessary to account for the special requirements of pulmonary surfactants. Standard methods noted in this chapter include organic chemical assays and purification methods, surface tension measurement methods, the Wilhelmy surface balance, and molecular biophysical methods (microscopic, spectroscopic, calorimetric) used to study surfactant films and multilayers. In addition, several techniques designed particularly for use with pulmonary surfactants, the pulsating bubble and captive bubble surfactometers, are described. These two instruments are highly useful in studying how adsorbed lung surfactant films lower surface tension during dynamic compression at physiologically relevant temperature, humidity, cycling rate, and area change. This information can then be used in conjunction with Wilhelmy balance and molecular biophysical studies, plus measurements of adsorption in the absence of diffusion and dynamic compression, to give a relatively complete picture of surface active behavior.

II. Overview of Analytical and Preparatory Methods Used with Lung Surfactants

The majority of lung surfactants are mixtures of chemical constituents, and an array of biochemical and organic chemical methods are used in analyzing and purifying them. Only a very brief overview of common analytical and preparatory methods in lung surfactant research is given here. For further details on methodologic descriptions, readers are referred to texts and reviews on biochemistry, analytical chemistry, and lung surfactant research (e.g., [175, 250, 436, 549, 918, 919, 942, 964, 1043]). Examples of preparatory and analytical methods applied to different surface active materials used in lung surfactant research are also discussed in Chapter 5.

A. Compositional Assays

Standard colorimetric assays for phosphate (e.g., [17, 122, 998]), protein (e.g., [545, 669]), and cholesterol (e.g., [231, 966, 1229]) have traditionally been used to measure the total phospholipid, protein, and neutral lipid contents of lung surfactants. Other organic chemical assays are applied as necessary to define compounds such as free fatty acids [506] or triglycerides [269]. Phospholipids are analyzed for headgroup composition by one- or two-dimensional thin layer chromatography [257, 542, 1085], and fatty chain distributions are determined by gas chromatography [665, 1082, 1214]. High-performance liquid chromatography (HPLC) can also be used to define the phospholipid composition of lung surfactants (e.g., [486, 541]). These standard assays are sufficiently accurate for most research applications involving lung surfactants, although validation and attention to specific artifacts are necessary. For example, interference from lipid must be minimized in spectrophotometric assays for protein by the use of sodium dodecyl sulfate (SDS) or other modifications. The accuracy of standard protein assays may also be lessened when applied to extremely hydrophobic materials such as surfactant proteins (SP)-B and SP-C or related synthetic peptides. Enzyme-linked immunosorbent assay (ELISA) tests for SP-A, SP-B, and SP-D are currently available (e.g., [504, 600, 997]).

B. Preparatory Methods

Research on lung surfactants utilizes specific preparatory methods in addition to analytical methods. Column (gel-permeation) chromatography using various solvents and types of packing can be used to separate subfractions of lipid and protein components from endogenous lung surfactant. For example, gel-permeation chromatography with chloroform:methanol on LH-20 packing is able to separate subfractions of mixed phospholipids and/or neutral lipids away from the hydrophobic surfactant proteins [383, 1144–1146], and a variety of other types of packing and solvent systems can also be used. Subsets of phospholipids such as zwitterionic phospholipids alone can similarly be isolated from lung surfactant by column methodology [383, 1144]. The purification and analysis of individual lung surfactant proteins SP-A, -B, -C, and -D are challenging technically and involve a variety of methods including solvent precipitation, column chromatography, HPLC, polyacrylamide gel electrophoresis (PAGE), amino acid analysis and sequencing, ELISA or other antibody-based analysis, and spectroscopic analyses of secondary and tertiary structure. The range of lipids, proteins, and other surface active materials used in lung surfactant research and their methods of preparation are described further in Chapter 5. The study of lung surfactant materials also involves a variety of physical methods for measuring surface tension, film behavior, and molecular biophysics as detailed in the remainder of this chapter.

III. Methods of Measuring Surface Tension

The surface tension of liquids varies in magnitude from a fraction of a mN/m for liquified gases to hundreds of mN/m for molten metals. For all liquids, surface tension is a function of temperature, pressure, and the presence of surface active solutes. Table 4-1 gives surface tension values for water as a function of temperature and also indicates surface tensions for several other liquids at room temperature. Water and mercury have comparatively high surface tensions that reflect contributions from hydrogen bonding and metal bonding, respectively, in addition to van der Waals and other intermolecular forces that contribute to the surface tensions of all liquids [991]. There are various experimental methods for

Table 4-1   Surface Tensions of Water and Other Liquids

Liquid	Surface tension (mN/m or dynes/cm)	Temperature (°C)
Water	73.49	15
	72.75	20
	71.97	25
	71.18	30
	70.38	35
	70.06	37
	69.56	40
Benzene	28.9	20
Acetic acid	27.6	20
Acetone	23.7	20
Carbon tetrachloride	26.8	20
Ethanol	22.3	20
n-Hexane	18.4	20
Mercury	485	20

Source: Data from Refs. 5, 766, 1158.

measuring surface tension based on its action as a force tending to minimize surface area.

Common methods for measuring surface tension include the capillary rise method, the drop-volume (drop-weight) method, the pendant drop and sessile drop methods, the ring method, the oscillating jet method, and the Wilhelmy method. Each of these methods is subject to experimental error, and significant effort has been devoted to the derivation of theoretical correction factors and experimental modifications to improve accuracy as detailed in texts on surface chemistry [5, 10, 86, 183, 295, 991]. The *capillary rise method* is based on the fact that the height to which a column of liquid rises in a capillary tube is directly proportional to its surface tension. The *drop-volume or drop-weight method* utilizes the mathematical relation between surface tension and the volume or weight of drops detached slowly from the tip of a capillary tube. The *pendant and sessile drop methods* allow surface tension to be determined based on solutions to equations relating the equilibrium shape of a liquid drop in air to surface tension. A similar principle is also applied in surface tension measurements using air bubbles in a liquid in instruments such as the pulsating bubble surfacto-meter and the captive bubble surfactometer as described later. The *ring method* (*du Nouy ring*) determines surface tension from the measured force pulling down on an immersed ring. The *oscillating jet method* is based on measure-ments of the shape of an oscillating jet of liquid in air, which can be related mathematically to surface tension, flow rate, and other parameters. The *Wilhelmy method*, which has been used extensively for surface tension measurements with lung surfactants, determines surface tension from the force pulling down on a rectangular plate (Wilhelmy slide) dipped into the liquid surface (Figure 4-1). When this method is incorporated in surface balance studies of surfactant films, the resulting apparatus is called a Wilhelmy surface balance (see following section).

The mathematical relationship describing the force on a slide dipped into a liquid surface as in Figure 4-1 was first derived by Wilhelmy in 1863 [1180]. This force is a sum of contributions from buoyancy, gravity, and surface tension. Buoyancy tends to push the slide up, while gravity and surface tension act to pull it in the downward direction. The contribution of surface tension to the downward force is dependent on the contact angle (θ), with the maximum downward force occurring when this angle is zero. A contact angle of zero means that the liquid completely "wets" the slide, a condition that is obtained experimentally by uniform roughening of the slide surface by sandblasting or other abrasives. The magnitude of the surface tension forces pulling on the slide is also dependent on its wetted perimeter and depth of immersion. However, the necessity to measure these dimensions can be bypassed by making measurements relative to the initial force on the slide when it is dipped into the pure subphase with no surfactant present.

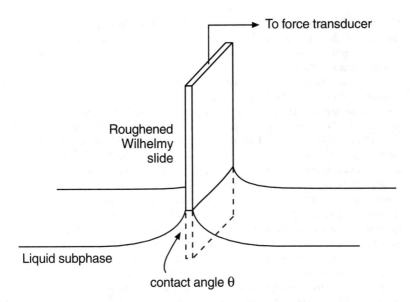

To force transducer

Roughened
Wilhelmy
slide

Liquid subphase

contact angle θ

Figure 4-1   The Wilhelmy slide method of surface tension measurement. The force pulling down on a thin slide dipped into a liquid can be used to measure surface tension. This method, defined by Wilhelmy in 1863, is commonly used in applications with lung surfactants. See text for details.

Mathematically, the net downward force F on the Wilhelmy slide in Figure 4-1 is given by a sum of contributions from surface tension, gravity, and buoyancy as follows [86, 183, 295, 991]:

$$F = 2\sigma(W + T)\cos\theta + (WTL)\rho_s g - (WTh)\rho_w g \qquad (4\text{-}1)$$

where W, T, and L are the width, thickness, and length of the slide, respectively; h is the height to which the slide is wetted by the subphase; σ is surface tension; θ is the contact angle; ρ_s is the density of the slide material; g is the acceleration of gravity; and ρ_w is the density of the subphase (water). In many applications, a modification of Equation (4-1) is used expressing the difference in force (ΔF) when the slide is dipped into the subphase without and with surfactant present. If the dimensions W, T, L, and h remain constant, this difference in force is given by:

$$\Delta F = 2(\sigma^\circ - \sigma)(W + T)\cos\theta \qquad (4\text{-}2)$$

where σ° is the surface tension of the pure subphase and σ is the surface tension of the subphase with surfactant present. Equation (4-2) can alternatively be rewritten in terms of surface pressure $\pi = \sigma^\circ - \sigma$ (Chapter 2) as:

$$\pi = \Delta F/2(W + T)\cos\theta \qquad (4\text{-}3)$$

Equations (4-1) to (4-3) are often simplified by neglecting the thickness T of the

Wilhelmy slide compared to its width W so that $W + T \doteq W$ (typical slide dimensions are ~1–2 cm in width, 2 cm in length, and thousandths of a millimeter in thickness). Further simplification occurs when complete wetting is present so that the contact angle θ is zero and $\cos \theta = 1$.

Artifacts from a nonzero or changing contact angle may be present in measurements of surface tension with the Wilhelmy method if slides are improperly cleaned or not adequately roughened. Achieving complete wetting of a Wilhelmy slide so that the contact angle is uniformly zero is particularly important in surface balance studies on cycled films described in the following section, because it minimizes the potential for changes in contact angle during the course of experimentation. Changes in contact angle alter the force on the Wilhelmy slide, leading to artifacts in data interpretation if this is assumed to be due to changes in surface tension. Artifacts from a nonzero and varying contact angle in studies with cycled surfactant films are typically most apparent in regions of the surface pressure–area isotherm where surface tension is high and where the liquid surface is advancing rather than receding over the slide. In Wilhelmy balance studies, surface tension is generally highest near the beginning of film compression and the end of film expansion. Also, the liquid surface making contact with the Wilhelmy slide is receding down the slide during compression as surface tension falls but advancing over the slide during expansion as surface tension rises. Sandblasting of platinum slides is one method that has been found to result in excellent wetting and minimization of contact angle artifacts in Wilhelmy measurements of surface tension on phospholipid and lung surfactant films [788, 790, 1058].

IV. Methods of Measuring the Surface Active Properties of Surfactants

A variety of different techniques are used to measure the surface properties of surfactants and surfactant films. Four experimental methods that are widely utilized with lung surfactants are the Wilhelmy surface balance, adsorption measurements with minimized diffusion, the pulsating bubble apparatus, and the captive bubble apparatus. The major features of these four techniques are described in subsequent sections, followed by a summary of molecular biophysical methods used to assess the interactions and aggregation of surfactant molecules in films and bilayers.

V. Wilhelmy Surface Balance

Instruments that measure surface tension–area (σ-A) or surface pressure–area (π-A) behavior in compressed surfactant films are termed *surface balances* [5, 10, 86, 183, 295, 991]. All such instruments share several design features. A rectangular or otherwise shaped *trough*, typically made of a hydrophobic material like Teflon, holds the liquid subphase. A surfactant film is formed at the air-liquid interface in this trough by spreading from an organic solvent or

---

Common Methods for Measuring the
Surface Active Properties of Lung Surfactants

Wilhelmy surface balance
Measures the surface tension lowering, respreading, and hysteresis properties of surfactant films compressed and expanded at the air-water interface. Uses a Wilhelmy slide to measure surface tension. Can be utilized to isolate effects occurring in the film and interfacial region.

Adsorption with minimized diffusion
Measures the adsorption of surfactants into the air-water interface from a liquid subphase that is stirred by a magnetic bar to minimize the mass transfer resistance from diffusion.

Pulsating bubble surfactometer
Measures the overall dynamic surface tension lowering of a surfactant dispersion at physiological cycling rate, area compression, temperature and humidity. A hypophase exchange modification also allows film behavior to be isolated. Can also be used in the static mode to measure adsorption.

Captive bubble apparatus
Measures the dynamic surface tension lowering ability of a surfactant dispersion or spread film at physiological cycling rate, area compression, temperature, and humidity. Can also be used in the static mode to measure adsorption with diffusion minimized by stirring.

---

by adsorption from the subphase. The film is then compressed and expanded by one or more barriers that confine the interface and generate a known change in surface area. The surface area during cycling is commonly expressed as either a simple percentage of the fully expanded trough area or as an inverse surface concentration in Å^2/molecule based on the amount of surfactant initially spread. While the surface is cycled, measurements of surface tension are made by a force-sensing device to define a σ-A or π-A isotherm at the desired temperature (Chapter 2) . Isotherm characteristics for a surfactant film are a function of its concentration, composition, cycling rate, temperature, humidity, and extent of compression (ratio of maximum area to minimum area during cycling).

The *Wilhelmy surface balance* is one of four biophysical methods empha-sized in Chapter 4 because of their widespread use in measuring the surface active properties of endogenous and exogenous lung surfactants. Several factors in lung surfactant film behavior favor the use of the Wilhelmy balance rather than the classical *Langmuir trough* introduced in the early part of the 20th century [624]. In the Langmuir surface balance, the liquid subphase extends over the trough walls, and a barrier lying across the top of the trough compresses the surfactant film while a second floating barrier attached to a torsion balance measures surface pressure. These design features do not lend themselves to the study of lung surfactant films that reach extremely low surface tensions during compression. At very low surface tension, the subphase in a Langmuir trough wets and overflows the walls, and film leakage under the barrier also occurs as the water level drops. The typical Wilhelmy balance used in lung surfactant studies has a liquid subphase that does not reach the top of the trough walls, and a recessed barrier system confines and compresses the air-water interface through the action of an external motor. A recessed dam-type barrier extending across the trough interior and sealing by pressure against the walls is common, but film leakage is most effectively reduced by a continuous ribbon barrier such as that designed by Tabak and Notter [1058] (Figure 4-2). A Wilhelmy slide hanging from a force transducer or related sensor is dipped into the subphase at the front of the trough to measure surface tension during cycling. Subphase volumes in Wilhelmy balances range from 50 to 1000 ml, much larger than those in pulsating or captive bubble experiments described later. Cycling rates are gener-ally in the range of 1–10 minutes per cycle, well below the range of those in normal respiration.

A major utility of Wilhelmy balance measurements for lung surfactants is that they can give information on behavior within the interfacial film itself separate from the influence of adsorption. To this end, films are typically spread directly at the air-water surface in a volatile spreading solvent (e.g., 9:1 or 95:5 v/v hexane:ethanol) that is immiscible with the subphase. A pause of 5–10 minutes is allowed for solvent evaporation prior to cycling of the film. The π-A isotherm for spread interfacial films indicates not only how they lower surface tension as a function of area during compression and expansion but also how film molecules respread on successive cycles of the surface [789, 796, 1057, 1093, 1145] (see Chapter 2). The relaxation of dynamic surface pressure with time can also be measured if cycling is halted at fixed area [321, 795, 1058, 1059]. Films formed by adsorption from surfactants dispersed in the subphase can also be studied in the Wilhelmy balance. Adsorbed films have physiological relevance for lung surfactants, but their initial composition and surface concen-tration are less defined than in the case of spread films. The presence of adsorbing, dispersed surfactant in the subphase also complicates analysis of interfacial respreading processes. Films formed in a Wilhelmy balance by either spreading or adsorption can be examined by molecular biophysical techniques for large and small scale structure and aggregation as described later in this chapter. Films can

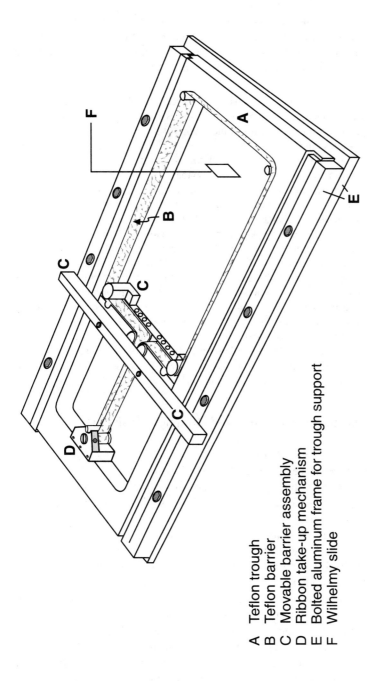

A Teflon trough
B Teflon barrier
C Movable barrier assembly
D Ribbon take-up mechanism
E Bolted aluminum frame for trough support
F Wilhelmy slide

Figure 4-2 Schematic of a Wilhelmy surface balance. The Wilhelmy balance shown here has a recessed Teflon ribbon barrier to compress and expand an interfacial surfactant film, minimizing the potential for film leakage during cycling. A rectangular Wilhelmy slide is dipped into the interface in front of the barrier to measure surface tension during cycling. An alternative barrier design (not shown) uses a dam-type barrier that presses against the trough walls. (Adapted from Ref. 1058.)

be studied *in situ* in the balance trough or deposited as Langmuir-Blodgett multilayers on a glass slide or other support dipped repeatedly through the surface prior to examination by electron microscopy, spectroscopy, X-ray diffraction, and other physical techniques.

The Wilhelmy surface balance is well adapted to study either single or multi-component films, including complex mixtures such as the majority of lung surfactants. However, data interpretation in terms of the actions of specific molecular components within the film becomes more difficult as the number of components increases. Similarly, data interpretation is most direct for Wilhelmy balance results for films formed by spreading at the air-water interface so that events occurring in the interfacial region are highlighted. Although residual solvent can affect measurements of this kind, the general validity of using spreading solvents for surfactant film formation is well documented in the surface chemistry literature [10, 295]. Wilhelmy balance studies on adsorbed films add additional considerations to data analysis, although they are clearly relevant for alveolar surfactant activity. A third, hybrid method of film formation sometimes used with pulmonary surfactants is to add surfactant dispersed in saline dropwise at the interface of the saline subphase. However, this technique yields ill-defined films representing an uncertain combination of adsorption and spreading that further limits data analysis and interpretations. Surface balance measurements on surfactant films, including not only π-A isotherms but also other film properties such as surface potential and surface viscosity, are discussed in detail in interfacial phenomena texts such as Gaines [295] and Adamson and Gast [10]. An additional array of molecular biophysical tools are also available that can be applied to study details of surfactant film behavior at interfaces as summarized at the end of this chapter.

VI. Adsorption with Minimized Diffusion Resistance

Adsorption is the process whereby surfactant molecules move from a liquid subphase into an interface to form a surface film (Chapter 2). Adsorption is defined in the absence of dynamic compression and is limited by the equilibrium spreading pressure of the surfactant substance in question. Experimental measurements of adsorption must rule out or account for the diffusive resistance to molecular transport in the liquid subphase. Diffusion is a comparatively slow process, particularly if aggregates of phospholipids or large molecules like proteins are involved. The mass transfer resistance from diffusion can be difficult to separate from adsorption in a stagnant subphase, but it can be minimized quite simply by stirring the subphase. Diffusion is still present in a thin boundary layer near the interface that can never be completely removed by stirring, but the influence of this layer is small and primarily affects data at short times. For most applications involving lung surfactants, measurements of surface tension as a function of time after addition of surfactants to a stirred subphase provide a sufficiently accurate assessment of adsorption behavior.

The most widely used apparatus for studying lung surfactant adsorption is based on the work of King and Clements [571, 572]. These investigators studied the adsorption of lung surfactant preparations dispersed in a aqueous subphase in a Teflon dish and agitated with a Teflon-coated magnetic stirring bar (Figure 4-3). A Wilhelmy slide connected to a force transducer and strip chart recorder was used to measure surface tension as a function of time. To approximate a clean interface with no adsorbed surfactant at the start of measurements, the surface was mechanically aspirated at time zero to remove the film that had already been formed by the dispersed surfactants. A less cumbersome alternative to aspiration was utilized by subsequent workers who added dispersed surfactants into an initially surfactant-free stirred subphase to initiate adsorption [783, 794]. Surfactant suspensions were injected beneath the interface or directly poured into the subphase from a container with a small surface area to minimize the possibility of transferring material from a preformed film. Surface pressure–time (π-t) or surface tension–time (σ-t) adsorption isotherms measured for lung surfactants with this kind of methodology have proved to be highly useful in characterizing

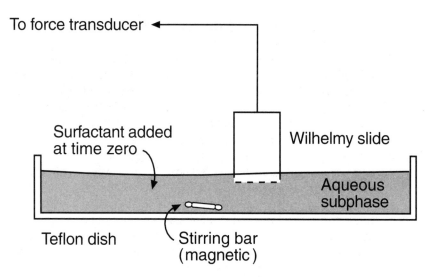

Figure 4-3 Apparatus for measuring adsorption. Adsorption of lung surfactants is often measured in a Teflon dish containing a subphase stirred by a magnetic bar to minimize the effects of diffusion. A surfactant suspended in saline or buffer is injected beneath the surface or poured into the subphase at time zero. Surface pressure or tension is measured as a function of time from the force on a hanging Wilhelmy slide. See text for details.

this important surface behavior. The captive and pulsating bubble surfactometers described below can also be used to make adsorption measurements.

Magnitudes of Surface Pressures in Adsorption versus Film Studies. Maximum surface pressures measured during adsorption can be significantly lower (minimum surface tension can be significantly higher) than found in dynamic film studies with the same surfactant in a Wilhelmy balance or bubble apparatus. No matter how rapidly and extensively a surfactant material adsorbs, it cannot generate surface pressures above its equilibrium spreading pressure in the absence of dynamic compression. Endogenous lung surfactant, for example, adsorbs to equilibrium surface tension values of about 22–23 mN/m at 37°C, equivalent to the equilibrium spreading pressure of fluid phospholipids. When adsorbed films of endogenous lung surfactant are compressed at physiological cycling rates in pulsating or captive bubble experiments, however, they generate much lower surface tension values of <1 mN/m (Chapters 7, 8). Other substances such as DPPC and related phospholipids generate low adsorption surface pressures because they have a high effective energy barrier for leaving subphase aggregates and entering the interface. However, if spread directly at the air-water interface, DPPC films have the ability to lower surface tension to <1 mN/m when compressed even at relatively slow rates in a Wilhelmy balance (Chapter 3).

VII.   The Pulsating Bubble Surfactometer

The pulsating bubble surfactometer developed by Enhorning and Adams [6] and Enhorning [236] is extensively used in lung surfactant research to assess the overall surface tension lowering ability of surfactants dispersed in the aqueous phase.[1] Measurements of surface tension lowering with the pulsating bubble apparatus reflect a sum of contributions from adsorption and dynamic film behavior at physiologically relevant cycling rate (20 cycles/min), area change (50%), temperature (37°C), and humidity (>95%). In the pulsating bubble surfactometer, a small air bubble is formed in an aqueous dispersion of surfactant held in a plastic sample chamber (Figure 4-4). The bubble, which communicates with ambient air, is then oscillated at a known rate between minimum and maximum radii of 0.4 and 0.55 mm, respectively, by means of a precision pulsator that moves liquid in and out of the sample chamber. The bubble can be monitored through a microscope to ensure that minimum and maximum sizes remain constant throughout an experiment. Pressure in the liquid during pulsation is measured by a transducer, and surface tension σ is calculated from the measured

---

1.   Dr. Enhorning has also developed a capillary surfactometer that determines surface activity based on the ability of surfactant materials to maintain airflow through a narrow tube with a liquid lining [241, 660].

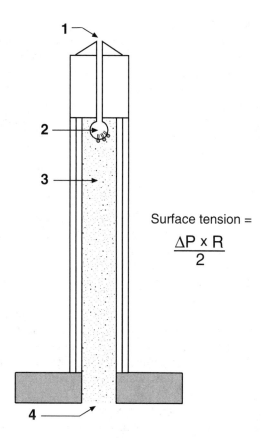

Surface tension =

$$\frac{\Delta P \times R}{2}$$

The sample chamber of the pulsating bubble surfactometer contains:

1 Capillary, open to air
2 Air bubble at tip of capillary tube
3 Surfactant suspension
4 Opening to pulsator unit and pressure transducer

Figure 4-4 Schematic of pulsating bubble surfactometer. The pulsating bubble surfacto-meter designed by Enhorning [236] measures overall surface tension lowering reflecting both adsorption and dynamic film behavior at physical conditions relevant for the lung. A tiny air bubble is pulsated at 20 cycles/min between radii of 0.55 and 0.4 mm (area reduction 50%) in a subphase containing dispersed surfactants. Pressure in the liquid is measured by a transducer and surface tension is calculated from the Laplace equation. Adsorption can also be measured in the absence of pulsation. (Adapted from Ref. 862.)

pressure drop (ΔP) and bubble radius (R) by the Laplace equation for a sphere: $\Delta P = 2\sigma/R$ [236] (the Laplace equation is discussed more fully in Chapter 6). A commercial version of the pulsating bubble surfactometer is made by General Transco (formerly Electronetics Corporation), Lancaster, NY.

The pulsating bubble apparatus has several advantages for studying endogenous and exogenous lung surfactants. It is relatively simple to use and requires only small amounts of surface active material. The interfacial geometry, cycling rate, area change, temperature, and humidity in pulsating bubble experiments are also analogous to those existing for alveolar surfactant during breathing. Surface activity measurements with this instrument also reflect contributions from both adsorption and dynamic film compression as is the case for surfactant in the lungs *in vivo*. Contact angle artifacts that can affect Wilhelmy slide measurements of surface tension are not present. The pulsating bubble apparatus is also well suited to study how the surface tension lowering ability of lung surfactants is compromised by plasma proteins, membrane lipids, fatty acids, and other inhibitory substances (Chapter 9). The sample chamber volume of 20 microliters allows surface activity studies on small samples of lavage or tracheal aspirates from fetal animals or premature infants. In addition to dynamic activity measurements, the bubble apparatus can be used in the static mode to measure adsorption. In this case, surface tension is determined as a function of time immediately after formation of an air bubble that is held at minimum (or maximum) radius in a surfactant dispersion. Adsorption measurements of this kind require less material than dish experiments with a stirred subphase, and dynamic activity can be studied at any point by initiating pulsation. However, because the subphase is not stirred, a diffusion resistance is present.

The accuracy and limitations of data analysis for the pulsating bubble surfactometer have been analyzed by several groups [378, 875]. One factor influencing measurements with this instrument is that the surfactant film in the sample chamber can migrate up the capillary tube during pulsation [875]. This can increase the film-covered surface area so that the area change induced by the pulsator is less than 50%, leading to elevated minimum surface tensions. If migration occurs, however, it is limited to early bubble pulsations and does not affect data at times longer than a few minutes [875]. Another factor relevant for pulsating bubble data analysis is the accuracy of the simple Laplace equation when the bubble deforms from a sphere to an oblate ellipsoid as surface tension decreases to low values [378]. Theoretical analysis shows that small air bubbles of the size in this apparatus have significant deformations only at very low surface tensions, so that the absolute error associated with the use of the Laplace equation is quite small (<0.5 mN/m) [378]. Interfacial theory also indicates that surface viscosity can potentially influence surface tension measurements during pulsation, but data at minimum and maximum bubble size are not affected because the time rate of change of radius is zero at these points [378]. Minimum and maximum surface tensions at these radii are the data most commonly reported in the literature for the pulsating bubble surfactometer.

Hypophase Exchange System for the Pulsating Bubble Apparatus. The pulsating bubble surfactometer can be modified with a hypophase exchange system that allows interfacial films to be isolated and examined during pulsation in the absence of adsorption [244, 449, 461]. This modification also permits interfacial films to be formed by adsorption in one subphase and then studied for interactions with other substances such as inhibitors introduced in a second subphase (see Chapter 9). The hypophase exchange system for the pulsating bubble apparatus utilizes two 22 gauge stainless steel tubes that form channels into the lumen of a specially designed sample chamber [244, 449]. One of these (inlet channel) opens at the top of the chamber near the site of bubble pulsation, and the other allows drainage from the lower part of the chamber. Hypophase exchange takes place at constant bubble size by equalizing inflow and outflow rates. To accomplish this, the two channels are connected externally to a Silastic tube divided by a migrating barrier (a spring-loaded motor-driven compressing wheel) that enlarges one section at the expense of the other while keeping total volume constant. At the start of the exchange process, regulatory valves are closed so that the inlet tube communicates with a syringe containing the new hypophase. The compressing wheel is then disengaged, moved to a position of maximal volume for the inlet tube, and reengaged after an external water level adjustment to equalize tube pressures on the two sides of the barrier. Valves are then opened so that the new hypophase passes through the bubble chamber in a volume 10 times that of the original hypophase while the bubble and surfactant film are maintained at equilibrium [244, 449].

VIII. Captive Bubble Apparatus

The captive bubble technique is conceptually related to the pulsating bubble method in that both utilize calculations based on relationships between interfacial tension and the shape of an air bubble in a liquid. In the case of the captive bubble apparatus developed by Schurch and co-workers (e.g., [953–955]), diameter and height measurements in an air bubble "captured" against a fixed agar gel support are used to calculate surface tension and surface area. Bubble compression and expansion rates up to 100 cycles/min against a fixed piston can be generated with the sample chamber mounted on a movable microscope stage [953, 954] (Figure 4-5). A modified version of the captive bubble apparatus generates cycling by pressure changes from an exterior reservoir [874, 875, 877]. In either case, the design of the system removes the possibility of film migration during cycling. Adsorption can also be examined in this apparatus by measuring bubble dimensions as a function of time in the absence of pulsation while the subphase is stirred to minimize diffusion. In a typical captive bubble experiment, an air bubble of 1–7 mm diameter is formed in a surfactant suspension in the sample chamber. Bubble shape is then monitored without external pulsation to define adsorption,

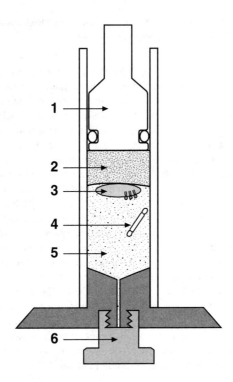

The captive bubble chamber, mounted on a movable stage of a microscope, contains:

1   Fixed piston
2   1% agarose gel to support bubble
3   Air bubble with interfacial surfactant film
4   Magnetic stirring bar
5   Surfactant suspension
6   Filling port plug at base of sample chamber

Figure 4-5   Schematic of captive bubble surfactometer. In the apparatus designed by Schurch and co-workers [953–955], an air bubble in a surfactant suspension is "captured" against an agar gel layer. Dynamic surface activity is measured by pulsating the bubble against a fixed piston or by means of external pressure variation (not shown). Adsorption can also be determined in the absence of pulsation with diffusion minimized by a stirring bar. Surface tension and bubble area and volume are computer calculated from bubble dimensions measured from recorded video images and used in theoretical solutions relating surface tension and bubble shape. See text for details. (Adapted from Ref. 953.)

and the adsorbed surface film is subsequently cycled to determine dynamic surface activity. Alternatively, the surfactant film can be spread directly at the air-subphase interface for dynamic activity studies [877]. Data acquisition and analysis for the captive bubble apparatus are more complex than with the Wilhelmy balance or pulsating bubble apparatus. The captive bubble is continuously monitored by one or more video cameras during cycling or adsorption experiments. Video images are recorded and stored on tape, and bubble diameter and height are determined from these recorded images on a single frame basis typically by means of computer image analysis. Dimensional data are then used to calculate surface tension, bubble area, and volume during adsorption or cycling based on the methods of Malcolm and Elliott [686] and Schoel et al [951]. The captive bubble apparatus can provide important information on the dynamic π-A or σ-A isotherm behavior of cycled surfactant films under conditions where film migration cannot occur. Also, because of its capability for subphase stirring, adsorption and film studies can be done without the complicating influence of diffusion.

IX. Molecular Biophysical Methods Used in Lung Surfactant Research

A variety of sophisticated physicochemical methods are used to study molecular behavior in surfactant films and multilayers. Several of the more important molecular biophysical techniques for lung surfactants have been developed in research on membrane biophysics, which also deals with the properties of phospholipid and protein molecules. Only a summary of molecular biophysical methodologies is given here, with more detailed coverage in texts, handbooks, and reviews on phospholipid biophysics, physical chemistry, and spectroscopy (e.g., [59, 300, 301, 339, 401, 479, 584, 708, 1007, 1033]).

A. Differential Scanning Calorimetry (DSC), Electron Spin Resonance (ESR) Spectroscopy, and X-Ray Diffraction

ESR and DSC are widely used to investigate the thermal and phase behavior of phospholipid bilayers and multilayers, while X-ray diffraction is utilized to determine bilayer dimensions and physical parameters. For lung surfactant applications, DSC and ESR are most useful in defining gel to liquid crystal transition temperatures to correlate with the surface tension lowering ability of phospholipid films (see Chapter 3), and these methodologies can also be used to assess the miscibility of phospholipids in films ([108, 301, 785, 985, 1007] for review). DSC utilizes measurements of the difference in heat input to a material sample in solvent relative to solvent alone to identify phase transitions as endothermic peaks on heating. ESR utilizes probes of defined electronic spin ("spin label probes") dispersed in a material to give information on local molecular environment and phase behavior. X-ray diffraction is based on analysis of scattering and diffraction patterns when a sample of interest is exposed to a powerful beam of X-rays. This method yields quantitative information on bilayer thickness and spacing in deposited surface films and stacked multilayers of phospholipids [301, 1007]. Bilayer dimensions determined by X-ray diffraction give information on whether phospholipids are in the gel or the liquid crystal phase and also define the degree of bilayer interdigitation if any is present (e.g., [566, 1006]).

B. Imaging Methods for Surfactant Films

Another group of biophysical methods important for lung surfactants are those involved with direct imaging of surface films. Over the past several decades, there has been a striking increase in the use of sophisticated imaging methodologies to study molecular behavior and aggregation in surfactant films, multilayers, and dispersions. Imaging techniques of this kind that have been applied to phospholipid and lung surfactant films or deposited multilayers include *Brewster angle microscopy* (BAM), *scanning tunneling microscopy* (STM), *atomic force microscopy* (AFM), *transmission and scanning electron microscopy, fluorescence microscopy or epifluorescence, and near-field scanning optical microscopy.*[2] BAM and fluorescence microscopy are used to define micron-scale structure and aggregation in surfactant films at the air-water interface. Near-field scanning optical microscopy with fluorescent probes can give resolutions as fine as 30–50 nanometers, while AFM and STM can elucidate structures on the molecular level. Small-scale structure can also be examined by transmission and scanning electron microscopy, often in conjunction with specialized techniques such as freeze-etching or freeze-fracture to help fix film and bulk phase aggregates for analysis. Current electron microscopes have practical limiting resolutions approaching 1 Å in crystalline systems, although the technical constraints of sample preparation lead to a higher practical limit of ~20Å for biological materials. The ongoing application of new, sophisticated imaging techniques to films of natural and synthetic lung surfactants is providing a more complete and quantitative understanding of phenomena like squeeze-out, collapse, and respreading.

BAM is based on the fact that for a pure Fresnel interface (approximated by pure water), the reflectivity of in-plane polarized light becomes zero at the Brewster angle (~53–54°) [431, 464, 1044]. The interface thus appears dark when illuminated by

---

2. For example, see Refs. 21, 174, 206, 431, 487, 602, 653-656, 758–761, 902–905, 931, 978, 1044, 1123 plus methodologic descriptions in Refs. 15, 300, 301, 401, 479, 584, 708, 1033.

plane-polarized laser light, except for a very low reflectivity due to surface roughness such as from capillary waves. The presence of a surfactant film at the interface increases the reflectivity, so that micron-scale regions differing in molecular density and orientation show up optically. Similar micon-level analysis of structure in surfactant films is feasible with fluorescence microscopy (epifluorescence), and submicron spatial resolutions can be achieved with near-field scanning optical microscopy in conjunction with fluorescent probes (e.g., [584, 708]). A fluorescent substance absorbs light at one wavelength and emits light at another longer wavelength. A variety of fluorescent probes are available for imaging studies, including those able to segregate in different regions of phospholipid layers or partition between the gel and liquid crystal phases. In the epifluorescence technique, probes of this kind are incorporated in spread or adsorbed surface films and visualized by fluorescence microscopy to elucidate micron-scale structure during cycling. Even finer resolution, in principle approaching molecular dimensions, can be achieved by electron micoscopy and by AFM and STM (e.g., [21, 300, 401, 479, 584, 1033]). AFM depends on sensing the force between the surface of interest and a sharp inert tip held in close proximity to it by a cantilever [21, 401]. The force is proportional to the deflection of the tip, and this can be measured, for example, from laser light reflected off the tip. The magnitude of the deflection can be of atomic dimensions, resulting in a high resolution image of the surface. STM also involves the use of a cantilever-held tip, but in this case a voltage is applied between the sample and the tip to generate a tunneling current that varies exponentially with distance [401]. A feedback control system determines the voltage necessary to keep this current constant, which is proportional to the deflection of the tip and can be converted into an image of the surface over which it travels.

C. Spectroscopic Methods for Studying Surfactant Films and Suspensions

A final group of molecular biophysical methods used in lung surfactant research are spectroscopic in nature. *Fourier transform infrared spectroscopy* (FTIR spectroscopy) and *nuclear magnetic resonance spectroscopy* (NMR spectroscopy) are perhaps the two techniques most widely used with lung surfactants, although a variety of other spectroscopic methodologies are also available (e.g., [59, 338, 479]). Spectroscopic methods are extremely powerful when used to analyze molecular properties and interactions in simple mixtures of small organic and inorganic molecules, but they are more difficult to apply to lung surfactants. The multicomponent composition typical of lung surfactants, and the relative structural complexity of glycerophospholipids and surfactant proteins, limits the ability of spectroscopic methods to discriminate molecular behavior and display component-specific effects. However, this can be addressed to some extent by comparing and contrasting spectroscopic data for separated subfractions of components and through studies with simplified model mixtures. Spectroscopic methods or related molecular-level analytical techniques have been applied with some success to elucidate molecular interactions in endogenous and exogenous lung surfactants.[3] A major emphasis of such studies has focused on molecular structural and orientational changes relevant for the surface active interactions of lung surfactant apoproteins or related peptides with phospholipids (e.g., Chapter 8).

The two general areas of experimental spectroscopy are *absorption and emission spectroscopy*. In absorption spectroscopy, the sample being analyzed absorbs energy from an external source after passage through a filter mechanism to restrict wavelength to the desired range (infrared, ultraviolet, etc.). In emission spectroscopy, the sample itself provides the radiant energy, so that an external source is not needed. Instruments used for either absorption or emission spectroscopy have additional common components including a fixed chamber or holder for the sample, a detector system that transduces radiant energy to a usable (usually electronic) form, and a signal processor/recorder system [59, 339]. FTIR spectroscopy utilizes light with wavelengths in the infrared range to excite vibrations within atoms, resulting in a signal containing information on specific atomic bonds or groups as a function of frequency. Modern IR spectrometers use Fourier transforms to process the signal and greatly enhance resolution. FTIR spectroscopy can be applied to study molecular behavior not only in the bulk phase but also directly in interfacial films (e.g., [208–211]). NMR spectroscopy is related in concept to ESR but involves resonance in the nuclear region rather than in electrons. In NMR spectroscopy, molecules with a nuclear magnetic moment such as ^1H, ^{13}C, or ^{31}P are aligned in a strong magnetic field, and are then excited by pulsed electromagnetic energy. As molecules relax from the excited state, they emit energy that can be expressed as a spectrum and analyzed to obtain information ranging from bond lengths to transition state characteristics to molecular motion in the local environment. New spectroscopic methods, as well as numerous modifications of current techniques, are continually being developed and applied in material science research. Examination of lung surfactant behavior using these methods, and the other molecular biophysical techniques noted above, is an active area of current research interest.

X. Chapter Summary

Chapter 4 has summarized methods for measuring the composition, surface tension, and surface active properties of lung surfactants. Four techniques given particular emphasis because of their extensive use in lung surfactant research were the Wilhelmy surface balance, adsorption measurements with diffusion

---

3. For example, see Refs. 36, 37, 106, 174, 210, 211, 298, 334, 528, 532, 535–537, 749, 775, 827, 828, 838, 839, 884, 891, 1002, 1111, 1112, 1122, 1138, 1154 plus methodologic descriptions in Refs. 59, 300, 339, 479, 584, 708.

minimized, and the pulsating and captive bubble surfactometers. Because of the multicomponent composition and complex phenomenology of lung surfactants, the most complete assessments of surface active behavior involve combining results found with several of these biophysical methods. In addition, a variety of sophisticated imaging, spectroscopic, and other molecular biophysical techniques are used to study interfacial films of lung surfactants and the interactions of different components within them.

Surface tension can be measured by several classical methods. The Wilhelmy method, widely used in lung surfactant research, determines surface tension from the force pulling down on a rectangular slide dipped in the liquid surface. The Wilhelmy surface balance uses this principle to measure surface pressure–area (π-A) or surface tension–area (σ-A) isotherms for surfactant films compressed and expanded at the air-water interface. These isotherms provide detailed information on film properties including dynamic surface tension lowering, respreading, compressibility, and hysteresis. Wilhelmy balance film studies are often supplemented with measurements of surfactant adsorption in the absence of dynamic compression. Adsorption can be measured in a dish with a stirred subphase to minimize diffusion, with results expressed as surface pressure or tension as a function of time at fixed temperature. The pulsating bubble surfactometer and captive bubble surfactometers can also be used to measure adsorption, but these instruments are even more useful in providing an overall assessment of surface activity that reflects both adsorption and dynamic film behavior at cycling rate (20 or more cycles/min), area change (50%), temperature, and humidity relevant for the lungs. Cycling rates in the Wilhelmy balance are typically significantly slower than normal respiration. Complementary studies using surface balance, bubble, and adsorption measurements cover the range of surface active properties important in lung surfactant function.

5

Lung Surfactant Materials, Research Complexities, and Interdisciplinary Correlations

I. Overview

Basic research on lung surfactants requires multidisciplinary correlations involving composition, molecular biophysics, surface activity, and physiological activity. Most endogenous and exogenous lung surfactants are complex mixtures whose composition and component molecular biophysics determine overall surface activity. The surface active properties and physiological actions of lung surfactants are similarly linked together. Activity assessments in lung surfactant research require the study of a spectrum of biologically derived and synthetic materials as summarized in this chapter. Many of these surface active materials are also relevant as components in exogenous lung surfactants. In addition to detailing materials used in lung surfactant research, this chapter briefly notes animal models of surfactant deficiency and dysfunction utilized in physiological activity studies.

II. Complexities in Lung Surfactant Research and the Need for Correlated Measurements

Lung surfactant research is subject to multiple factors that limit interpretations based on any single methodology or analysis. Correlated data from complementary methods and analyses are thus necessary to optimize the consistency and validity of interpretations. Complicating factors in lung surfactant research can be grouped arbitrarily as *compositional, biophysical, and anatomical and physiological complexities* (Table 5-1). *Compositional* assessments for endogenous lung surfactant are hampered by the difficulty of sampling the alveolar interface or type II cell without contamination from nonsurfactant lipids and proteins

Table 5-1 Examples of Complexities Affecting Research on Lung Surfactants

---

Compositional complexities
   Complex composition of most endogenous and exogenous lung surfactants
   Difficulty of directly sampling the alveolar interface *in situ*
   Widespread biological distribution of lipids and proteins as chemical classes
   Necessity to determine lung surfactant composition in a functional context
Biophysical complexities
   Importance of dynamic as well as equilibrium behavior
   Multiple surface properties are relevant for biophysical function
   Functional biophysics involves interacting components
   Complex thermal and hydrated phase behavior of phospholipids
   Complex molecular structure and multicomponent nature of most lung surfactants
   limit spectroscopy and related molecular biophysical methods
Anatomical and physiological complexities
   Intricate, interdependent structure of the alveolar network
   Difficulty of precise morphological measurements in lungs
   Difficulty of defining the extent of the alveolar hypophase
   Complexity of pulmonary tissue and vascular biomechanics
   Dependence of gas exchange on multiple factors in addition to lung surfactant activity

---

The many complexities affecting lung surfactant research lead to a necessity for correlated inter-disciplinary characterizations. See text for details.

present throughout biological tissues and fluids. Compositional analyses are further complicated by the large number of functional components making up endogenous lung surfactant. The majority of exogenous lung surfactants are also complex mixtures of interacting components. Understanding the roles and inter-actions of these components requires the study of multiple surfactant materials, ranging from biologically derived surfactants to synthetic phospholipids, lipids, proteins and peptides, and so on. The composition of all lung surfactants is best defined in a functional context that links measurements of composition, surface activity, and physiological activity (Chapter 8). *Biophysical* research on lung surfactants has its own set of complications. The surface activity of most lung surfactants reflects a sum of dynamic and equilibrium behaviors generated by interactive system components. Biophysical assessments are further complicated by the metastable phase and aggregation behavior of phospholipids. The use of spectroscopic and other sophisticated physical methods to define molecular behavior is also hampered by the nontrivial structure and complex composition of surfactant lipids and proteins. *Anatomical and physiological* complexities also affect research on pulmonary surfactants. The intricate, interdependent structure of the alveolar network complicates theoretical analyses relating mechanics and surface activity. The complex mechanical behavior of lung tissue and the pulmonary vasculature also limits the precision of theory, as does the dependence of gas exchange on multiple factors in addition to lung surfactant function. All of these factors makes it essential to correlate findings from compositional, biophysical, and physiological research on lung surfactants; i.e., a multidisciplinary perspective is necessary.

III. Multidisciplinary Approaches in Lung Surfactant Research

Pulmonary surfactant research can be arbitrarily subdivided as involving 1) physical and chemical studies *in vitro*, 2) biological studies in lung cells and tissue, 3) lung functional and mechanical studies in animals, and 4) clinical studies (Figure 5-1). Each of these categories incorporates multiple scientific disciplines. *Physical and chemical studies in vitro* involve the disciplines of biophysics, organic and physical chemistry, and interfacial phenomena to study lung surfactant composition, surface activity, and component molecular interactions. Theoretical analyses in this category also include principles of thermodynamics, interfacial phenomena, and mechanics. *Studies in lung cells and tissue* involve the application of cell and molecular biology and biochemistry to examine the synthesis, secretion, and recycling of lung surfactant and its functional components. *Mechanical and functional studies* in animal lungs *in vivo* or *in situ* incorporate the disciplines of physiology, mechanics, and pharmacology, among others, to define the activity of endogenous and exogenous pulmonary surfactants in a more comprehensive manner than feasible or desirable in patients. *Clinical studies* involve applications from medicine, pathology, pharmacology, biostatistics, and epidemiology in randomized controlled trials and pilot studies that define the therapeutic effects of exogenous surfactants in human disease. The involvement of so many scientific disciplines in lung surfactant research leads to a particularly broad and complex literature, only a fraction of which is cited in the alphabetized references at the end of this book.

IV. Overview of Materials Used in Lung Surfactant Research

A variety of lung-derived and synthetic materials ranging from single compounds to complex multicomponent mixtures are utilized in research on endogenous and exogenous lung surfactants (Table 5-2). Many of these materials are studied not only in terms of their own intrinsic activity but also to help elucidate the roles and importance of specific components and molecular structural features in the activity of more complex mixtures. Understanding the source and composition of lung surfactant materials, and their distinctions and similarities, is

Multidisciplinary Approaches in Lung Surfactant Research

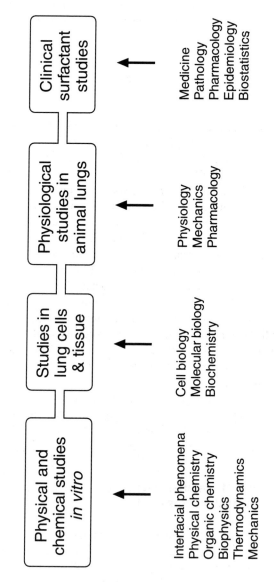

Figure 5-1 Multidisciplinary research approaches for studying pulmonary surfactants. Research on endogenous and exogenous lung surfactants involves a variety of scientific disciplines. Experience has shown that correlated experiments and analyses across these disciplines are necessary for understanding lung surfactant function and for developing effective surfactant replacement therapy.

Table 5-2 Surface Active Materials Used in Basic Research on Lung Surfactants

I. Lung-derived surfactant materials
  Endogenous (whole, natural) lung surfactant
    Obtained by bronchoalveolar lavage
    Obtained in isolated lamellar bodies
    Processed from tissue
  Aggregate subpopulations (subtypes) of endogenous surfactant
  Organic solvent extracts of endogenous surfactant
  Organic solvent extracts of lung tissue
  Surfactant apoproteins purified from endogenous surfactant
  Recombinant surfactant apoproteins
  Separated subfractions of endogenous surfactant lipids
    Total phospholipids plus neutral lipids
    Total phospholipids
    Subsets of phospholipid classes
II. Synthetic surface active materials
  Synthetic glycerophospholipids
  Synthetic phospholipid/phosphonolipid analogs
  Synthetic peptides
  Synthetic polymers
  Fatty acids and alcohols
  Di- and triglycerides
  Other surface active chemicals

Recombinant surfactant apoproteins are arbitrarily included in lung-derived materials although they are produced *in vitro*; the category of "synthetic" materials also includes some compounds purified from biological sources rather than synthesized *de novo*. Clinical exogenous surfactants are not specifically tabulated but also fit within or contain components from the listed categories of materials (Chapter 14).

essential for understanding their activity. The broad category of "lung-derived" surfactants in Table 5-2 encompasses a spectrum of materials that vary in composition and behavior: endogenous lung surfactant, organic solvent extracts of lavaged lung surfactant, organic solvent extracts of lung tissue, purified lung surfactant lipid subfractions, purified surfactant proteins, and so on. Within each of these categories there are finer distinctions. Endogenous lung surfactant prepared from different animals by different methods, for example, is not biochemically identical although compositional commonalities do exist. Major features of the different kinds of surfactant materials in Table 5-2 are described below.

A. Endogenous Lung Surfactant (Whole Surfactant, Natural Surfactant)

Endogenous lung surfactant can be harvested from mammalian lungs by a variety of methods ([319, 569, 790, 829, 942, 1105] for review). Endogenous surfactant obtained by different methods and from animals of different species and ages does vary in composition and activity. However, all active preparations of endogenous surfactant share the compositional features of having a high content of phospholipids rich in DPPC and other phosphatidylcholines, as well as containing specific biophysically important surfactant proteins: SP-A, B, and C. A small content of neutral lipid, primarily cholesterol, is also typically present (the composition and component molecular biophysics of endogenous lung surfactant are detailed in Chapters 6 and 8). Although the composition of endogenous lung surfactant has been clarified over decades of research, variabilities in surfactant material obtained by different methods and purification schemes has complicated interpretations in this area. Two important methods, bronchoalveolar lavage and lamellar body isolation, used to obtain endogenous lung surfactant are described below.

Surfactant from Broncholalveolar Lavage

The simplest and most direct method of obtaining endogenous lung surfactant is to wash it from the alveolar spaces with normal saline (0.15 M NaCl) instilled into the intact airways. This process is called bronchoalveolar lavage (BAL) (derived from the French: lavage = washing). BAL is the most widely used technique for obtaining isolates of endogenous surfactant.[1] A modification of BAL used in early research was to instill saline via the intact pulmonary vasculature to wash out alveolar surfactant (e.g., [93]). Alveolar surfactant exists primarily as aggregates, which can be readily sedimented by centrifugation from BAL (Figure 5-2). Recovered BAL fluid is immediately centrifuged at a low gravitational force to remove any cells (150–200 × g), followed by centrifugation of the supernatant at higher speed to pellet endogenous surfactant. A gravity force of 10,000–12,000 × g is commonly employed to sediment endogenous surfactant [460, 541, 732, 783, 791, 1214], but sedimentation at gravity forces as low as 1,000–2,000 × g is sometimes used to isolate larger aggregates only. If applied with care to fresh, intact lungs, BAL yields a suspension of alveolar surfactant in saline with minimal contamination from cell membrane lipids and other lung tissue components. Moreover, since obtaining lung surfactant from BAL fluid involves very little processing, the ratio of components originally present in

---

1. For example, Refs. 286, 287, 296, 460, 541, 569, 570, 732, 744, 783, 791, 1032, 1177, 1214.

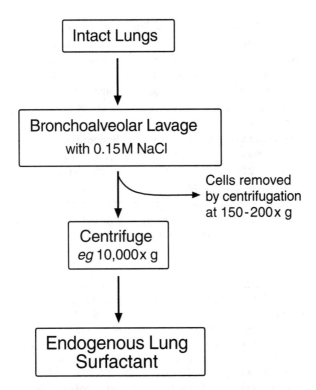

Figure 5-2 Endogenous lung surfactant obtained by bronchoalveolar lavage. The simplest and most direct method of obtaining alveolar surfactant is to wash it from the airspaces of fresh, intact lungs by bronchoalveolar lavage with normal saline. Surfactant aggregates are then pelleted by simple centrifugation. Endogenous lung surfactant can also be processed by more elaborate procedures from lung tissue. See text for details.

alveolar surfactant is not significantly altered. However, even BAL isolates of lung surfactant can be contaminated with cellular material if lungs are damaged or hemorrhagic, or if cellular lysis and surfactant degradation have occurred prior to or during lavage.

Surfactant in Isolated Lamellar Bodies

Endogenous surfactant can also be obtained by isolating lamellar bodies from animal lung tissue homogenates (e.g., [305, 391, 397, 863, 1020]). Purification procedures for lamellar bodies are much more complex than surfactant isolation in BAL and require multiple steps of conventional and density gradient centrifugation. Analysis of lamellar body contents has been very helpful in understanding surfactant biosynthesis in the type II cell, as well as in defining relationships between cellular surfactant and alveolar surfactant. Lamellar bodies that have been isolated with minimal disruption as verified by electron microscopy and the presence of specific marker enzymes are found to have a phospholipid content very close to that in alveolar surfactant [391, 397, 863, 1020]. However, the potential for contamination by nonsurfactant constituents is nontrivial in any preparation processed from lung tissue. A variety of studies of lung surfactant composition have used homogenized lung tissue as a starting point and attempted to enrich lung surfactant components by differential and density gradient centrifugation and/or physicochemical processing (e.g., [281, 570, 571, 829, 834, 942]). All such studies are affected to some extent by the inability to fully separate cellular lipids from surfactant lipids once the two are mixed at the molecular level. The content of surfactant proteins and their compositional ratios relative to lipids also can be affected by the more elaborate processing required when lung surfactant is obtained from homogenized tissue as opposed to being isolated by lavage.

Example of Methods for Isolating Lamellar Bodies. A representative experimental procedure for lamellar body isolation from Spalding et al [1020] is illustrated in Figure 5-3. In brief, animal lungs are removed, trimmed, and minced, followed by homogenization in cold 0.25 M sucrose. A series of conventional and density gradient centrifugation steps at 4°C then specifically isolate lamellar bodies. The tissue homogenate in 0.25 M sucrose is first centrifuged at $1,100 \times g$ for 10 min (Sorvall RC 2-B, SS-34 rotor), the pellet discarded, the supernatant centrifuged at $4,300 \times g$ for 10 min, the pellet again discarded, and the supernatant centrifuged at $21,000 \times g$ for 20 min. The pellet from this last centrifugation is then resuspended in 0.25 M sucrose, centrifuged again at $21,000 \times g$, the supernatant discarded, and the pellet resuspended in 0.25 M sucrose (1 ml/gm of minced lung). This material is layered at 8.5 ml/gradient on 30 ml of 0.60 M sucrose and the discontinuous gradients centrifuged at $82,000 \times g$ for 60 min (Beckman L3-50, SW 27 rotor). Lamellar bodies locate in a distinct interfacial band, which is removed, diluted with cold distilled water to a sucrose concentration of 0.25 M, and centrifuged a final time at $12,000 \times g$ for 12 minutes to collect the lamellar bodies in a white pellet [1020]. Other lamellar body preparation procedures are similar in concept but differ in the specific number and conditions of conventional and density gradient steps [305, 391, 397, 863]. Lamellar body isolates are typically characterized by electron microscopy and marker enzyme analysis to verify their integrity and purity.

B. Aggregate Subfractions (Subtypes) of Endogenous Surfactant

Subpopulations of aggregates can be isolated from aqueous suspensions of endogenous surfactant by differential or density gradient centrifugation [67, 350, 352, 379, 685, 872, 873, 1200]. These lung surfactant aggregate subfractions exhibit differences in surface activity, with larger aggregates typically having both a higher apoprotein content and better surface tension lowering ability. In a general sense, lung surfactant aggregates reflect the proclivity of phospholipids

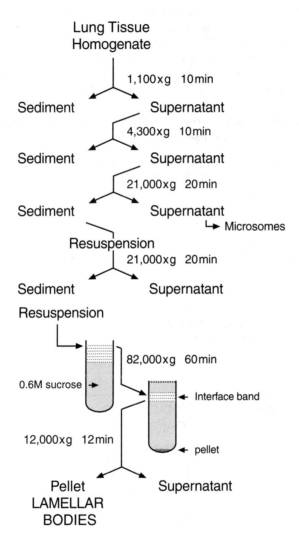

Figure 5-3 Isolation of lamellar bodies from type II cells. Lamellar bodies can be isolated from type II cells in lung tissue by elaborate protocols involving multiple steps of conventional and density gradient centrifugation. Final isolates can be examined by electron microscopy and marker enzyme analysis to assess integrity and purity. (Redrawn from Ref. 1020.)

to organize as multilamellar structures in water (Chapter 3). In addition, aggregate subfractions, also called *surfactant subtypes*, may represent alveolar surfactant in different functional stages (e.g., freshly secreted surfactant vs material expelled from the surface vs material being processed for reuptake) [350–352, 1199, 1201] (Chapter 6). Lung surfactant aggregates are also found to be altered in size and density or activity during lung injury, contributing to surfactant dysfunction or inactivation [64, 348, 349, 379, 640, 871, 1114] (Chapter 9).

C. Organic Solvent Extracts of Endogenous Lung Surfactant (Lung Surfactant Extracts)

When an aqueous suspension of endogenous lung surfactant is extracted into appropriate organic solvents, hydrophilic protein is removed while hydrophobic components are retained. Such preparations are called organic solvent extracts of lung surfactant (Figure 5-4). A number of different hydrophobic solvent systems can be used for extraction, but the most common is 2:1 chloroform:methanol [88, 273]. Chloroform:methanol extracts of lavaged alveolar surfactant, at least in principle, contain the hydrophobic lipids and proteins of the surfactant system

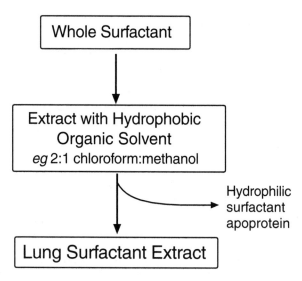

Figure 5-4 Organic solvent extracts of endogenous lung surfactant. Aqueous suspensions of whole lung surfactant can be extracted into hydrophobic organic solvents such as 2:1 chloroform:methanol to remove hydrophilic surfactant apoprotein. Lung surfactant extracts containing only surfactant lipids and hydrophobic apoproteins display high surface activity. Additional extraction with solvents such as acetone or ethyl acetate to deplete cholesterol and neutral lipids can also be done. See text for details.

in close approximation to the natural ratio. There is no evidence that extraction with this solvent system, per se, alters the compositional ratios of the hydrophobic constituents of lung surfactant. Chloroform:methanol extracts of lavaged lung surfactant have very high surface activity and form the basis for several clinical exogenous surfactants (Chapter 14). Additional extraction and processing can be applied to chloroform:methanol extracts of lung surfactant to modify their composition. For example, extraction with acetone or with ethyl acetate can remove or deplete neutral lipids such as cholesterol relative to phospholipids. Organic solvent extracts of endogenous lung surfactant as a group are sometimes called "lipid extract surfactants" or "lung surfactant extracts." Although useful as general categorizations, the former designation does not account for the functionally important hydrophobic apoproteins that are present, and neither discriminates the significant compositional variations that can exist between extract preparations.

Use of the Term "Extract" in Lung Surfactant Research. In the current lung surfactant literature, the term "extract" is generally reserved to describe surface active material that has been extracted into hydrophobic organic solvents. However, in the early lung surfactant literature, the terminology "lung extract" was widely used to describe *any* surface active material extracted from lungs. This nonspecific terminology did not necessarily discriminate between whole surfactant extracted in saline and surfactant extracted with organic solvents. This distinction is now understood to be highly important. Whole surfactant in saline contains the complete mix of surfactant lipids and proteins, including hydrophilic protein (predominantly SP-A) as well as hydrophobic surfactant proteins SP-B and SP-C. Hydrophilic surfactant proteins are removed by extraction with hydrophobic organic solvents such as chloroform:methanol (e.g., [792, 1177, 1214]). The resulting organic solvent extracts contain only phospholipids, neutral lipids, and the hydrophobic surfactant proteins. At a further level of distinction, hydrophobic organic solvent extracts of this kind may differ significantly in the details of their composition depending on the specific solvents and surfactant preparative methodology used as noted above. This is particularly true if lung surfactant is processed from tissue as opposed to being isolated by lavage.

D. Organic Solvent Extracts of Lung Tissue

If homogenized lung tissue as opposed to lung surfactant is extracted into hydrophobic organic solvents, the resulting material is a lung tissue extract (Figure 5-5). Organic solvent extracts of homogenized lung tissue by necessity contain a mixture of alveolar and intracellular surfactant plus cell membrane lipids and other nonsurfactant components. The significant difference in lipid content between alveolar surfactant and lung tissue homogenates was widely recognized by early lung surfactant researchers (e.g., [286, 744, 942]). In general, lung tissue contains lower percentages of total and disaturated phosphatidylcholine, and higher percentages of phosphatidylethanolamine and sphingomyelin,

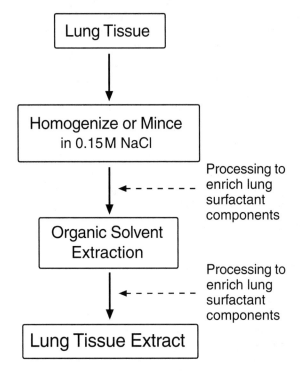

Figure 5-5 Organic solvent extracts of lung tissue. Lung tissue homogenates contain nonsurfactant constituents as well as alveolar and intracellular surfactant. Lipids and other hydrophobic components from cells and tissue coextract in hydrophobic organic solvents along with surfactant lipids and hydrophobic proteins. Further physical or chemical processing of lung tissue extracts can be applied to enrich the final product in lung surfactant components. See text for details.

compared to alveolar surfactant. Unprocessed organic solvent extracts of homogenized lung tissue have much lower surface activity than endogenous surfactant due to their high content of fluid cellular lipids and other tissue components. However, additional physicochemical processing either prior to or following extraction can be used to enrich tissue extracts in lung surfactant components (Figure 5-5). Lung tissue extracts can also be supplemented with synthetic additives to increase their surface activity.

E. Purified Lung Surfactant Proteins

Purified isolates of the biophysically active surfactant proteins SP-A, SP-B, and SP-C are widely utilized in research on endogenous and exogenous lung surfactants. SP-D, although not involved in lung surfactant biophysics, is also isolated and studied for its biological effects. SP-A is generally purified from lung surfactant by precipitation in organic solvent to separate it from hydrophobic lipids and proteins, plus additional steps such as column chromatography to isolate it from plasma proteins (e.g., [66, 92, 363, 422, 927, 1046, 1130]). SP-B and SP-C are typically purified from lung surfactant by a combination of extraction into hydrophobic organic solvents and gel-permeation chromatography and/or HPLC (e.g., [179, 180, 317, 383, 853, 1071, 1143, 1174, 1216]). SP-D can be isolated from the supernatant of centrifuged lung lavage by several methods including barium sulfate extraction or EDTA extraction plus gel permeation chromatography or affinity chromatography (e.g., [610, 613, 844, 845]). Surfactant apoprotein purification is nontrivial biochemically. All the surfactant proteins form oligomers, and SP-B and SP-C are particularly challenging to process because of their hydrophobicity, overlapping oligomeric molecular weights, and strong associations with lipid. Many isolates of "pure" SP-B and SP-C studied for interactions with phospholipids are in actuality not completely pure and also vary widely in their content of associated lipid. Nonetheless, research using purified preparations of SP-A, SP-B, and SP-C has provided crucial information on the molecular biophysics of these materials and their surface active interactions with phospholipids. The structure and biophysical properties of lung surfactant apoproteins are detailed in Chapter 8.

Examples of Purification methods for SP-A, SP-B, SP-C. SP-A is typically isolated from whole surfactant by a combination of centrifugation, column chromatography, and/or organic solvent extraction [92, 363, 422, 927, 1046, 1130]. Butanol or other hydrophobic solvents such as ethanol:ether (3:1) precipitate SP-A away from hydrophobic proteins and lipids, although some SP-A–associated lipid remains. SP-A can then be separated from coprecipitated plasma proteins by column chromatography (e.g., using DEAE-cellulose with a linear gradient of 0.1–0.5 M NaCl [1046] or Sephadex G-150 or G-200 with 0.05 M sodium borate and 0.1% SDS [574]). Further purification can be done by washing in buffer where SP-A is insoluble (e.g., 5 mM Tris/0.15 M NaCl/octyl-β-D-glucopyranoside at pH 7.4), resuspension of insoluble protein in 5 mM Tris where SP-A is soluble, dialysis, and removal of Tris-insoluble material by centrifugation (e.g., [422]). SP-A can also be isolated from lung surfactant by gel electrophoresis and isoelectric focusing [1175]. In reduced form, purified SP-A migrates at 28–36 kDa with glycosylated and unglycosylated isoforms on polyacrylamide gel electrophoresis with sodium dodecyl sulfate (SDS-PAGE) and silver staining. The hydrophobic apoproteins SP-B and SP-C are most directly and easily isolated from lavaged whole surfactant that has been extracted with hydrophobic solvent such as 2:1 chloroform:methanol (C:M) to remove hydrophilic material (organic solvent extracts of processed lung tissue can also be used). The extract containing hydrophobic surfactant components is then subjected to gel permeation chromatography on LH-20, LH-60, Silica C8, or related packing using an elution solvent such as 2:1 or 1:1 (v/v) C:M to separate SP-B/C away from surfactant lipids (e.g., [180, 383, 853, 1071, 1143, 1174]). Column fractions are typically monitored by ultraviolet adsorption and/or SDS-PAGE with tricine buffers to define molecular content [944]. Fractions containing hydrophobic apoproteins can also be processed with additional gel chromatography on LH-60 or by HPLC to purify SP-B and SP-C individually. For example, initial column fractions containing primarily SP-B or SP-C can be pooled and applied to a clean LH-60 column, eluted with C:M:HCl, and appropriate fractions pooled and dialyzed to remove acid from the final pure isolates. If necessary, additional purification of SP-B or SP-C can be accomplished by HPLC, for example, with the column equilibrated with 95% methanol:5% water or methanol:trifluoroacetic acid prior to apoprotein injection in C:M. Both SP-B and SP-C avidly retain associated lipid, and isolates of these apoproteins vary in lipid content depending on specific processing and dialysis. Final apoprotein purity is typically verified by SDS-PAGE, silver staining, and N-terminal amino acid (AA) sequencing. Protein concentrations are determined either by colori-metric assays or by quantitative AA analysis. ELISA tests are available for all of the surfactant apoproteins with the exception of SP-C [361, 463, 504, 600, 997]. In addition, biochemical methodology is available to purify specific oligomers of the hydrophobic surfactant proteins [37, 1174] and to modify apoprotein isolates such as by deacylation of SP-C using trimethylamine or carbonate buffer treatment [179, 841, 1142].

F. Recombinant Apoproteins

Recombinant surfactant proteins are studied in basic research on lipid-protein interactions in lung surfactants and are also under active investigation as components in synthetic exogenous surfactants. Genetic engineering technology makes it feasible to produce and study recombinant versions of all of the human surfactant apoproteins.[2] Although SP-A [367, 1130, 1131], SP-B [1208], SP-C [425, 947, 1113], and SP-D [166] can all be produced by recombinant techniques, primary attention in research has focused on the two hydrophobic surfactant proteins, particularly SP-C. Recombinant human SP-C in several forms is currently under commercial development as a component in exogenous surfactants for use in RDS and ARDS [184, 374, 375, 425, 638, 968]. One widely used form of recombinant SP-C (Byk Gulden Pharmaceutical, Germany) has phenylalanine substituted for cysteines in positions 4 and 5 of the human sequence and isoleucine substituted for methionine in position 32 [184, 374, 375, 497, 638]. Recombinant SP-C without these substitutions is also available following expression in bacteria [425], and a palmitoylated form of recombinant SP-C has also been reported [1113]. Research on recombinant SP-C and other human sequence recombinant proteins and related studies involving synthetic hydrophobic peptides are active areas in current exogenous surfactant research and development (Chapter 14).

---

2. For example, Refs. 160, 166, 184, 360, 367, 374, 375, 425, 497, 535, 638, 947, 968, 1045, 1113, 1130, 1131, 1208.

Example Methods for Producing Recombinant Surfactant Proteins. Standard genetic engineering techniques are followed to generate recombinant surfactant apoproteins, although purifications are complicated particularly for SP-B and SP-C by their hydrophobicity, lipid-binding affinity, and other physicochemical properties. Recombinant versions of the two hydrophobic apoproteins are typically produced in bacteria, while recombinant SP-A and SP-D are produced in mammalian cells due to their more extensive intracellular processing. Bacteria used to produce SP-B and SP-C are generally strains of *Escherichia coli* (*E. coli*) [425, 947, 1208], while Chinese hamster ovary (CHO) cells are commonly utilized for production of SP-A and SP-D [166, 1130, 1131]. The genes or reconstructed complementary DNA (cDNA) for the human surfactant proteins are introduced into the designated bacteria or mammalian cells by means of specific vectors (e.g., plasmid pTrp233, pCR-Bac, or pKK223-3 in bacteria or simian virus SV40 enhancer in mammalian cells). Vectors are typically inserted with control sequences of DNA and genes for resistance to antibiotics that can be used to identify appropriately transfected organisms. Cells or bacteria are cultured with various agents used to stimulate DNA transcription and protein expression (e.g., 3-β-indoleacrylate, isopropyl-β-D-thiogalactoside, methionine sulfoximine). Clones producing the desired surfactant protein are identified, isolated, and then cultured specifically. Product proteins are obtained from culture supernatants or by centrifugation after cell or bacterial lysis and purified by techniques such as affinity chromatography [166, 1130, 1131] or HPLC [425]. Surfactant proteins can be produced directly or as part of larger fusion proteins that are subsequently cleaved to provide the desired product. Specific chemical modifications such as the covalent attachment of acyl groups to some forms of recombinant SP-C can subsequently be applied *in vitro*.

G. Column-Separated Subfractions of Lung Surfactant Lipids

Column chromatographic methods permit subfractionation and isolation of surfactant lipids as well as surfactant proteins [383, 1144–1146, 1215, 1217]. Activity studies on defined subsets of surfactant lipids isolated in their endogenous ratio provide a valuable adjunct to research on model synthetic mixtures for clarifying component roles and interactions in lung surfactant. As in the case of the hydrophobic apoproteins above, separations of lung surfactant lipids typically begin with lavaged surfactant extracted into chloroform:methanol or other hydrophobic solvents to remove hydrophilic protein. The use of surfactant obtained by lavage rather than processed from tissue is generally preferable to minimize the content of nonsurfactant lipids. The lung surfactant extract is fractionated using gel permeation chromatography with LH-20, LH-60, or C8 packing and a hydrophobic elution solvent. The fractions in which specific surfactant components appear are a function of column packing, elution solvent, flow rate, and so on. Small volume fractions of column output are typically collected and analyzed for composition, followed by pooling of appropriate fractions to obtain subsets of components such as total phospholipids plus neutral lipids (N&PL), total phospholipids alone (PPL), zwitterionic phospholipids alone (mPPL), phospholipids

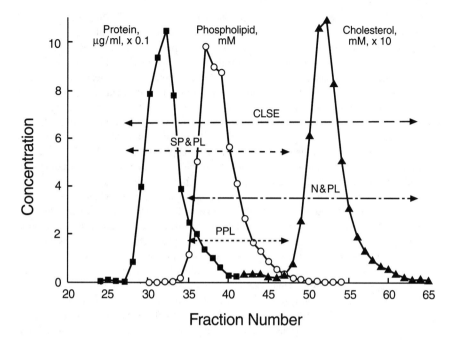

Figure 5-6 Column-separated subfractions of lung surfactant lipids, phospholipids, and proteins. Subfractions of lipid and protein components can be separated from calf lung surfactant extract (CLSE) or related materials by gel permeation chromatography with hydrophobic elution solvents like acidified chloroform:methanol. These subfractions can then be used to examine the biophysics of subsets of surfactant components at or near their endogenous ratio as a complement to studies on synthetic model mixtures. N&PL, total phospholipids plus neutral lipids; PPL, total phospholipids; SP&PL, phospholipids plus mixed hydrophobic proteins. Additional subfractions such as zwitterionic phospholipids alone can be obtained by varying column conditions and elution solvents. See text for details. (Redrawn from Ref. 383.)

plus hydrophobic proteins (SP&PL), and so on (Figure 5-6). Recombined surfactant formed from the totality of column-separated components is found to have surface activity equivalent to the parent surfactant extract, indicating that the separation process does not significantly alter surface activity [383, 1215, 1217].

Example of Methods for Purifying Lung Surfactant Lipids. Whole surfactant obtained by centrifugation of bronchoalveolar lavage at $10,000-12,000 \times g$ is extracted with chloroform:methanol [88], and the extract is loaded on a column packed with hydroxypropyl dextran (LH-20, Pharmacia-LKB) or related packing [383, 1144–1146]. Sample sizes and flow conditions in separations vary depending on the specific surfactants, columns, packing, and elution solvents used (e.g., a representative sample size and flow rate for a 15×1.5 cm LH-20 column is ~20–100 μmoles phospholipid eluted at 4 ml/hr). Examples of elution solvents include chloroform:methanol:0.1 N HCl (C:M:HCl) in volume ratios of 1:1:0.1 or 2:1:0.1 or C:M without acid in ratios of 2:1 or 1:1. Elution with acidified C:M is generally used to obtain the complete mix of purified surfactant phospholipids (PPL), while unacidified C:M allows separation of anionic phospholipids from zwitterionic phospholipids [383, 1144–1146]. Columns can be run either at room temperature or at low temperature (e.g., 4°C) to minimize lipid degradation in the presence of acid. Fractions of column eluent are generally collected at fixed intervals (e.g., every 7.5 minutes) and phospholipid-containing fractions localized by phosphate assay or by monitoring ultraviolet absorption. Appropriate fractions are then pooled to generate different subsets of surfactant lipids. If acidic solvent is used, final pooled fractions can be extracted with 2:1 C:M to remove acid. A second column pass can also be used as necessary to further purify phospholipids away from hydrophobic surfactant proteins. Final pooled fractions of specific components are typically analyzed for phospholipid content [17, 122, 998], protein content [545, 669], and cholesterol content [231, 966] to assess purity. Phospholipid class content is defined by thin layer chromatography [257, 1085].

H. Synthetic Glycerophospholipids

Synthetic glycerophospholipids are widely studied alone and in model mixtures to investigate component roles and interactions in endogenous lung surfactant. Synthetic glycerophospholipids are also important constituents in many clinical exogenous surfactants. A variety of highly pure synthetic phospholipids are available commercially (sources in the United States include Avanti Polar Lipids, Inc; Calbiochem-Novabiochem; Fluka Chemika-BioChemika; the Sigma Chemical Co., and ICN Pharmaceuticals). Phospholipids and their headgroup and chain molecular structures are described in Chapter 3. In addition to glycerophospholipids synthesized in the laboratory, the terminology "synthetic" phospholipids is commonly used in the lung surfactant literature to include phospholipids purified from nonlung sources such as eggs or soybeans. Phosphatidylcholines purified from eggs or soy (egg-PC, soy-PC), for example, are complex mixtures containing PC molecules with different fatty chains. Mixtures of molecular species within other phospholipid classes such as PG and PI can also be obtained from eggs, soy, or related sources. Heterogeneous phospholipid mixtures of this kind are sometimes used as convenient (but very approximate) substitutes for the actual mixture of phospholipids in endogenous lung surfactant.

I. Synthetic Phospholipid Analogs

Synthetic phospholipid and phosphonolipid analogs with specific structural differences from biological phospholipids are another group of surface active compounds used in lung surfactant research.[3] Synthetic phospholipid analogs, either synthesized *de novo* or prepared by reaction from biological glycero-phospholipids, are helpful in studying the molecular basis of surface behaviors such as rapid film respreading or adsorption [657, 658, 1091–1093]. Analog compounds can also be used to investigate structural contributions to phospholipid phase behavior [566, 659, 673, 1006, 1093] and have potential utility as novel constituents in synthetic exogenous surfactants [207]. Examples of analog compounds that have been studied for structure-activity behavior include phospholipids and phosphonolipids with ether or amide linkages instead of ester linkages between the fatty chains and glycerol backbone plus additional substitutions in the region of the phosphate moiety and N-headgroup (e.g., Figure 5-7). A variety of other synthetic analogs of glycerophospholipids can also be made. Structural changes in the region of the chain-backbone linkage and phosphate group not only affect surface activity but also reduce molecular susceptibility to cleavage by phospholipases. The ability to resist degradation by phospholipases may enhance the utility of active synthetic phospholipid analogs as components in exogenous surfactants for use in inflammatory lung injury [207, 651, 1092, 1093] (Chapter 14).

J. Synthetic Peptides

Synthetic peptides are widely used in lung surfactant research both to study the molecular biophysical basis of apoprotein function and to serve as components in exogenous surfactants.[4] Modern chemical technology allows the *de novo* synthesis of peptides containing any of the biologically relevant amino acids. These amino acids have known frequencies of occurrence in α-helices, β-sheets, and reverse turns in the secondary structures of proteins (Table 5-3), and this can be exploited in designing peptides with specific folding and conformation characteristics. Other known physicochemical properties of amino acids such as their hydrophobicity/hydrophilicity can also be incorporated as desired in peptides. Peptides with varying degrees of homology to all of the biophysically important surfactant proteins have been synthesized and used in lung surfactant research (see citations in footnote). The majority of these peptides contain hydrophobic amino acids and have some degree of structural relevance for SP-B or SP-C. Full length and regional peptides covering the entire primary 78–80 AA sequence of

---

3. For example, Refs. 207, 566, 651, 657–659, 673, 1006, 1091–1093.
4. For example, Refs. 18, 19, 36, 106, 139, 142-144, 160, 249, 334, 532, 535, 653, 654, 656, 666, 667, 679, 711, 713, 730, 775, 895, 939, 1064, 1118, 1137, 1138, 1140, 1154, 1155, 1225.

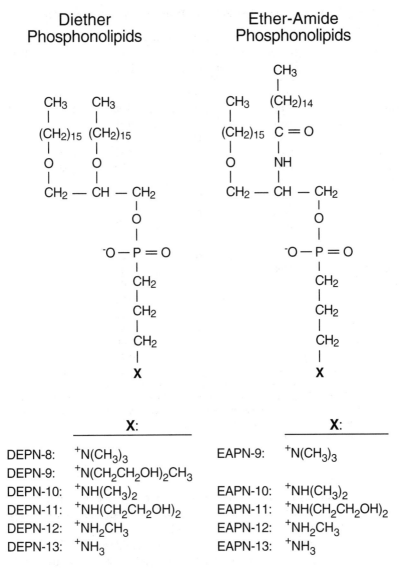

Figure 5-7 Molecular structures of selected synthetic phospholipid analogs used in lung surfactant research. Compounds shown are diether and ether-amide phosphonolipids (DEPN or EAPN series) synthesized by Turcotte, Notter, and co-workers [651, 657–659, 673, 1091–1093]. Synthetic analogs of this kind can be used to study structure-activity correlates for phospholipid-like molecules or as novel constituents for synthetic exogenous surfactants. DEPN compounds have two identical palmitylether chains; EAPN compounds have one palmitylether chain and one palmitoylamide chain. Analogs 10–13 in both series are shown in the protonated form existing near neutral pH. Many other synthetic phospholipid analogs are also available for research applications. (Redrawn from Ref. 657.)

Table 5-3 Amino Acids and Their Relative Occurrence in the Secondary Structures of Proteins

Amino acid	3-letter abbreviation	1-letter abbreviation	Alpha helix	Beta sheet	Reverse turn
Alanine	Ala	A	1.29	0.90	0.78
Arginine	Arg	R	0.96	0.99	0.88
Asparagine	Asn	N	0.90	0.76	1.28
Aspartic acid	Asp	D	1.04	0.72	1.41
Cysteine	Cys	C	1.11	0.74	0.80
Glutamine	Gln	Q	1.27	0.80	0.97
Glutamic acid	Glu	E	1.44	0.75	1.00
Glycine	Gly	G	0.56	0.92	1.64
Histidine	His	H	1.22	1.08	0.69
Isoleucine	Ile	I	0.97	1.45	0.51
Leucine	Leu	L	1.30	1.02	0.59
Lysine	Lys	K	1.23	0.77	0.96
Methionine	Met	M	1.47	0.97	0.39
Phenylalanine	Phe	F	1.07	1.32	0.58
Proline	Pro	P	0.52	0.64	1.91
Serine	Ser	S	0.82	0.95	1.33
Threonine	Thr	T	0.82	1.21	1.03
Tryptophan	Trp	W	0.99	1.14	0.75
Tyrosine	Tyr	Y	0.72	1.25	1.05
Valine	Val	V	0.91	1.49	0.47

Amino acids are typically grouped as hydrophobic or nonpolar (A, F, I, L, M, P, V, W); polar but uncharged (C, G, N, Q, S, T, Y); basic or positively charged near neutral pH (H, K, R); acidic or negatively charged near neutral pH (D, E).
Source: Adapted from Ref. 1043.

human SP-B are available, as are regional and full length SP-C peptides. Examples of several N-terminal human sequence SP-B peptides are illustrated in Figure 5-8. Also studied are derivatized SP-B and SP-C peptides with specific amino acid substitutions and deletions, plus SP-C peptides with acyl moieties of varying length attached covalently at selected locations. Less specific synthetic peptides are also used in lung surfactant research, including KL4, a 21 amino acid peptide containing repeating subunits with one lysine (K) and four leucine (L) residues that is combined with DPPC, palmitoyl-oleoyl-PG (POPG), and palmitic acid as a synthetic exogenous surfactant [142–144, 730] (Chapter 14). Examples of other nonspecific hydrophobic synthetic peptides include synthetic homopolymers of valine, phenylalanine, leucine, or other hydrophobic amino acids [1118] and synthetic amphipathic α-helical peptides [711–713, 1225].

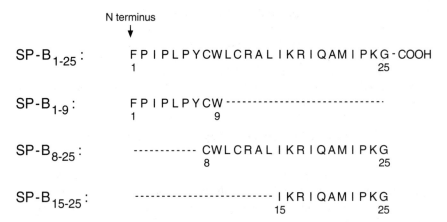

Figure 5-8 Examples of synthetic human sequence N-terminal SP-B hydrophobic peptides. Lung surfactant research utilizes synthetic peptides structurally related to all three biophysically active apoproteins as well as less specific hydrophobic peptides. Shown as examples are several synthetic human SP-B peptides incorporating amino acids 1–25, 1–9, 8–25, and 15–25 numbered from the N-terminal phenylalanine (see Chapter 8 for complete amino acid sequences of human surfactant apoproteins). (Adapted from Ref. 334.)

Examples of Peptide Synthesis Methods. Peptides in lung surfactant research are commonly synthesized by a modification of the solid-phase methods defined by Merrifield (e.g., [726, 727]). Amino acids are added sequentially to the N-terminal end of the growing peptide, which is anchored at the C-terminal end to an insoluble matrix or resin such as on the surface of polystyrene beads. To obtain specific coupling, the N-terminal of the incoming amino acid is blocked, such as with tertiary butyloxycarbonyl (t-boc) [727] or Fastmoc (F-moc) reagents [253]. Conversely, the carboxyl group of the incoming amino acid is activated to facilitate reaction with the unprotected N-terminus of the resin- coupled growing peptide utilizing reagents such as dicyclohexylcarbodiimide (DCC) or *O*-benzotriazol-1-yl-*N,N,N',N'*-tetramethyluronium hexafluorophosphate (HBTU). After each new amino acid is incorporated, the matrix is washed to remove excess reactants and uncoupled products, and the now protected N-terminus of the growing peptide is deprotected (e.g., by dilute acid, which removes t-boc but leaves peptide bonds intact). The next amino acid, with N-terminus protected, is then added and synthesis proceeds. Peptide synthesis utilizing commercial peptide synthesizers (e.g., Perkin Elmer-Applied Biosystems Model 430A or 431A) is typically on a 0.1–1 mmole scale. Crude peptides are released from the resin using anhydrous hydrogen fluoride or trifluoroacetic acid and purified further by reverse phase HPLC (e.g., with a Vydac C4 column using a water-acetonitrile or water-acetonitrilea:propanol (1:1) gradient containing 0.1% trifluoroacetic acid). Peptide composition is verified by quantitative AA analysis, and molecular weights are confirmed by mass spectrometry. Peptide oligomerization and secondary structure with and without lipid present can be assessed by circular dichroism and spectroscopic methods including FTIR and IR spectroscopy. A variety of modifications can be incorporated in peptide synthesis. For example, fatty acid deriviatized SP-C peptides can be synthesized utilizing selective deprotection of cysteine residues. Peptides with a variety of specific amino acid substitutions relative to native apoproteins can also be made to investigate the importance of specific charge and conformation characteristics in activity. Isotopic substitutions such as ^{13}C for ^{12}C in selected amino acids can also be incorporated to follow residue- specific changes in secondary structure during interactions with lipids.

K. Fatty Acids and Alcohols, Triglycerides, and Miscellaneous Surfactants

The majority of the surfactant materials above either are present in endogenous lung surfactant or are related structurally to its lipid and protein components. A variety of other surface active chemicals are also utilized in lung surfactant research. Examples include a number of synthetic components added to exogenous surfactants to enhance adsorption and spreading. As described in Chapter 14, palmitic acid (C16:0) and the neutral triglyceride tripalmitin are specific additives along with synthetic DPPC in the clinical exogenous surfactants Survanta and Surfactant-TA, while the C16:0 alcohol hexadecanol and the nonionic detergent polymer tyloxapol are constituents of Exosurf.

L. Clinical Exogenous Surfactants

The clinical exogenous surfactants currently used to treat RDS or ARDS in humans are detailed in Chapter 14. All of these surfactants contain materials from the categories of materials in Table 5-2. These preparations include ALEC (Britannia Pharmaceuticals, Redhill, Surrey, UK), Alveofact (Thomae GmbH, Biberach, Germany), bLES (bLES Biochemicals Inc., Ontario, Canada), Curosurf (Chiesi Farmaceutici, Parma, Italy), Exosurf (Glaxo-Wellcome, Research Triangle Park, NC, USA), Infasurf (ONY, Inc., Amherst, NY, USA), Surfactant-TA (Tokyo Tanabe, Tokyo, Japan), and Survanta (Abbott Laboratories, North Chicago, IL, USA). Two additional clinical exogenous surfactants, KL4 and recombinant SP-C surfactant (Byk Gulden Pharmaceutical, Konstanz, Germany), are now undergoing clinical evaluations of efficacy in patients, and additional clinical exogenous surfactants are also under development.

V. Categories of Animal Models for Studying the Physiological Activity of Lung Surfactants

In addition to a utilizing a range of materials, research on lung surfactants involves a variety of experimental methods including physiological assessments in animal models (e.g., [452, 455, 620, 643, 784, 913] for review). Animal research is of crucial importance in developing and optimizing surfactant replacement therapy for RDS and ARDS as detailed in Chapter 11. As an overview, animal models in lung surfactant research can be subdivided into those primarily involving surfactant deficiency and those primarily involving surfactant dysfunction or inactivation (Table 5-4). This categorization is not absolute, and animal models of surfactant deficiency may also contain surfactant dysfunction and vice versa. Animal models of surfactant deficiency are primarily relevant for RDS, while animal models of surfactant dysfunction and lung injury are most applicable for ARDS. Within these two broad categories, further subdivision allows specific features of different models to be delineated. Some animal models of surfactant deficiency, for example, focus primarily on pressure-volume (P-V) mechanics in small premature animals or surfactant-depleted adult animals. Other animal models of surfactant deficiency emphasize functional assessments of respiratory physiology. Many of these latter models involve larger premature animals where detailed measurements of lung function during mechanical ventilation are technically feasible and are made over a significant, although still acute, time scale. Examples of commonly measured lung functional variables are arterial partial pressure of oxygen (PaO2), arterial partial pressure of carbon dioxide (PaCO2), arterial-alveolar oxygen ratio (a/A ratio) and alveolar-arterial oxygen difference (A-a DO2). Also commonly determined are ventilator pressures and other clinically-relevant ventilator-associated variables such as ventilator rate, fraction of inspired oxygen (FiO2), mean airway pressure (MAP), peak inspiratory pressure (PIP), positive end-expiratory pressure (PEEP), inspiratory/expiratory (I/E) ratio,

Table 5-4 Categorization of Animal Models Involving Lung Surfactant Deficiency or Dysfunction

I. **Animal models of surfactant deficiency**
A. *Models emphasizing measurements of P-V mechanics*
Excised adult rat lungs depleted in surfactant by multiple lavage
Small premature fetal animals such as 27-day gestation rabbits
B. *Models emphasizing pulmonary functional measurements*
Premature large animals such as lambs, baboons, monkeys, canines
Adult animals depleted in surfactant by *in vivo* lavage (also used as an animal model of lung injury)
II. **Animal models of surfactant dysfunction in acute lung injury**
A. *Models of acute lung injury in adult animals*
Multiple lung injury initiators including hyperoxia, sepsis, vagotomy, aspiration, fatty acid infusion, viral infection, antibody- or toxin-induced lung injury, and *in vivo* lavage
B. *Models of acute lung injury in full-term newborn animals*
Multiple lung injury initiators similar to those used in adult animals above

Details on the tabulated animal models of surfactant deficiency and dysfunction and their use in research on surfactant therapy for RDS and ARDS-related acute lung injury are given in Chapter 11.

and so on. Lung functional and mechanical measurements are made not only in animal models of surfactant deficiency but also in animal models of surfactant dysfunction and acute lung injury. Models of acute lung injury utilize both adult and newborn animals and are commonly categorized by how the lung injury is caused. Many of the animal models of lung injury in Table 5-4 are relevant for clinical ARDS in adult or pediatric patients. The most meaningful assessments of exogenous surfactant activity utilize studies in multiple animal models, with physiological findings correlated with data on surface activity and composition. The use of animal models in studying surfactant replacement therapy for RDS and ARDS is covered in detail later (Chapter 11).

VI. Chapter Summary

Research on endogenous and exogenous lung surfactants is complicated by a variety of factors. Most lung surfactants are complex mixtures of biophysically interacting components with nontrivial molecular structures that are difficult to analyze spectroscopically or model theoretically. The cellular metabolism of pulmonary surfactant is also complex and is influenced by multiple mediators during lung development, growth, and injury. Pulmonary structure, mechanics, and function are similarly complicated and involve a variety of features and phenomena in addition to pulmonary surfactant and its activity. Correlated basic research across a range of disciplines is necessary in order to understand lung surfactants and their activity. This basic science understanding then has to be translated and integrated with clinical research on the efficacy of surfactant-based therapies for respiratory disease.

A broad range of materials is utilized in lung surfactant research. These include endogenous surfactant obtained by lavage or processed from lung tissue, organic solvent extracts of lavaged surfactant, organic solvent extracts of lung tissue, purified lung surfactant lipids and apoproteins, recombinant apoproteins, and a spectrum of synthetic phospholipids, peptides, phospholipid analogs, and other surface active compounds. Clinical exogenous surfactants for treating diseases of lung surfactant deficiency and dysfunction are also studied. Meaningful assessments of lung surfactants and their activity require that composition, surface activity, and physiological activity be viewed in combination rather than in isolation. In addition to biochemical and biophysical assessments, this leads to the study of animal models where physiological activity can be examined more comprehensively and quantitatively than in patients. Animal models in lung surfactant research can be categorized as those involving primarily surfactant deficiency or surfactant dysfunction. The former are typically used to study surfactant replacement interventions in RDS, while the latter involve acute lung injury and are most relevant for ARDS.

6

Discovery of Endogenous Lung Surfactant and Overview of Its Metabolism and Actions

I. Overview

This chapter gives an overview of the pulmonary surfactant system and its composition, surface behavior, and physiological actions. Coverage describes why surface tension and surfactants are important in the lungs and how the Laplace equation shows that surface active agents are necessary to reduce the work of breathing and to stabilize alveolar inflation and deflation. The discovery of lung surfactant is discussed, and its biochemical components are summarized. Alveolar type II epithelial cells and their crucial roles in the synthesis, secretion, and recycling of lung surfactant are also covered. The biophysical behaviors and surface properties responsible for the physiological activity of endogenous lung surfactant are also described. The direct effects of all lung surfactants on respiratory function follow from their specific adsorption and dynamic surface tension lowering properties as discussed here. Further details on the component-based surface active biophysics and physiological activity of endogenous and exogenous lung surfactants are given in subsequent chapters.

II. Pulmonary Surface Area

A network of multiple small communicating alveoli is required to provide a pulmonary gas exchange membrane of sufficient area to sustain life. Alveolar sizes and numbers vary with age and animal species but the lungs maintain a basic characteristic: their total surface area is hundreds of times larger than a single large sphere of equivalent volume. The total internal surface area in mammalian lungs at full expansion is of order 1 m^2 per kilogram of body weight. For a 70 kg human, pulmonary surface area is about 70 m^2 at total lung capacity [155, 1165].

This area is equivalent to that of a singles badminton court [742, 1081]. Human lungs contain approximately 3×10^8 alveoli with diameters ranging from about 75 to 300 microns in adults [155, 1165]. Figure 6-1 shows surface areas calculated for a pair of lungs, each containing from one to 150,000,000 spherical alveoli. More precise calculations based on measured alveolar morphometrics, tissue volumes, and packing are possible, but the estimates in Figure 6-1 illustrate the tremendous increase in surface area associated with a network of millions of small alveoli.

The common observation in the physiological literature that the surface area of the human lungs is equal to that of a tennis court is a slight overstatement. While an internal surface area of 70 m² is certainly impressive, it is actually only equal to that of a singles badminton court and is smaller than even one side of a singles tennis court. A full two-sided singles badminton court measures 17 ft × 44 ft (5.18 m × 13.41 m) [742, 1081], a surface area of 69.5 m². In contrast, one side of a singles tennis court measures 27 ft × 39 ft (8.23 m × 11.89 m²) [742, 1081], covering an area of 97.8 m².

Calculated lung surface area with different numbers of alveoli

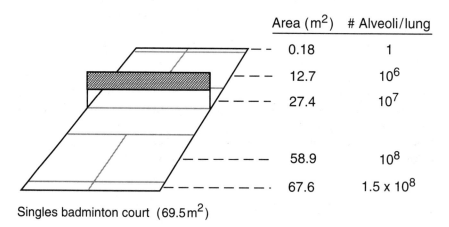

Area (m²)	# Alveoli/lung
0.18	1
12.7	10^6
27.4	10^7
58.9	10^8
67.6	1.5×10^8

Singles badminton court (69.5 m²)

Figure 6-1 Pulmonary surface area and alveolar number. Surface area is approximated on an imaginary singles badminton court for a pair of lungs, each containing 1, 1 million, 10 million, 100 million, or 150 million spherical alveoli. Total uncorrected alveolar volume is constant in all calculations (2.57 liters/lung). The tremendous increase in surface area as alveolar number increases is apparent. Area calculations for lungs with more than one alveolus have an arbitrary correction factor of 0.7 to account for packing limitations, septal tissue volume, and other phenomena that reduce volume and surface area *in vivo*.

III.   The Alveolar Hypophase and Surface Tension

Much of the internal surface of the mammalian lungs is covered by a thin liquid lining or "hypophase" (e.g., [303, 304, 306, 307, 832, 1167–1169, 1171]). The air-liquid interface associated with this hypophase generates surface tension forces. Surface tension is not a particularly strong force on a per area basis, but the very large wetted surface to volume ratio in the lungs leads to total surface tension forces of appreciable magnitude. These surface tension forces have a significant impact on the stability and mechanics of alveoli. As detailed in subsequent sections, the action of lung surfactant in moderating and varying surface tension as a function of alveolar size is essential for normal respiration.

> The existence of a thin liquid layer over a major portion of the alveolar surface in mammalian lungs has been documented in a variety of studies, although its volume, extent, and continuity are not known exactly. Electron microscopic studies indicate that the hypophase within each alveolus is variable in thickness and tends to be thickest in the septal corners [303, 304, 306, 307, 788, 1167–1169]. A continuous hypophase of uniform thickness is not required for rationalizing the importance of surface forces in the lungs or for theories of lung surfactant activity (Chapter 7). The overall volume of the alveolar hypophase is small since its average thickness is very thin. However, the hypophase still qualifies as a bulk liquid phase relative to molecular dimensions. For example, an aqueous layer of thickness 0.1–0.5 μm (1,000–5,000 Å) as observed in many portions of the alveolar network is more than adequate as a bulk liquid phase in molecular dimensions but contains a volume of only 0.1–0.5 ml/kg body weight if pulmonary surface area is 1 m²/kg. This volume is small compared to typical pulmonary gas volumes and does not adversely affect respiration or gas exchange.

IV.   Importance of Surface Tension Forces
     in Lung Mechanics

The importance of surface tension forces in pulmonary mechanics was demonstrated in 1929 by von Neergaard [1124]. Von Neergaard measured the pressure-volume (P-V) characteristics of animal lungs inflated with air and with isotonic saline. He showed that the pressure needed to inflate or maintain the lungs at fixed volume was greater when they were filled with air than with saline (Figure 6-2). Inflating or maintaining the lungs at fixed volume with air requires work against two kinds of forces: tissue forces and surface tension forces. Saline removes the liquid-air interface that generates surface tension forces, leaving only lung tissue forces. The difference between P-V curves for air and saline thus refects the contribution of surface tension forces to pulmonary mechanics. Analysis of this difference by von Neergaard [1124] indicated that surface tension was responsible for a majority of the force of static lung recoil. The primary contribution of

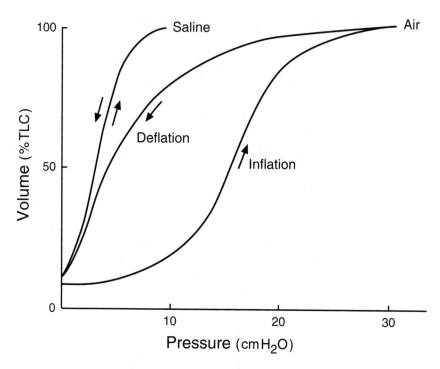

Figure 6-2 Differing pulmonary mechanics for air and liquid filling. The pressure necessary to inflate the lungs to any volume, or to maintain a given volume during deflation, is greater for air filling compared to saline filling. The difference between P-V behavior for air and saline reflects surface tension forces (see text). Surface tension forces contribute the majority of the static work of breathing. Note that the close approximation of air and saline P-V deflation curves at low lung volumes is consistent with very low pulmonary surface tension.

surface tension forces to pulmonary P-V mechanics during inflation and deflation with air has since been verified in a large number of experimental and theoretical studies (see Chapter 7 for further details).

V. Alveolar Stability and the Laplace Equation

The finding that surface tension forces dominate quasi-static pulmonary mechanics has direct consequences for the stability of the alveolar network during breathing. This follows from the Laplace equation, which relates the pressure drop (ΔP) across a spherical interface at equilibrium to its radius (R) and surface tension (σ) as:

$$\Delta P = 2\sigma/R \qquad\qquad (6\text{-}1)$$

Equation (6-1) predicts that, at equilibrium, a smaller alveolus requires a higher pressure to inflate or maintain it compared to a larger alveolus (Figure 6-3). If two spherical alveoli of different size but equal surface tension are connected, the smaller will tend to collapse into the larger. Similarly, if air at uniform pressure is added to such a system, the larger alveolus will expand preferentially over the smaller. Normal lungs do not exhibit this behavior. Although interconnected alveoli with a distribution of sizes are present, smaller airsacs are comparatively stable during expiration and are recruited effectively during inspiration. This implies that alveolar surface tension is not constant but instead varies as a function of alveolar radius during breathing.

   The Law of Young and Laplace. The Laplace equation for a spherical interface in Equation (6-1) is a special case of the more general law of Young and Laplace that holds for an interface of arbitrary shape defined by two principal radii of curvature R_1 and R_2 [5, 10, 183]:

$$\Delta P = \sigma(1/R_1 + 1/R_2) \qquad\qquad (6\text{-}2)$$

For a sphere, the two radii of curvature R_1 and R_2 are both equal to the radius R, and Equation (6-2) simplifies to the spherical Laplace equation: $\Delta P = 2\sigma/R$. For a cylinder, the axial radius of curvature is infinite and the law of Young and Laplace reduces to $\Delta P = \sigma/R$. This latter relationship is also sometimes referred to as the Laplace equation, but in this case holding for a cylindrical interface such as in a pulmonary airway or a blood vessel. The law of Young and Laplace in its general and simplified forms has a variety of applications in physics and biology. Aside from being used to analyze surface forces and their contributions to pulmonary mechanics and function, these range from helping to explain the vapor pressure and evaporation behavior of liquid droplets to describing cardiac and blood vessel mechanics. Equations (6-1) and (6-2) are derived from thermodynamics and thus incorporate a presumption of equilibrium. Analysis of the mechanics of nonspherical alveoli with Equation (6-2) leads to conclusions about stability and inflation identical to those in Figure 6-3, i.e., that alveolar surface tension in the normal lungs must vary as radius changes during breathing. Theories of stability taking into account these and other factors including alveolar septal interdependence and specialized fibrous architecture are noted in Chapter 7.

   Physical Nature of Surface Tension–Generated Pressure Differences. As opposed to osmotic pressure that arises from chemical potential differences and is not itself a physical pressure, surface tension–generated pressure differences from the law of Young and Laplace are directly measurable hydrostatic entities. A graphic demonstration of how the curvature of a liquid surface affects pressure is shown by vapor pressure variations in small liquid drops of different sizes. Thermodynamic theory related to that defined by the law of Young and Laplace predicts that the vapor pressure of a liquid in a droplet will vary inversely with the curvature of the droplet surface and proportionally to its surface tension [5, 10]. The magnitude of this effect is also dependent on the molar volume of the liquid. For water droplets at 20°C, calculations

Alveolar Instability by Laplace's Law

Two idealized alveoli, one larger than the other, require different
internal pressures at equilibrium if their surface tensions are equal

Small alveolus: Radius R_S
Surface tension σ
Internal pressure P_S
Outside pressure P_O

Large alveolus: Radius $R_L > R_S$
Surface tension σ
Internal pressure P_L
Outside pressure P_O

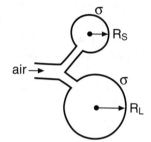

From Laplace's Law:

Pressure drop across small alveolus: $P_S - P_O = 2\sigma/R_S$

Pressure drop across large alveolus: $P_L - P_O = 2\sigma/R_L$

Comparison for $R_L > R_S$ gives: $P_S - P_O > P_L - P_O$

For constant outside pressure P_O: $\boxed{P_S > P_L}$

*This system is unstable: the small alveolus will tend to
collapse while the large one overinflates*

Figure 6-3 Instability of different-sized alveoli from the Laplace equation. The Laplace
equation shows that two spherical alveoli with different radii but equal surface tension
cannot coexist at equilibrium at uniform pressure. The smaller alveolus requires a higher
internal pressure to maintain its size and will tend to collapse into the larger. Conversely,
the larger alveolus will tend to overexpand at the expense of the smaller under a uniform
applied pressure. This stability problem can be avoided if surface tension is lowered and
varied as a function of alveolar size by pulmonary surfactant.

based on a constant surface tension and verified by experiment indicate that the actual vapor pressure over the curved droplet surface (vp′) is greater than that over a plane water surface (vp), approximately as follows [5]:

Water droplet radius (cm)	vp′/vp
10^{-4}	1.001
10^{-5}	1.011
10^{-6}	1.114
10^{-7}	2.95

Physical pressure differences stemming from surface tension forces have an important impact on pulmonary mechanics and stability as described in this chapter and in Chapter 7.

VI. Pulmonary Surface Tension Is Much Less Than That of Water

In addition to indicating that a variable surface tension is needed to stablize alveolar mechanics, the Laplace equation predicts that the average surface tension in the lungs is much less than that of water (70 mN/m at 37°C). If alveolar surface tension were 70 mN/m, the interfacial pressure drop necessary to inflate or maintain alveoli open against surface tension would be too large to be consistent with measured pressures in the pulmonary interstitium and intrapleural space

---

Rationale for the Existence of Pulmonary Surfactant

The lungs have a huge internal surface area (~1 m²/kg body weight), much of which is lined by a thin alveolar hypophase.

Surface tension forces at this air-hypophase interface dominate pulmonary static P-V mechanics.

Lung surfactant moderates alveolar surface tension forces and is essential for normal respiration.

By lowering and varying surface tension during breathing, lung surfactant reduces respiratory work and stabilizes alveoli against collapse and overexpansion.

(Figure 6-4). Interstitial pressures in the normal lung are found to be only slightly subatmospheric in the range of –1 to –10 cm H_2O (e.g., [832, 1171]). The much larger subatmospheric pressures calculated in Figure 6-4 would not only require an abnormally increased work of breathing but also generate a significant hydrostatic driving force for fluid filtration from pulmonary capillaries into the interstitium. Arguments such as these about the magnitude of pulmonary surface tension helped form the rationale suggesting the existence of surface active agents

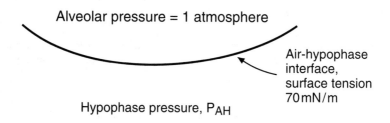

Alveolar pressure = 1 atmosphere

Air-hypophase interface, surface tension 70 mN/m

Hypophase pressure, P_{AH}

Alveolar Radius (microns)	Hypophase Pressure P_{AH} (cm H_2O)
10	-143
25	-57
50	-29
100	-14

These subatmospheric hydrostatic pressures are unreasonably large compared to normally measured interstitial and intrapleural pressures

Figure 6-4 Alveolar hypophase pressures calculated for the surface tension of water. Hypophase pressures calculated from the Laplace equation are shown for a spherical alveolar interface with radii of 10, 25, 50, and 100 microns and an assumed surface tension of 70 mN/m (water at 37°C). Calculated hypophase pressures are much more subatmospheric than measured pulmonary interstitial pressures of –1 to –10 cm H_2O (–1 to –7 torr) [832, 1171]. This indicates that alveolar surface tension is actually much lower than water, as verified by direct experimental measurements [952, 956, 958, 959] See text for details.

(surfactants) in the lungs in the 1950's. It has since been verified by direct experimental measurements using oil spreading techniques that surface tension in the lungs is indeed much less than that of water throughout the breathing cycle and approaches very low values at volumes near functional residual capacity [952, 956, 958, 959].

VII. Discovery of Pulmonary Surfactant

The presence of surface active molecules in the mammalian lungs was conclusively demonstrated in 1955 by Pattle [831], who studied the stability of air bubbles in foam squeezed from lung slices. By analyzing the lifetime of these bubbles, Pattle was able to show that surfactants capable of achieving very low surface tension were present. This result was extended by Clements [135], who demonstrated that washings from minced animal lungs reduced surface tension to low values (<10 mN/m) when compressed in a modified Wilhelmy surface balance. The composition of lung surfactant, and its source of production from type II pneumocytes, remained the subject of research for some time. However, the direct demonstration that surface active material of high activity was present in the lungs was itself significant for understanding pulmonary function and dysfunction.

The existence of pulmonary surfactant explained the stability and relative ease of inflation found in normal lungs. By lowering surface tension at all volumes, lung surfactant reduces the work of breathing and the pressure differences leading to instability (Figure 6-5). The surfactant film in each alveolus also varies surface tension in concert with radius during breathing. During expiration, alveolar radius decreases and surface tension is reduced in the compressed lung surfactant film at the air-hypophase interface. During inspiration, alveolar radius increases and surface tension rises in the expanded surface film. The ratio of surface tension to radius is thus more constant within any alveolus, and throughout the lung, in the presence of lung surfactant. From the Laplace equation, this minimizes pressure differences in different sized alveoli during breathing. The tendency for small alveoli to collapse at end-expiration is reduced, and the uniformity of alveolar recruitment and expansion is improved during inspiration. Other factors such as specialized fibrous tissue and the interdependent, septal-sharing design of the alveolar network also enhance stablility (Chapter 7). However, lung surfactant plays a paramount role in maintaining normal pulmonary mechanics. Within a short time after the existence of lung surfactant was established, its deficiency was being linked to the atelectasis and increased work of breathing found in premature infants with hyaline membrane disease [34, 356] (Chapter 10).

Schematic view of surfactant action

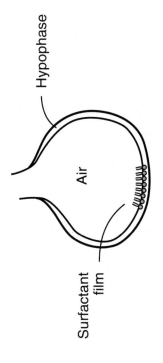

Hypophase

Air

Surfactant film

Conceptually:

When alveolar radius is small ⟶ the surfactant film is compressed and surface tension is small

When alveolar radius is large ⟶ the surfactant film is expanded and surface tension is larger

The ratio of surface tension to radius is more uniform in each alveolus and throughout the lung, stabilizing P-V behavior from Laplace's law.

Figure 6-5 Idealized lung surfactant action in an alveolus. Lung surfactant reduces surface tension in concert with alveolar size. When alveolar radius becomes small during expiration, the surfactant film is compressed and surface tension is reduced to low values. As radius increases during inspiration, the surface film expands and surface tension rises. This leads to a more uniform ratio of surface tension to radius during breathing, minimizing pressure inequalities from the Laplace equation as discussed in the text.

VIII. Overview of the Composition of Endogenous Lung Surfactant

The composition of endogenous lung surfactant is detailed more fully in Chapter 8 and is also described in a variety of review articles and book chapters.[1] Endogenous lung surfactant, also called natural surfactant or whole surfactant, is a complex mixture of lipids and proteins. Representative values for the composition of lavaged lung surfactant are approximately 85–90% phospholipid, 6–8% biologically active protein, and 4–7% neutral lipid by weight (Table 6-1). Lung surfactant composition varies somewhat among animals of different species and age, but certain basic compositional features are always present. The most abundant single component of lung surfactant is the disaturated phospholipid dipalmitoyl phosphatidylcholine (DPPC), which makes up about one third of total phospholipids. DPPC is not common in cell membranes and was recognized as an important surfactant component soon after the discovery that surface active agents were present in the lungs [102, 103]. In addition to DPPC, lung surfactant contains a spectrum of unsaturated and saturated phospholipids. Phosphatidylcholine (PC) is by far the most prevalent class, accounting for about 80% of total lung surfactant phospholipids when DPPC is included. Other phospholipid classes are present in much smaller amounts: phosphatidylglycerol (PG), phosphatidylinositol (PI), phosphatidylserine (PS), phosphatidylethanolamine (PE) and sphingomyelin (SPH). PC, PE, and SPH are zwitterionic near neutral pH, while PG, PI, and PS are anionic (Chapter 3).

There are four known lung surfactant proteins, imaginatively named surfactant proteins (SP)-A, B, C, and D. The first three of these proteins have crucial biophysical significance in lung surfactant function. SP-A makes up about 5% of lavaged lung surfactant on a weight basis and is isolated in intimate association with surfactant lipids during centrifugation. SP-B and SP-C together account for approximately 1.5% by weight of typical lavaged surfactant isolates and also are tightly associated with surfactant lipids. SP-D is less abundant than SP-A and is found largely in the lipid-depleted supernatant of centrifuged lavage rather than sedimenting with the majority of surfactant lipids and SP-A, B, and C [610, 613]. SP-D is also not present with packaged intracellular surfactant in type II cell lamellar bodies [169, 1129]. The existence and functional importance of specific proteins in lung surfactant was an area of active controversy through the 1970's. Even after the existence of functionally important proteins ("apoproteins") in endogenous surfactant was accepted in the 1980's, additional research was

---

1. Reviews of the composition and characteristics of lung surfactant lipids and proteins include, for example, Refs. 61, 148, 163, 362, 365, 419, 423, 529, 568, 614, 722, 786, 798, 865, 925, 936, 1105, 1162.

Table 6-1 Representative Composition of Lavaged
Endogenous Lung Surfactant

85–90% phospholipids
  DPPC plus other saturated and unsaturated phospholipids
  PC is by far the most prevalent phospholipid class
  PG, PI, PS, PE, Sph are also present
6–8% biophysically-active apoproteins
  SP-A
  SP-B
  SP-C
4–7% neutral lipids
  Primarily cholesterol

Tabulated values are representative only (weight percent). Compositional values for lavaged surfactant vary with species and age and with
the populations of aggregates obtained by centrifugation. Additional
variation is present for surfactant processed from tissue (Chapter 5).
Details of lung surfactant composition and component biophysics are
given in Chapter 8. PC, phosphatidycholine; PG, phosphatidylglycerol; PI, phosphatidylinositol; PS, phosphatidylserine; PE, phosphatidylethanolamine; Sph, sphingomyelin.

necessary to establish their precise molecular characteristics and behavior. Varying designations were initially used for lung surfactant proteins in the literature,
partly reflecting the different purification and characterization methods used in
early studies. The now standard nomenclature of SP-A, SP-B, and SP-C was
proposed by Possmayer [861] in 1988. SP-D was named according to the same
system when it was subsequently identified [844, 845]. In addition to phospholipids and proteins, lung surfactant isolates contain small amounts of neutral
lipids, primarily cholesterol. The molecular biophysical roles of different lung
surfactant components in surface active function are detailed in Chapter 8.

IX. The Pulmonary Surfactant System:
Lung Surfactant and the Type II Pneumocyte

Lung surfactant and the alveolar type II epitheliel cell are sometimes called the
pulmonary surfactant system (Figure 6-6). As described in the following sections, lung surfactant is synthesized, processed, packaged, secreted, and recycled
by type II pneumocytes. It is stored in characteristic lamellar body organelles
in the cytoplasm prior to secretion into the alveolar hypophase. Lung surfactant exists in the alveolar hypophase in the form of heterogeneous phospholipid-
rich aggregates including tubular myelin that range in size from nanometers to

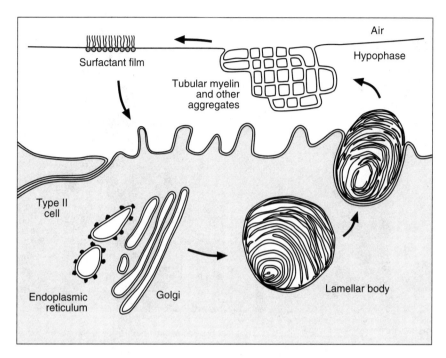

Figure 6-6 Schematic overview of the pulmonary surfactant system. Lung surfactant is synthesized in alveolar type II epithelial cells as a complex mixture of lipids and proteins. It is stored in lamellar body organelles and secreted into the alveolar hypophase, where it forms heterogeneous phospholipid-rich aggregates including tubular myelin. Lung surfactant adsorbs from these aggregates to form a film at the air-hypophase interface, which acts to lower and vary surface tension during breathing. Surfactant in the hypophase is eventually taken up back into the type II pneumocyte for recycling. See text for details. (Adapted from Ref. 319.)

microns. Lung surfactant adsorbs from these aggregates to form a film at the air-water interface. This film, compressed and expanded during breathing, is the ultimate site where surface tension is lowered and varied. After performing their physical function, surfactant molecules eventually leave the interface and are cleared from the alveolar space. An important part of this process involves reuptake by endocytosis back into type II cells, where surfactant components are recycled to reduce the need for *de novo* synthesis. An electron micrograph of a rat lung fixed by vascular perfusion showing the location of the alveolar surfactant film, and the associated hypophase and epithelium, is shown in Figure 6-7.

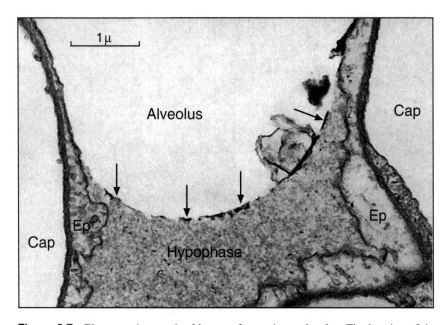

Figure 6-7 Electron micrograph of lung surfactant in an alveolus. The location of the interfacial film of surfactant in a rat alveolus is indicated by arrows. The alveolar hypophase, lumen, and epithelium are also visible, along with two capillaries. The lungs were fixed by vascular perfusion to maintain their structural features prior to examination by electron microscopy. Approximate magnification ×17,000. (From Ref. 1168 as reproduced in Ref. 942.)

X. Alveolar Type I and Type II Epithelial Cells

There are approximately 40 different cell types in the human lungs, but the alveolar epithelial lining contains only two: alveolar type I and type II epithelial cells. *Type I epithelial cells* have a long, thin cytoplasm and serve as the basic structural support cell for the epithelium. They cover about 90% of the alveolar surface but account for less than 10% of distal lung cells [158, 376].[2] Human type I cells have an average volume of about 1800 μm^3 and cover an area of about 5100 μm^2 [158]. The flattened cytoplasm of type I cells has a thickness <0.2 μm over large portions of the alveolar surface, providing a structural lining with very short diffusion distances for gas exchange between alveolar air and capillary blood. *Type II epithelial cells* (*type II pneumoctyes*) are cubiodal in shape and

---

2. Distal lung cells include type I and type II epithelial cells, capillary endothelial cells, interstitial cells, and macrophages.

cover about 10% of the alveolar surface while making up 10–15% of the cells in the distal lung [158, 159, 376] (Figure 6-8). Type II cells are more numerous than type I cells in the alveolar lining (about ⅔ of the epithelial cells), but they cover much less surface area. Human type II pneumocytes have a volume of approximately 900 μm^3 [158]. Alveolar type II pneumocytes are the primary cells of lung surfactant metabolism, and they also play other important roles in the normal and injured lung including being stem cells for the alveolar epithelium as noted below.

Methods Used to Isolate and Culture Type II Cells. Much of the understanding about alveolar type II epithelial cells and their various functions comes from studies on isolated cells. Type II pneumocytes can be isolated from the lungs by several strategies involving dissociation from the epithelium and extracellular matrix followed by density gradient centrifugation, flow cytometry, or other methodology to enrich the final purity and yield of type II cells (e.g., [212, 698] for review). Prior to dissociation, vascular perfusion and bronchoalveolar lavage with normal saline or buffers are employed to minimize the presence of blood cells and other free contaminating lung cells. Much effort has been directed at developing conditions to accomplish enzymatic dissociation and purification of type II cells while minimizing cell damage and optimizing yields (e.g., see [212] for comprehensive review of specific enzymes, concentrations, digestion times, enrichment methods, and other conditions used in type II cell isolation). The most common enzymes used for dissociation include trypsin and elastase, alone or together, with DNAase often added to reduce cell clumping; collagenase is also sometimes utilized. Enzymatic dissociation of type II cells was initially accomplished by Kikkawa and co-workers in 1974–5 [564, 565], folowed, for example, by Mason et al [695], Fisher and Furia [261], and Pfleger [849] in 1977. Finkelstein and co-workers [259, 260] were among those demonstrating the potential for cellular biochemical damage during dissociation and defining low protease concentration conditions to minimize it. After enzymatic dissociation, cells are harvested from the lung by lavage or mincing plus filtration. Differential sedimentation by density gradient centrifugation (e.g., Ficoll, Percoll, albumin gradients) or related techniques is then used to enrich type II cells and deplete contaminating cells. Type II cells have sizes and densities not greatly different from several other kinds of lung cells, particularly macrophages. Various strategies including particle ingestion have thus been used to deplete these latter cells in final type II cell isolates [212]. Type II cells are often studied in suspension immediately after isolation to avoid artifacts from subsequent culture. Density gradient purification methods have the ability to generate yields of type II cells sufficient for metabolic and other biochemical studies. Cell yields are not as large for type II cell isolation by laser flow cytometry [631, 632], although this technique can give highly pure cell populations. The culture of type II cells is quite difficult, since as epithelial cells and stem cells they tend to lose their differentiated characteristics and can become altered significantly when cultured over time. A variety of cellular enzymes, membrane receptors, and synthesis products, along with morphological criteria, have been used as markers to identify "normal" type II cells in culture. Dobbs [212] and Rannels and Rannels [885], for example, review strategies used to maintain or promote the differentiated phenotype of type II cells in culture.

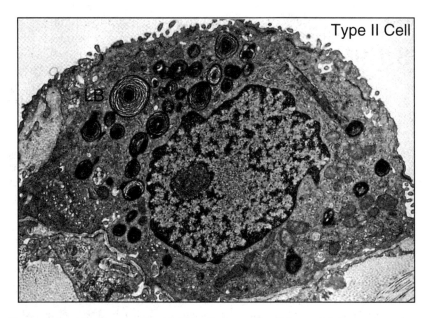

Figure 6-8   Alveolar type II epithelial cell. A type II cell from adult human lung shows characteristic cytoplasmic lamellar body organelles under electron microscopy. These organelles contain stored pulmonary surfactant. Type II pnuemocytes are the primary cells of lung surfactant metabolism. They are also stem cells for the alveolar epithelium and play other roles in pulmonary biology. Magnification ×8,000. (From Ref. 698 with permission.)

A.   Nonsurfactant Functions of Type II Cells

Although surfactant-related functions of type II cells are emphasized here, these cells have multiple roles in pulmonary biology and pathobiology ([212, 218, 258, 698, 1195] for review). Alveolar type II epithelial cells respond to and produce multiple cytokines, growth factors, and other mediators and products during lung growth, development, inflammation, injury, and repair. Examples of the many mediators produced by type II cells include transforming growth factors (TGF) α and β; interleukin (IL)-6, 8, 11; tumor necrosis factor (TNF) α; various complement components; monocyte chemoattractant protein (MCP)-1; and granulocyte-macrophage colony stimulating factor (GM-CSF) . Type II cells also respond to these mediators plus a host of other self-produced or externally elaborated factors including keratinocyte growth factor (KGF), epidermal growth factor (EGF), hepatocyte growth factor (HGF), fibroblast growth factor (FGF), IL-1, and many more. Some of the specific mediators that affect and help regulate surfactant metabolism in type II cells are noted in the following section. In addition to lung surfactant, type II cells produce a variety of products including extracellular matrix components like fibronectin, collagen, proteoglycans, and many others. Type II cells and their products and responses affect not only the alveolar epithelium and extracellular matrix but also the pulmonary microvasculature, airways, and interstitium. The role of type II cells as stem cells for the alveolar epithelium is particularly important. In many forms of lung injury, type I cells are damaged, causing type II cells to proliferate and dedifferentiate to repopulate the epithelium (e.g., [96, 1194]). Type II cell alterations during lung injury, whether direct or indirect, can add a component of surfactant deficiency into the pathophysiology of acute lung injury (Chapter 10).

XI. Overview of Lung Surfactant Synthesis and Metabolism in Type II Cells

The synthesis and metabolism of lung surfactant lipids and proteins in the type II pneumocyte are detailed in a number of comprehensive review articles and book chapters.[3] The type II pneumocyte was not positively identified as the cell responsible for pulmonary surfactant synthesis for some time after the initial demonstration that surfactant material was present in the lung. The alveolar type II cell is now known to be the primary site of synthesis and metabolism for all of the phospholipid and protein components of active lung surfactant. Neutral lipids including cholesterol are also synthesized in type II cells, although some of the cholesterol in lung surfactant may be plasma-derived. In addition to new synthesis, surfactant components are taken up by the type II cell from the alveolar lumen and reutilized in synthesis pathways or directly incorporated along with newly synthesized surfactant into lamellar body stores. A summary of surfactant synthesis, secretion, and recycling by type II cells is given below.

A. Intracellular Sites of Surfactant Synthesis and Processing in the Type II Cell ([61, 365, 419, 1106] for review)

Lung surfactant phospholipids and proteins are synthesized in the endoplasmic reticulum and are processed and transferred via the Golgi system to the lamellar bodies. The typical type II cell can have 100–150 membrane-bound lamellar body organelles that vary in size but are roughly 1 μm in diameter (cf. Figure 6-8). The lamellar bodies are the final storage location for lung surfactant prior to secretion. Under electron microscopy, these organelles are filled with distinctive multilamellae and concentric swirls formed by phospholipid bilayers. Electron-

---

3. See, for example, Refs. 61, 148, 163, 365, 419, 420, 423, 515, 520, 529, 614, 722, 865, 922, 925, 1105, 1106, 1162, 1199, 1201.

dense amorphous material, presumptively proteinaceous in nature, is also observed in lamellar bodies. SP-A, SP-B, and SP-C have been specifically demonstrated in lamellar bodies [280, 806, 808, 856, 1134, 1163], while SP-D is absent [169, 1129]. The intracellular movement of phospholipids in the type II cell involves specific transfer proteins for phosphatidylcholines and for anionic phospholipids. These transfer proteins may participate in segregating lung surfactant phospholipids from cell membrane phospholipids within the type II cell, although details of this process are unclear. Surfactant proteins SP-A, SP-B, and SP-C are transported from the Golgi to the lamellar bodies via multivesicular bodies [124, 1126, 1127]. In addition to transport by this pathway, some post-translational protein processing occurs at this time. For example, at least some proteolytic processing of SP-B is thought to occur in a related endosomal/ lysosomal compartment [424, 1126], possibly mediated by a cathepsin-like protease [1160]. There is still uncertainty about many of the intracellular sites and mechanisms for post-translational processing of surfactant proteins and their incorporation with surfactant phospholipids. However, it is clear that the majority of intracellular processing for surfactant phospholipids and proteins occurs prior to the lamellar body level. Lamellar bodies, for example, do not have the full complement of enzymes required for *de novo* synthesis or remodeling of phospholipids. Nonetheless, some final processing of lung surfactant phospholipids and/or proteins may occur within these organelles.

B. Synthesis of Surfactant Phospholipids
([61, 148, 365, 698, 922, 925, 1106] for review)

Lung surfactant phospholipids are synthesized in type II pneumocytes by pathways similar to those found in other cells and tissues, but with modifications and regulation to give a high content of disaturated and monoenoic phosphatidylcholines. The synthesis of all classes of diacyl glycerophospholipids requires phosphatidic acid formed by acylation of glycero-3-phosphate. Diacylglycerol derived from phosphatidic acid is then acted on by specific transferases to generate compounds with specific headgroups such as PC, PG, PI, and PE. Lung surfactant phosphatidylcholines are synthesized *de novo* almost exclusively by the CDP-choline incorporation pathway common to other cells. A good deal is known about the regulatory aspects of lung surfactant PC production in type II cells, with choline-phosphate cytidylyltransferase being one of the most important regulatory enzymes. About 25–45% of disaturated PC in lung surfactant has been reported to be synthesized via the choline incorporation pathway utilizing dipalmitoyl-*sn*-glycerol derived from dipalmitoylphosphatidic acid. However, the majority of disaturated PC in lung surfactant is thought to be formed through a specialized deacylation-reacylation mechanism (remodeling pathway) that converts monoenoic PC to disaturated PC. The final mix of lung surfactant phosphatidylcholines retains a significant content of monoenoic species, with only small amounts of compounds containing two or more double bonds. The anionic phospholipids PG and PI in lung surfactant are synthesized from CDP-diacylglycerol by well-defined pathways involving specific transferase enzymes. Less is known about the regulatory details of PG and PI synthesis compared to PC synthesis, including the extent of enzymatic specificity toward particular molecular species. The content and biophysical function of specific phospholipids in lung surfactant are discussed in Chapter 8.

C. Molecular Genetics and Synthesis of Surfactant Proteins in Type II Cells ([362, 418, 420, 423, 529, 614, 722, 925, 1162] for review)

All four surfactant proteins are synthesized in type II cells. However, only SP-C appears to be exclusively synthesized in this location, and all the remaining surfactant proteins have been shown also to be present or synthesized in other cellular locations (see following section). The molecular genetics of surfactant protein synthesis has been extensively studied and characterized. The gene for human SP-A is located on the long arm of chromosome 10 [107, 266]. Three different genes including one pseudogene for SP-A in humans have been reported, but the protein products vary only slightly. The primary translation product for human SP-A is a preproprotein containing 228 amino acid residues[4] plus a 20 amino acid signal peptide. This preproprotein then undergoes processing including signal peptide cleavage, proline residue hydroxylation, disulfide bond formation, glycosylation, acetylation, and carboxylation. The mature active form of SP-A is an oligomer containing 18 monomers organized as six sets of triplets [366, 1130]. The gene for SP-D is located on the long arm of chromosome 10 near that of SP-A [171]. SP-D has a number of structural and synthetic similarities to SP-A. SP-D is synthesized as an ~43 kDa preproprotein that is subsequently modified in an analogous fashion to SP-A, including oligomerization to a dodecamer containing four sets of triplet monomers [168, 671, 846]. The gene for human SP-B is located on chromosome 2 [234, 855]. Human SP-B is synthesized as a preproprotein with an approximate molecular mass of 43 kDa containing 381 amino acids including an N-terminal signal peptide [317, 422, 513]. The prepro form of SP-B undergoes extensive post-translational modification including removal of glycosylated sequences to generate a final active hydrophobic peptide [233, 317, 422, 513, 1126, 1160, 1163, 1205]. The final active form of human SP-B is 78–79 AA in length and has a molecular mass of 8.7 kDa [178, 533].

---

4. See Chapter 8 (Tables 8-5, 8-6) for primary amino acid sequences of human SP-A, SP-B, and SP-C. The molecular structure and biophysical behavior of surfactant apoproteins are also detailed in Chapter 8.

Intracellular processing of SP-B occurs prior to the level of the lamellar bodies, and only the small mature form is found in these organelles [1163]. The two genes for human SP-C are located on chromosome 8 [265, 315]. Human SP-C is synthesized as a proprotein 191–197 AA in length that does not contain a signal peptide and lacks potential glycosylation sites [267, 316, 1156]. This proprotein is subsequently processed to the final active human hydrophobic SP-C peptide that contains 35 AA and has a molecular mass of 4.2 kDa including two cysteine-linked palmitic acid moieties [179, 531, 534].

Extra-Alveolar Surfactant Protein Synthesis and Nonbiophysical Surfactant Functions ([172, 362, 614, 696, 857] for review). In addition to synthesis in type II cells, lung surfactant apoproteins have been found in alveolar macrophages and in bronchiolar cells including Clara cells (e.g., [32, 167, 170, 543, 650, 850, 1129, 1134, 1198]). Surfactant proteins in alveolar macrophages may at least in part reflect material taken up by phagocytosis, but mRNA consistent with synthesis of all surfactant proteins except SP-C has been found in nonciliated bronchiolar cells (Clara cells). The extra-alveolar distribution of surfactant proteins, particularly SP-A and SP-D, is consistent with biologic roles for these proteins outside lung surfactant biophysics. Both SP-A and SP-D are members of the collectin family of collagenous lectins important in host defense [172, 696]. SP-A, for example, is thought to participate in host defense in the alveoli and airways by enhancing macrophage-mediated phagocytosis of bacteria, viruses, and particulates [689, 1078, 1107, 1108]. SP-A also has important nonbiophysical roles at the alveolar level in helping to regulate lung surfactant secretion and reuptake by type II cells [214, 611, 896, 1015, 1202, 1203, 1212]. SP-D is also thought to participate in host defense and has been shown to affect gram-negative bacterial aggregation, to interact with bacterial lipopolysaccharides and viruses, and to facilitate macrophage function [411, 603, 604, 1109]. Studies have also shown that phospholipid components of lung surfactants can affect inflammatory cells and the production of inflammatory mediators [13, 23, 24, 857, 1005, 1210]. Potential roles of lung surfactant proteins and other surfactant components in host defense are described in detail elsewhere ([172, 362, 614, 696, 857] for review). In addition, all or some of the surfactant apoproteins may be involved in helping to regulate other metabolic, developmental, or growth-related processes in the lung. However, in terms of direct effects on pulmonary mechanics and stability, it is the surface active biophysics of the mix of lung surfactant lipids and apoproteins produced by type II pneumocytes that is of primary relevance as emphasized in coverage here.

D. Regulation of Type II Cell Maturation and Lung Surfactant Synthesis ([49, 148, 347, 420, 722, 864, 925, 1105, 1162] for review)

A variety of hormones, growth factors, cytokines, and other mediators influence and regulate type II cell maturation and lung surfactant synthesis. Messenger RNA for the two hydrophobic surfactant proteins appears in the human lungs as early as the end of the first trimester [650, 1178], while mRNA for SP-A is not present at 16–20 weeks and appears early in the third trimester [50, 1014]. A similar pattern is followed in all mammalian species, with mRNA for the hydrophobic proteins present prior to midgestation, mRNA for SP-A appearing after about 65–85% gestation, and SP-D appearing last [148, 722]. Phospholipids are present in the lungs of humans and animals very early in development, but lung surfactant phospholipid levels increase significantly along with protein over the last third of gestation [148]. Surfactant metabolism in type II cells is responsive to a variety of hormones including glucocorticoids, thyroid hormones, estrogens, androgens, and insulin. Type II cell maturation and function are also affected by growth factors such as epidermal growth factor (EGF), platelet-derived growth factor (PDGF), insulin-like growth factor (IGF), transforming growth factor (TGF)-β, and tumor necrosis factor (TNF)-α, among others. Additional mediators including interferon-γ and other cytokines also are known to influence surfactant synthesis. The details of specific receptor molecules and signal transduction pathways for type II cell maturation and surfactant synthesis are still not fully defined. The regulatory actions of glucocorticoid steroids on type II cells have received particular attention because of the early recognition that these substances could stimulate fetal lung maturation and increase surfactant levels (e.g., [199, 648, 649]). A large body of work has now verified the potent infuence of glucocorticoids on type II cell function *in vitro* and in the developing fetus ([49, 148, 864] for review). The maternal administration of steroids in the 24–72 hours prior to premature delivery is now a standard obstetric therapy to enhance fetal lung maturity [154, 173, 648] and can be synergistic with surfactant therapy [526, 547]. However, it is the balance between multiple regulatory mediators rather than just the actions of glucocorticoids alone that ultimately determines normal type II cell maturation and surfactant metabolism. The important influence of estrogens and androgens on surfactant-related type II cell functions, for example, is reflected in the higher rate of surfactant-deficient lung disease found in male vs female fetuses. Similarly, the regulatory importance of insulin is shown by the increased incidence of surfactant-deficient disease in infants of diabetic mothers.

E. Secretion of Lung Surfactant
([419, 515, 698, 865, 922, 925, 1105, 1199, 1201] for review)

Lung surfactant is secreted by exocytosis from type II cell lamellar body organelles into the alveolar hypophase. Estimates of basal secretion rates ranging from about 5 to 40% of lamellar body surfactant per hour have been reported. Newborn animals typically have higher rates of secretion on a weight-normalized basis compared to adults. *De novo* synthesis rates for many of the lipid and protein components of lung surfactant are lower than secretion rates, leading to the necessity for significant recycling of alveolar surfactant components (see following section). As reviewed for example by Jobe [515], the overall rate at which phosphatidylcholine moves from new synthesis through lamellar body storage and secretion is relatively slow in adult, newborn, and preterm animals. Radio-labeled palmitic acid starts to become incorporated into lung surfactant phos-phatidylcholine only minutes after intraveneous injection [512, 523], but peak alveolar levels in adult and newborn animals do not appear for 12–30 hours or even longer depending on the species studied [499, 512, 519, 521, 523, 525]. Preterm animals have an equivalent or longer delay between PC synthesis and secretion [496, 498, 521, 524]. Secretion itself appears to be regulated largely at the local level, and direct nerve-mediated control has not been demonstrated. Hyperventilation and mechanical stretch are strongly stimulatory to surfactant secretion (e.g., [772, 815, 1193]), although details of receptor proteins and signal transduction pathways are still uncertain. Chemical mediators known to stimulate secretion in type II cells *in vitro* include catecholamines and other β-adrenergic agents, ATP, protein kinases A and C, cAMP, phorbol esters, prostaglandins, and leukotrienes (e.g., [105, 213, 308, 731, 815, 854, 938, 1148]). SP-A is strongly inhibitory to lung surfactant secretion [214, 611, 896], and whole surfactant and surfactant lipids have also been reported to have an inhibitory effect on surfactant secretion from type II cells *in vitro* [214].

F. Surfactant Clearance, Reuptake, Recycling, and Catabolism
([61, 365, 419, 423, 515, 520, 865, 1199, 1201] for review)

Reuptake by endocytosis into type II cells is the major mechanism by which surfactant is removed from the alveolar lumen. The great majority of this surfactant is reutilized directly or indirectly to augment cellular surfactant stores rather than being lost from the alveolar compartment. Only about 10–15% of alveolar surfactant appears to be taken up into macrophages. Most of this surfactant is presumably degraded rather than reutilized, and this pathway prob-ably accounts for much of the loss from the alveolar compartment over time. A small fraction of 2–5% of alveolar surfactant is also thought to be cleared to the airways. Surfactant metabolism within the alveolar lumen is not well under-stood, although both biophysical and chemical processing in the hypophase have been proposed (see following section on surfactant subtypes). The recycling of alveolar surfactant phospholipids and apoproteins within the type II cell itself is well established. Some surfactant components are transported to the lamellar bodies without degradation and are combined intact with newly synthesized surfactant, while others are catabolized to products that are incorporated into synthesis pathways. The term "recycling" is sometimes used in the literature to refer to only the first of these processes and at other times to refer to both. Newborn and adult lungs differ in their balance of reutilization mechanisms for at least some lung surfactant components.

Turnover times from the alveolar lumen ranging from about 1 to 24 hours for surfactant lipids and proteins in animals have been reported.[5] These turnover rates reflect reuptake and reutilization in type II cells as well as loss from the alveolar compartment as noted above. Newborn animals have significantly slower weight-normalized rates of loss from the alveolar compartment (lumen and type II cell together) compared to adult animals, primarily as a result of a higher rate of recycling. The percentage of alveolar phosphatidylcholine that is recycled as opposed to being lost from the alveoli is of the order of 90% in many newborn animals, with lower efficiencies of 25–50% in adults and in preterm animals (e.g., [509, 515, 899]). On a per weight basis, full-term newborn animals tend to have larger intracellular and alveolar surfactant pools, a much higher rate of recycling, and a lower rate of synthesis relative to adult animals. Preterm animals generally have surfactant pools at or below the lower end of the pool sizes found in adults. Turnover rates for alveolar surfactant are component dependent as well as age dependent. The alveolar turnover of surfactant apoproteins is rapid, for example, with the majority of the cleared material entering recycling pathways and not lost from the lungs [56, 57, 493, 497, 502, 934, 1015, 1099, 1212]. Receptor-mediated binding may be involved in the uptake of at least some surfactant apoproteins (e.g., [934]). However, the details and regulation of type II cell uptake mechanisms for most lung surfactant components are still uncertain. SP-A has been shown to enhance the uptake of surfactant phospholipids into type II cells [1015, 1202, 1203, 1212], and the hydrophobic surfactant proteins may also influence this process [134, 897].

Recycling of Components in Exogenous Surfactants. Less is known about the type II cell uptake and recycling of specific components in exogenous lung surfactants compared to endogenous surfactant. However, recycling of phospholipids, proteins, and other components in exogenous surfactants by type II cells is known to occur [30, 393, 496, 497, 502, 508, 515, 520, 524, 1060]. For example, a significant portion of the DPPC in Survanta (Abbott/Ross Laboratories) apparently becomes incorporated into type II cell surfactant after instillation into preterm lambs, since only about 20–40% is recoverable in alveolar wash during the 24-hour postinstillation period and relatively little is lost from the lung [496]. DPPC in synthetic exogenous surfactants such as Exosurf (Glaxo-Wellcome, formerly Burroughs-Wellcome) and ALEC (Britannia Pharmaceuticals) has also been shown to be taken up from the alveolar lumen into type II cells [30, 393]. Tripalmitin in Survanta is recovered in a similar pattern to phosphatidylcholine, while palmitic acid in this preparation is rapidly cleared and almost completely incorporated into lung phospholipids [496, 1060]. As in the case of endogenous surfactant, the overall pulmonary lifetimes of many components in exog-

---

5. For example, see Refs. 56, 57, 389, 493, 496, 497, 499, 502, 508, 509, 510, 512, 519, 523–525, 685, 1099, 1202, 1212.

enous surfactants are longer in newborn animals than in adult animals due at least partly to higher rates of type II cell uptake and recycling. The entry of components in exogenous surfactants into endogenous recycling pathways can be an important indirect mechanism whereby long-term physiological benefits can be achieved by an instilled surfactant preparation independent of its surface active properties. Even exogenous surfactants that have relatively low surface activity can ultimately benefit lung function by combining with and supplementing intracellular surfactant pools, although their initial acute effects on respiration may not be large. Another indirect mechanism by which exogenous surfactants can improve lung function is by combining with components of type II cell-derived surfactant in the alveolar lumen. The influence of both of these indirect mechanisms can affect respiratory function and long term outcomes in patients treated with exogenous surfactants (Chapter 12).

G. Example Calculations of Metabolic Rates and Surfactant Pools in Newborn and Adult Animals

Data for the uptake and half-life of radiolabeled lung surfactant phospholipids can be analyzed to yield approximate cellular and alveolar pool sizes and metabolic flux rates. Although such calculations are far from precise theoretically and depend on data that are difficult to measure and exhibit significant variability, they do give an indication of the balance of processes involved. As an example showing the balance of synthesis, recycling, and loss in newborn vs adult animals, Jobe [515] compared steady-state estimates of these processes for lung surfactant phosphatidylcholine calculated from data in 3 day old and adult rabbits [509, 510, 512].[6] On a per kg body weight basis, newborn rabbits had a total PC synthesis rate of 0.46 µmole/hr, a lamellar body pool size of 38 µmoles, a secretion rate of 7.7 µmole/hr, an alveolar pool size of 77 µmoles, a reuptake (recycling) rate of 7.3 µmole/hr, and a loss rate of 0.46 µmole/hr (recycle efficiency of >90%). On the same per kg body weight basis, adult rabbits had a PC synthesis rate of 1.5 µmole/hr, a lamellar body pool size of 16 µmoles, a secretion rate of 2 µmole/hr, an alveolar pool size of 11µmoles, a reuptake rate of 0.5 µmole/hr, and a loss rate of 1.5 µmole/hr (recycle efficiency of 25%). Estimates of total phosphatidylcholine pool sizes (type II cells plus lavagable alveolar surfactant) of order 75–125 µmole/kg in newborns and 15–30 µmole/kg in adults have also been obtained in other animal species (e.g., [61, 515] for review). Preterm animals with surfactant deficient lung disease have estimated phosphatidylcholine pool sizes at or below the lower limit in adult animals. Infants with respiratory distress syndrome have been estimated to have alveolar phosphatidylcholine pools of 1–10 µmole/kg [8, 395].

---

6. Disaturated phosphatidylcholine (DSPC) and DPPC account for approximately 55–60% and 40%, respectively, of this total surfactant phosphatidylcholine (see Chapter 8).

XII. Lung Surfactant Microstructure and Subtypes

Lung surfactant exists in the alveolar hypophase as a heterogeneous population of phospholipid-rich aggregates ranging in size from nanometers to microns. Subpopulations of aggregates obtained by standard or density gradient centrifugation from suspensions of lavaged endogenous lung surfactant are sometimes termed surfactant "subfractions" or "subtypes."[7] Surfactant aggregates separated on the basis of size or density vary in their surface activity and content and ratio of phospholipids and proteins. Large aggregate subfractions are typically found to have higher surface activity than small aggregate subtypes [350, 379, 685, 872, 873, 1115, 1200]. The basis for this appears to be at least partly related to a greater content of surfactant apoproteins in large aggregates [685, 872, 1115, 1200]. It has been proposed that surfactant subtypes may reflect alveolar surface active material in various stages of function. Large aggregate subtypes, for example, may contain recently secreted material that is most active in adsorption and dynamic surface tension lowering. In contrast, smaller aggregate subtypes may reflect "spent" surfactant that has left the interface after repetitive cycling, i.e., a form of biophysical processing prior to reuptake of alveolar surfactant by type II cells for subsequent recycling. Chemical processing of alveolar surfactant by enzymes such as serine proteases is also thought to occur [64, 350–352]. Abnormal alterations in lung surfactant subtypes may also contribute to lung surfactant inactivation during lung injury [64, 348, 349, 379, 640, 871, 1114] (Chapter 9). The biophysics and biochemistry of alveolar surfactant subtypes, and their physiological significance, are the subject of ongoing research.

XIII. Tubular Myelin and Lung Surfactant Adsorption

One unique aggregated microstructure in endogenous lung surfactant is tubular myelin.[8] As described in Chapter 3, this characteristic microstructure appears under electron microscopy as a cross-hatched network of phospholipid bilayers organized by and containing specific surfactant proteins. SP-A, SP-B, and calcium are necessary to order phospholipids into the tubular myelin microstructure; SP-C is not required [1053, 1184]. Tubular myelin is important biophysically in endogenous lung surfactant because it has been associated with rapid adsorption to the air-water interface (e.g., [685, 791, 872]). However,

---

7. For example, see Refs. 64, 348–352, 379, 640, 685, 871–873, 1114, 1115, 1200 plus reviews in Refs. 61, 515, 1199, 1201.

8. For example, Refs. 413, 685, 791, 872, 1034, 1036, 1037, 1053, 1182–1184, 1200.

other aqueous phase aggregates of phospholipids and surfactant proteins also exhibit rapid adsorption. For example, chloroform:methanol extracts of lavaged lung surfactant lack SP-A and do not form tubular myelin but can be suspended in water to exhibit adsorption equivalent to that of whole surfactant [783, 791, 794, 1116] (Figure 6-9). Organic solvent extracts of processed lung tissue that lack SP-A and do not form tubular myelin also adsorb rapidly to the air-water interface [1062], as do a variety of mixtures of synthetic phospholipids and hydrophobic surfactant proteins or peptides ([560, 784, 792, 798] for review).

XIV. Summary of Physiological Actions of Lung Surfactant

The primary actions of lung surfactant in respiratory physiology are to reduce the work of breathing, stabilize small alveoli against collapse and large alveoli against overdistension, and to moderate interstitial pressures so as to decrease the hydrostatic driving force for edema fluid to filter out of pulmonary capillaries. These actions are crucial for normal gas exchange. Their conceptual rationale based on Laplace's law has been given earlier (e.g., Figures 6-4, 6-5 and accompanying text). A direct and explicit demonstration of the physiological actions of lung surfactant is provided by the pathology observed in surfactant-deficient premature infants with RDS. These infants exhibit increased work of breathing, decreased compliance, prominent atelectasis with reduced functional residual capacity, impaired gas exchange, and diffuse interstitial edema [606, 1061, 1075] (Chapter 10). Similar pathological features have also been documented in multiple animal models of prematurity and RDS (Chapter 11). The activity of lung surfactant in stabilizing alveoli has also been directly demonstrated in morphological experiments (e.g., [39, 42, 739, 1009]), and its action in reducing edema by moderating interstitial pressures has been veri-

---

Major Physiological Actions of Lung Surfactant

Reduces the work of breathing (increases lung compliance)
Increases alveolar stability against collapse
Improves uniformity of alveolar inflation
Reduces the pressure driving force for pulmonary edema

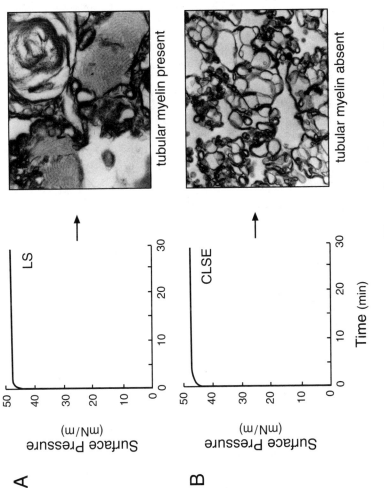

Figure 6-9 Adsorption of whole and extracted calf lung surfactant related to tubular myelin. Lung surfactant (LS) lavaged from calves adsorbs rapidly in the presence of calcium, and electron microscopy demonstrates tubular myelin and other phospholipid-rich aggregates in its complex microstructure (A, ×17,500). CLSE, obtained from LS by chloroform:methanol extraction, has equivalent adsorption but does not form tubular myelin (B, ×17,500). Phospholipid concentration 0.25 mg/ml for both LS and CLSE. (Data from Ref. 791.)

fied by pressure and fluid measurements in surfactant-sufficient and surfactant-deficient lungs [14, 767].

Additional Physiological Actions of Surfactants in the Airways. Although the primary physiological actions of type II cell–produced surfactant are at the alveolar level, surfactants can also affect physical behavior in other areas of the lung such as the airways. It has suggested that surfactant in the liquid lining layer of small airways would help to stabilize them against closure by lowering transmural pressure differences and by reducing liquid bridging [237, 245, 544, 681, 812] (see Chapter 7). Surface active material in the airways, including exogenous surfactants deposited there during replacement therapy, could also enhance clearance mechanisms by promoting surface transport toward the central airways or the epithelium [354, 723, 957] or by interacting with mucociliary clearance mechanisms [196, 745, 933]. Surfactant present endogenously in the airways is unlikely to be identical in composition and activity to surfactant secreted from alveolar type II cells. However, airway surfactant could include some cleared type II cell–derived material in addition to phospholipids and surfactant proteins synthesized in airway cells [32, 167, 170, 312, 313, 650, 850, 1129, 1134, 1198].

XV. Summary of the Biophysical Properties of Active Lung Surfactant

It is possible on conceptual grounds to define a qualitative set of surface properties necessary for lung surfactant to achieve its physiological effects on respiratory function and mechanics. These properties include the ability to adsorb rapidly at the air-water interface to form a surface film. In addition, this surface film must be capable of varying surface tension and reducing it to low values during compression, while also being able to respread effectively at the interface during expansion. The rationale for this set of surface behaviors is as follows (see Chapter 2 for general discussion of surface film behavior). Lung surfactant *in vivo* is initially secreted from type II pneumocytes into the alveolar hypophase. This secreted surface active material must adsorb to the air-water interface and form a surface film, since this is the site where surface tension lowering actually occurs. The adsorbed surfactant film must decrease surface tension to low values during dynamic compression in order for the work of breathing to be reduced appropriately. The film must also vary surface tension with surface area during cycling, since this is required by the Laplace equation in order to stabilize different-sized alveoli. Finally, the lung surfactant film must "respread" effectively during expansion to allow molecules squeezed out of the interface during compression to reenter and reintegrate back into the film. Respreading, along with continued adsorption from the bulk phase, ensures that sufficient interfacial surfactant is available to lower surface tension during subsequent cycles of expiration and

---

Surface Properties of Functional Lung Surfactant

Adsorb rapidly to the air-water interface
Reach low minimum surface tension during dynamic film compression
Vary surface tension with surface area during cycling
Respread well on successive cycles of the surface

---

inspiration. No single constituent of endogenous surfactant exhibits all of these surface properties. Instead, the components of endogenous lung surfactant interact biophysically to achieve the desired overall system behavior. Exogenous surfactants must also exhibit a similar set of surface behaviors in order to most effectively influence lung function and mechanics.

Quantitative Surface Activity Limits on Functional Lung Surfactants. In addition to the qualitative surface behaviors above, a variety of specific criteria for functional lung surfactants have been proposed in the literature (e.g., quantitative limits on film compressibility, area-dependent surface tension lowering, adsorption and spreading rates). Highly specific limits on required surface active properties for functional lung surfactants have attractions, but they are difficult if not impossible to validate rigorously given the limitations of theory (Chapters 5, 7). It has been suggested that physiologically effective lung surfactants must achieve minimum surface tension values of 10 dynes/cm or less [136–138], and this is probably conservative. Suspensions of lavaged endogenous lung surfactant, or organic solvent extracts prepared from it, have been shown to lower surface tension to <1 mN/m under dynamic cycling in multiple biophysical studies *in vitro* ([320, 784, 788, 798, 1105] for review). Surface tensions approaching this extremely low limit have also been measured in intact lungs at small volume [952, 956, 958, 959]. Empirical indices such as the stability index (SI) [137] and the recruitment index (RI) [797] have also been defined to help quantitate physiologically relevant surface tension lowering characteristics. Both these indices are defined so that lung surfactants can have physiological activity without lowering surface tension to <1 mN/m but predict that their effectiveness will increase as minimum surface tension decreases. The most complete assessments of the surface active properties of lung surfactants depend on complementary and detailed measurements of adsorption rate, equilibrium surface pressure, and surface tension lowering and respreading under dynamic cycling at physiologically relevant physical conditions. Biophysical methods for such measurements are discussed in Chapter 4, and theories of lung surfactant activity are detailed in Chapter 7.

XVI. Correlating Composition, Surface Activity, and Physiological Activity in Lung Surfactants

Subsequent chapters present an integrated view of the composition, surface activity, and physiological effects of endogenous and exogenous lung surfactants. The intrinsic connection between molecular composition, surface activity, and physiological activity in lung surfactants makes correlating these factors essential. The component-specific biophysics of endogenous and exogenous lung surfactants is central to their physiological and therapeutic effects. The chemical constituents in lung surfactants interact biophysically to generate surface active properties, which in turn determine physiological activity. Experiments and theories relating surface activity to lung mechanics and alveolar stability are described in Chapter 7. The chemical components and molecular biophysics underlying the functional composition of endogenous lung surfactant are detailed in Chapter 8. The characteristics and mechanisms involved in lung surfactant dysfunction or inactivation are described in Chapter 9 and the role of surfactant deficiency and dysfunction in human lung disease in Chapter 10. Correlations between surface and physiological activity for lung surfactants in animals and humans in the context of surfactant replacement therapy for RDS and ARDS are then detailed in Chapters 11–14.

Correlating Composition and Activity in Endogenous and Exogenous Lung Surfactants

Composition, surface activity, and physiological activity are inherently linked for any lung surfactant material.

Endogenous lung surfactant is a functional mixture; no single component has all the surface properties of the system as a whole.

Most exogenous lung surfactants are also functional mixtures of biophysically interacting components.

Correlating the component-based surface activity and physiological effects of lung surfactants is essential for understanding these materials as emphasized in subsequent chapters.

XVII. Chapter Summary

Much of the huge internal surface area of the lungs of air-breathing animals is covered by a thin liquid lining or "hypophase" that generates appreciable surface tension forces. The Laplace equation shows that if pulmonary surface tension were constant, small alveoli would collapse during expiration and larger alveoli would overinflate during inspiration. The Laplace equation also shows that pulmonary surface tension must be much less than that of water to be consistent with measured interstitial pressures and the normal work of breathing. Lung surfactant, discovered in the 1950's, lowers and varies surface tension as a function of area to allow normal respiratory mechanics and function. Endogenous surfactant is a complex mixture primarily containing phospholipids and three biophysically active proteins: SP-A, SP-B, and SP-C. A fourth surfactant protein, SP-D, does not influence biophysical function. The alveolar type II epithelial cell is the primary cell of lung surfactant metabolism. The components of lung surfactant are synthesized, processed, stored, secreted, and recycled by alveolar type II epithelial cells, as described in this chapter. The type II cell also plays a variety of roles in pulmonary biology and lung injury outside its crucial roles involving lung surfactant.

Lung surfactant acts physiologically to reduce respiratory work, stabilize small alveoli against collapse, make alveolar inflation more uniform, and reduce the hydrostatic driving force for pulmonary edema. Endogenous surfactant exists in the alveolar hypophase in complex phospholipid-rich aggregates including tubular myelin. To directly affect pulmonary mechanics and function, lung surfactant materials must adsorb rapidly to form an interfacial film that can lower and vary surface tension effectively during breathing. Films of lavaged alveolar surfactant can reach extremely low surface tensions <1 mN/m when compressed at rapid physiologic rates (Chapters 7, 8). In addition to adsorbing and lowering surface tension effectively, lung surfactant films must respread during cycling to maintain adequate amounts of interfacial material over time. No single component of endogenous lung surfactant has all these surface properties, and its overall activity reflects a sum of contributions from its lipid and protein components. The component biophysics and functional composition of lung surfactant are detailed in Chapter 8.

7

Analyses of Surfactant Activity and Its Contribution to Lung Mechanics and Stability

I. Overview

This chapter describes theoretical and experimental research relating the surface tension–area (σ-A) behavior of lung surfactant to pulmonary pressure-volume (P-V) mechanics. Also discussed are theories and experiments examining the role of lung surfactant in alveolar stability and theoretical work on the adsorption and dynamic film properties of phospholipids and lung surfactants. Coverage focuses on an overview of basic concepts and significance, without extensive mathematical detail. The thermodynamic relationship between the surface activity of lung surfactant and pulmonary pressure-volume work was recognized in the 1950's and has since been examined in a variety of studies. Research has also modeled surface activity theoretically and correlated surface active properties with effects on mechanics. An extensive body of theory and experiment has also addressed the contributions of surface and tissue forces to pulmonary biomechanics, as well as the importance of lung surfactant in alveolar stability as discussed here.

II. Relationships Between Surface Tension–Area and Pulmonary P-V Behavior

Theories and experiments of varying sophistication have addressed the relationship between alveolar surface tension forces and pulmonary P-V mechanics. As described in Chapter 6, the importance of surface tension forces in the mechanical behavior of the lungs was initially shown by von Neergaard in 1929 [1124]. The difference in P-V behavior found when lungs are inflated with air and saline provides a direct demonstration of the quantitative significance of forces at the liquid-air interface (surface tension forces). Air-filled lungs and saline-filled lungs both have tissue forces, but only the former have surface tension forces. Radford [880] was the first to show that the difference between P-V curves for air and saline could be analyzed from thermodynamics to obtain information about pulmonary surface area and surface tension. Radford's analysis was done just prior to the discovery of pulmonary surfactant and assumed that surface tension in the lungs was constant [880]. Subsequent workers modified this to account for variable surface tension so that surface tension–area (σ-A) behavior could be inferred from P-V mechanical measurements.

Initial theoretical treatments relating alveolar σ-A behavior and quasi-static[1] P-V mechanics were published by Mead et al [714], Brown [104], Radford [881], Clements et al [136, 137], and Gruenwald et al [358] soon after the demonstration that surface active material was present in the lungs [135, 831]. Despite the use of simplifying assumptions such as independent alveoli whose area changed in direct proportion to the two-thirds power of volume, these studies established the essential link between lung surfactant film properties and P-V mechanics ([41, 138] for review). Theory indicated that σ-A hysteresis in cycled pulmonary surfactant films was a major factor in the P-V hysteresis found when lungs are inflated and deflated with air [714]. Inflation-deflation hysteresis is almost completely abolished when lungs are inflated with saline so that surface tension forces are not present (e.g., [40, 41, 43, 1009]). Theory also allowed the calculation of σ-A isotherms from P-V data that verified that alveolar surface tension decreased with surface area during deflation and increased with surface area during inflation (Figure 7-1). This agreed with σ-A behavior measured in surface balance studies on films of pulmonary washings, although the correspondence of theory and experiment was not quantitatively exact. Early studies relating σ-A and P-V behavior are reviewed by Clements and Tierney [138] and Bachofen et al [41].

The mathematical relationship between pulmonary surface tension–area changes and quasi-static P-V mechanics follows from thermodynamics. As analyzed by Radford [880] and extended by Brown [104] and Bachofen et al [41], the differential component of respiratory work due to surface tension (dW) can be expressed as:

$$dW = P_s\, dV = \sigma\, dA \qquad (7\text{-}1)$$

where P_s is the component of pulmonary retractive pressure due to surface tension, V is volume, σ is surface tension, and A is surface area. The surface tension–related component of pressure P_s is determined as the difference in pressure measured in the

---

1. Surface tension forces are related most directly to quasi-static pulmonary P-V mechanics. Dynamic pulmonary mechanics and compliance are in addition strongly affected by flow-related pressure differences (e.g., see Refs. 503, 837 for review).

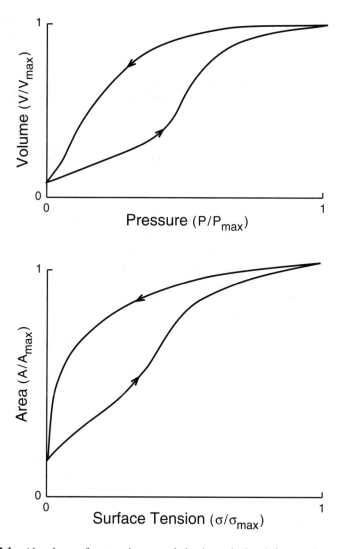

Figure 7-1 Alveolar surface tension–area behavior calculated from pulmonary P-V mechanics. *Top,* Dimensionless P-V curve measured in air-filled lungs. *Bottom,* Dimensionless surface tension–area (σ-A) diagram calculated from P-V curves for air-filled and saline-filled lungs. The calculated σ-A isotherm exhibits low minimum surface tension, variation of surface tension with area, and compression-expansion hysteresis. These features agree at least qualitatively with surface balance measurements on lung surfactant films. Curves shown are representative only. See text for details.

air-filled and saline-filled lungs at the same volume. As an approximation in initial studies, pulmonary surface area was assumed to be proportional to the two-thirds power of volume to give:

$$A = kV^{2/3} \tag{7-2}$$

where k is a proportionality constant called the shape factor. Equation (7-2) can be differentiated to relate dA and dV and the result substituted into Equation (7-1) to give:

$$\sigma = (3/2k)P_s V^{1/3} \tag{7-3}$$

The shape factor k is constant by assumption and can be determined if σ and P_s are known at any volume. This volume is often chosen as the full inflation volume (V_{max}), so that k is given by:

$$k = 3/2(V_{max})^{1/3}P_{s,max}/\sigma_{max} \tag{7-4}$$

where $P_{s,max}$ is the retractive pressure due to surface tension at V_{max}. The surface tension σ_{max} is generally assumed to be ~40–50 mN/m based on surface balance studies with lung surfactant films or on the known surface tension of tissue fluid. Once k is determined from Equation (7-4), a σ-A curve can be calculated from Equations (7-2) and (7-3) from pulmonary P-V data measured for air and saline filling. Alternatively, k can be eliminated from the analysis by combining Equations (7-3) and (7-4) to give:

$$\sigma/\sigma_{max} = (P_s/P_{s,max})(V/V_{max})^{1/3} \tag{7-5}$$

Equation (7-5) can then be used to plot σ/σ_{max} against dimensionless area A/A_{max} as calculated from normalized P-V data [41]. Although the above analysis contains approximations and is clearly not exact, it illustrates the direct linkage between alveolar σ-A changes and pulmonary P-V mechanics.

After initial studies relating alveolar σ-A changes to P-V mechanics were published, extensive theoretical and experimental research continued to analyze the mechanistic basis and relative magnitude of pulmonary surface tension and tissue forces.[2] Although results from such studies differ somewhat in detail, their consensus clearly indicates that alveolar surface tension and its variation with area have a significant influence on P-V mechanics. The majority of analyses also display features consistent with the surface behavior of endogenous lung surfactant as found in surface balance studies (e.g., [7, 135, 136, 287, 1032, 1157]). At low lung volumes, where alveolar surface tension is known to be low from direct oil spreading measurements [952, 956, 958, 959], theory indicates that the contribution of surface forces to quasi-static P-V behavior is small. As lung volume increases, the contribution of surface forces to mechanics becomes much

---

2. For example, Refs. 39, 40, 42–44, 268, 278, 293, 294, 304, 307, 340, 468, 472, 567, 1009, 1025–1028, 1171, 1187–1190.

**Direct Relation between Surface Tension–Area Behavior
and Pulmonary P-V Mechanics**

Quasi-static pulmonary P-V mechanics for air and saline filling
can be analyzed thermodynamically to infer alveolar surface
tension–area (σ-A) behavior.

Theory and experiment agree that surfactant-generated variations
in surface tension are important in P-V mechanics.

The difference in P-V behavior when lungs are inflated vs deflated
with air is primarily due to σ-A hysteresis in the surfactant film
during cycling.

Surface forces have little effect on normal mechanics at low
volumes where surface tension is low due to the action of lung
surfactant. Surface forces make a larger contribution as volume
increases, and tissue forces are limiting at high volumes.

---

larger. This agrees with the concept that the surfactant film at the alveolar
interface reduces surface tension appreciably during compression to low area and
varies it during cycling. At high lung volumes, tissue forces including those
associated with the chest wall and the stretched diaphragm ultimately become
dominant in restricting further expansion.

III. Experiments Correlating Surface Active
Properties with Effects on Mechanics

Although theory provides a framework within which data and interpretations
involving surface activity and mechanics must be consistent, it does not remove
the need for direct experimental measurements. Neither surface active properties
nor P-V mechanics can be predicted from the other with sufficient accuracy.
The precision of theory is limited by the mechanical and geometric complexity
of the lungs as well as by the compositional and biophysical complexity of the
surfactant system (Chapter 5). The importance of specific surface active behaviors in lung surfactant physiological function, however, can be demonstrated
experimentally. For example, it can be shown that surfactants that differ in
functionally relevant surface properties generate different mechanical responses
when instilled into surfactant-deficient lungs. A widely used experimental model
for studying such effects is excised rat lungs that have been depleted in endoge-

nous surfactant by repetitive lavage [73, 74, 382, 452, 492, 500, 1144].[3] In this model, each lung acts as its own control. Normal quasi-static P-V deflation mechanics are measured immediately after excision, followed by saline broncho-alveolar lavage to remove endogenous surfactant and measurement of surfactant-deficient mechanics. An exogenous surfactant is then instilled and its effectiveness in returning P-V deflation mechanics toward normal is assessed. Rigorous protocols for P-V measurements including degassing and stress relaxation procedures for each curve can be used to ensure mechanical data that are reproducible and free from leakage and gas-trapping artifacts [73, 74].

The physiological importance of several surface behaviors including rapid adsorption and the generation of low minimum surface tension can be demonstrated in the excised rat lung model. Five surfactants with known differences in surface active properties were instilled into the surfactant-deficient lungs: CLSE, 2:1 (w/w) CLSE:cholesterol, Survanta, DPPC, and Triton X-100 [73, 382, 784, 786]. These surfactants differ significantly in their ability to adsorb to the air-water interface, to lower and vary surface tension during dynamic compression, and to respread on successive cycles of the interface (Table 7-1). When instilled into surfactant-deficient rat lungs at a uniform dose, they are found to improve P-V mechanics in the following order: CLSE > Survanta > Triton X-100 ~ 2:1 CLSE:cholesterol ~ DPPC (Figure 7-2). This order of effectiveness allows inferences to be drawn about the importance of different surface behaviors in the physiological activity of lung surfactants.

CLSE, a chloroform:methanol extract of lung surfactant lavaged from calves, contains all of the lipids and hydrophobic proteins of endogenous surfactant but has hydrophilic apoprotein removed (Chapter 5). Like the alveolar surfactant from which it is derived, CLSE is able to reduce surface tension to <1 mN/m during dynamic cycling and also exhibits rapid adsorption and the ability to respread effectively during cycling (Table 7-1). The physiological correlate of these properties is that CLSE restores P-V mechanics almost to normal when instilled into surfactant-deficient rat lungs at the designated dose (curve 1, Figure 7-2). Survanta, an organic solvent extract of processed bovine lung tissue supplemented with synthetic DPPC, tripalmitin, and palmitic acid, has lower activity than CLSE in several surface properties (Table 7-1) and has an intermediate effect on P-V mechanics (curve 2). When large amounts of cholesterol (33% by weight) are added to CLSE to inhibit its surface tension lowering ability, the resulting mixture has little beneficial effect on mechanics (curve 4). The increased minimum surface tension of 2:1 CLSE:cholesterol (19 mN/m) compared to CLSE alone is the primary surface property difference between these

---

3.   The excised rat lung model and other animal models used in physiological assessments in lung surfactant research are described in detail in Chapter 11.

Table 7-1 Biophysical Properties of Different Surfactants Studied in Excised Lungs in Figure 7-2

Surfactant	Minimum surface tension (dynamic) (mN/m)	Adsorption surface tension (mN/m)	Postcollapse respreading (cycle 2/1)	Variation of surface tension (qualitative)
CLSE	<1	22	>0.9	Good
2:1 CLSE:cholesterol	19	22	>0.9	Good
Survanta	4	28	>0.9	Good
Triton X-100	27	27	>0.95	Poor
DPPC	20 (bubble) <1 (spread)	68	<0.4	Excellent

Minimum surface tensions are measured with the pulsating bubble surfactometer (37°C, 20 cycles/min) except spread film minimum for DPPC is from Wilhelmy balance studies at 37°C; adsorption surface tensions are plateau equilibrium values; respreading ratios (higher ratio means better respreading and 1 is perfect respreading) are representative from Wilhelmy balance studies at 37°C; variation of surface tension with area is qualitatively listed as poor, good, or excellent since specific values depend on multiple experimental variables.
Source: Data from Refs. 73, 284, 382, 796, 967, 1145.

two mixtures (Table 7-1). This comparison thus indicates that the ability to reach very low surface tensions during dynamic compression is important functionally in lung surfactants.

The functional importance of adsorption is indicated by contrasting the behavior of CLSE and DPPC in the excised lungs (curve 1 vs curve 4, Figure 7-2). DPPC adsorbs and spreads very poorly compared to CLSE (Table 7-1). Because of its poor adsorption, suspensions of DPPC are not able to form an interfacial film of sufficient concentration to lower surface tension effectively when studied on the oscillating bubble apparatus. DPPC, however, does have the intrinsic ability to reduce surface tension to very low values <1 mN/m if spread directly in a film at the air-water interface by means of a spreading solvent (Table 7-1). The poor adsorption of DPPC and its poor respreading correlate with its inability to improve P-V mechanics when instilled as a suspension in lavaged excised lungs (curve 4, Figure 7-2). However, rapid adsorption and the ability to respread do not result in high physiological activity unless a low minimum surface tension is also reached. Both 2:1 CLSE:cholesterol and Triton X-100 adsorb and spread much better than DPPC, but they do not reach low minimum surface tension during dynamic compression and give minimal improvements in P-V mechanics (curves 3, 4). The direct relevance of the surface active proper-

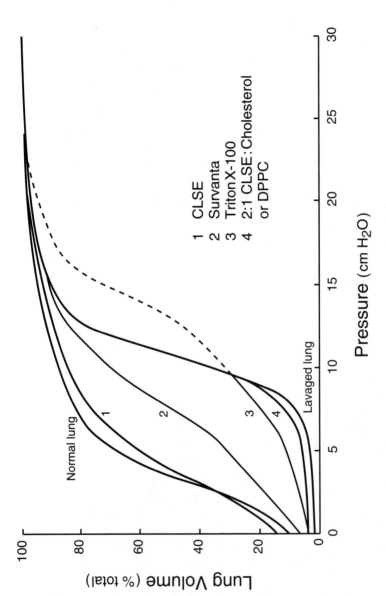

Figure 7-2 Effects of surfactants with differing biophysical properties on P-V mechanics in excised rat lungs at 37°C. Each surfactant in Table 7-1 was instilled into rat lungs made surfactant-deficient by a standardized lavage protocol. Normal P-V deflation mechanics were measured for the freshly excised lung prior to lavage. Deflation curves 1–4 show the response of the lavaged, surfactant-deficient lungs to instillation of 25 mg of (1) CLSE; (2) Survanta; (3) Triton X-100; (4) 2:1 CLSE:cholesterol or DPPC. Instillation volume was uniform at 2.5 ml. (Data from Refs. *73, 382, 784, 786*.)

ties of lung surfactants for physiological function is discussed further in subsequent chapters.

Stability Index and Recruitment Index. Two empirical indices designed to reflect the importance of lung surfactant surface tension–area properties in alveolar mechanics are the stability index (SI) [137] and the recruitment index (RI) [797]. The SI parameter introduced by Clements et al [137] is defined as:

$$SI = 2(\sigma_{max} - \sigma_{min})/(\sigma_{max} + \sigma_{min}) \qquad (7\text{-}6)$$

where σ_{max} and σ_{min} are the maximum and minimum surface tensions, respectively, measured for the surfactant film. The functional form of SI emphasizes that the stablilizing influence of a surfactant on alveolar mechanics involves both how much it lowers surface tension and how much it changes surface tension during cycling. As σ_{min} approaches 0, the variation of surface tension relative to σ_{max} has less influence on SI. The magnitude of SI can range from zero (a totally ineffective surfactant which has $\sigma_{max} = \sigma_{min}$) to two (a surfactant that has a $\sigma_{min} = 0$ regardless of σ_{max}). The RI parameter defined by Notter et al [797] emphasizes the role of lung surfactant in recruiting alveoli during inhalation in addition to stablilizing them at end-expiration:

$$RI = (\sigma_{min} + \Delta\sigma_{IE})/\sigma_{min} \qquad (7\text{-}7)$$

where $\Delta\sigma_{IE}$ is the rise in surface tension (fall in surface pressure) in the initial region of the expansion curve after compression to low surface tension. Conceptually, this initial rise in surface tension promotes alveolar recruitment [797]. As an alveolus begins to inflate and the surfactant film expands, surface tension increases. From the Laplace equation, this increase in surface tension increases the pressure needed for further inflation, leading to the recruitment of other alveoli. Because this effect is proportional to the percentage increase in surface tension, which depends on the value reached at end-compression, σ_{min} is also incorporated in RI. By definition, RI increases as surfactants generate a larger relative initial rise in surface tension and becomes infinite if σ_{min} actually reaches zero. Both SI and RI predict that surfactants will have the greatest effect on alveolar stability and recruitment if they achieve very low minimum surface tensions during cycling [137, 797]. The fact that endogenous lung surfactant is able to reach actual minimum surface tensions <1 mN/m under rapid, physiologic dynamic compression is described more fully below.

IV. Theories and Experiments on Low Minimum Surface Tensions in Dynamically Compressed Lung Surfactant Films

Because of its importance for physiological activity, the minimum surface tension reached by films of phospholipids and lung surfactants during dynamic compression has received significant attention. Because surface tension is a thermodynamic quantity, there has been some controversy about the interpretation of surface tension measurements in monolayers that are studied under nonequilib-rium conditions. A major part of the physiologically relevant surface activity of lung surfactants is generated during dynamic cycling. As discussed earlier in Chapters 2 and 3 and also below, dynamic cycling allows films of some surface active agents to achieve surface tension lowering substantially below equilibrium limits. These low nonequilibrium surface tensions are not only dynamically generated but also are typically associated with high degrees of film compression where monolayer stability limits can be exceeded. However, film behavior under such conditions can in many cases be viewed as sufficiently *metastable* that conceptual difficulties with thermodynamic arguments and interpretations do not arise [301, 785]. Minimum surface tension values near or below 1 mN/m in dynamically compressed films containing a substantial content of rigid dis-aturated phospholipids like DPPC have been shown to be stable over time scales in excess of normal rates of breathing [236, 321, 378, 795, 955, 1059]. Whether the reality of these very low interfacial forces is best interpreted as surface tension or as some other equivalent but nonthermodynamic quantity is effectively moot as far as the lungs are concerned. Physiologically, minimizing these forces will be equally significant in influencing alveolar stability and the work of breathing.

A. Factors Affecting Minimum Surface Tension in Surface Films

The minimum surface tension in any surfactant film is a function of multiple variables including cycling rate, temperature, extent of compression, and surfac-tant concentration (Chapter 2). Compression rate as noted above is a major factor affecting minimum surface tensions in films of phospholipids and lung surfac-tants, with lower values typically reached as cycling rate increases. Increases in temperature, on the other hand, tend to fluidize surface films so that higher minimum surface tension values are generated. Minimum surface tension is also affected by the extent of surface area compression. Surfactant films com-pressed to small areas and high surface concentrations generally reduce sur-face tension to lower values. Meaningful assessments of minimum surface tensions for lung surfactants require knowing temperature, extent of compres-sion, cycling rate, and initial film concentration. Minimum surface tension can also be affected by whether adsorbed films or spread films are studied. In this case, the bulk phase concentration of surfactant as well as the surface concentration becomes an important variable affecting minimum surface ten-sion. Moreover, surfactants such as DPPC that adsorb poorly can have greatly elevated minimum surface tensions when film formation is by this mechanism as opposed to direct deposition at the air-water interface in a volatile spread-ing solvent.

B. Ability of Interfacial Films of DPPC and Endogenous Lung Surfactant to Achieve Minimum Surface Tensions <1 mN/m Under Dynamic Compression

Phospholipids have maximum equilibrium spreading pressures equivalent to surface tensions of 22–23 mN/m at 37°C [22, 301, 852, 1059], providing an effective limit on the equilibrim surface tension lowering of endogenous lung surfactant during adsorption. Under dynamic compression at physiologic rates, however, films of endogenous surfactant and disaturated phospholipids such as DPPC are able to reduce surface tension to values significantly below equilibrium limits.[4] This behavior relates to the ability of DPPC and related rigid gel phase phospholipids to achieve metastable, dynamically generated surface pressures in the compressed film as detailed in Chapter 3 (see Figures 3-6, 3-7, 3-8, and associated text). Although minimum surface tensions of "zero" are sometimes reported for lung surfactant films, no surfactant film precisely reaches this value. A stable gas-liquid interface always has a finite, positive surface tension. Given this caveat, however, surface tension can indeed be very small, and this is the case with the most active lung surfactants. Experimental evidence is unequivocal that extremely low surface tensions ≤1 mN/m are actually reached by films of DPPC and whole or extracted endogenous lung surfactant under rapid interfacial cycling at physiologically relevant rates and area compressions at 37°C (e.g., see citations in footnote 4).

The ability of lung surfactant films to reach physically real surface tensions ≤1 mN/m has been questioned on conceptual grounds (e.g., [53]). However, rigorous analysis of deformed air bubbles in suspensions of lung surfactant during dynamic cycling directly demonstrates the existence of surface tension forces of this extraordinarily low magnitude [378, 953, 955]. Figure 7-3 shows an example of oblate spheriodal deformation that occurs even in very small air bubbles upon pulsation to 50% area at a rate of 20 cycles/min in a suspension of calf lung surfactant extract at 37°C. This deformation occurs despite the tiny size of the air bubble in question (radius of order 0.4 mm) and indicates that surface tension forces are very small. Calculations from interfacial theory demonstrate that the air bubble in Figure 7-3 has a surface tension of order 1 mN/m [378]. This low surface tension was found to be stable over 30–60 seconds when bubble pulsation was halted at minimum area. Surface tensions ≤1 mN/m have also been measured explicitly when concentrated films of DPPC are spread from organic solvents and cycled at the air-water interface in a Wilhelmy balance even at slow rates of 1–10 min/cycle at 37°C [416, 796, 1093]. Moreover, oil spreading experiments in intact

---

4. For example, Refs. 236, 321, 378–380, 415, 416, 719, 796, 872–875, 877, 951, 953–955, 1093, 1145, 1214.

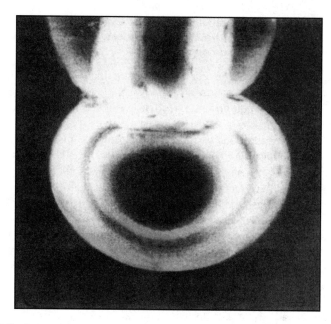

Figure 7-3 Low minimum surface tension in a pulsated air bubble in lung surfactant. A tiny air bubble pulsated to 50% area at a rate of 20 cycles/min in a suspension of calf lung surfactant extract deforms to an oblate spheroid as shown in the photograph. The deformed bubble shape is stable over 30–60 seconds when cycling is halted at end-compression. Analysis using rigorous interfacial theory shows that the actual surface tension in the deformed bubble is of order 1 mN/m. See text for details. (From Ref. 378.)

lungs indicate that surface tensions of 1–3 mN/m are actually reached by lung surfactant *in situ* at low lung volumes at 37°C [952, 956, 958, 959].

   Molecular Mechanisms of Dynamic Surface Tension Lowering in Phospholipid-Rich Films. The ability of films with a significant content of disaturated phospholipids like DPPC to achieve nonequilibrium states during dynamic compression is related to the molecular characteristics of the rigid gel phase (Chapter 3). If the temperature of such films is raised above the gel to liquid crystal transitions of the major components, the ability to achieve low, nonequilibrium surface tensions during dynamic compression is either lost or seriously impaired. Films of fluid liquid crystal phase phospholipids cannot generate minimum surface tensions substantially below equilibrium limits during dynamic compression [416, 560, 785, 796]. However, the specific molecular mechanisms that allow surface films of gel phase disaturated phosphatidylcholines and related phospholipids to reduce surface tension well below equilibrium during dynamic compression are not well understood. The nonequilibrium surface states achieved by these films presumptively have increased packing density and molecular order conducive to reduced interfacial tension. Regardless of the underlying molecular mechanisms, the ability of disaturated phospholipid and lung surfactant films to reduce interfacial tension to very low values is well established as described above. The physiological importance of this behavior is also clear, and theoretical understanding suggests that the greatest activity in lungs is associated with surfactants reaching the lowest minimum surface tension values (e.g., [136–138, 797, 798]).

V. Theoretical Modeling of Lung Surfactant Adsorption and Dynamic Film Behavior

A large body of thermodynamic and interfacial theory has been applied to the adsorption and film behavior of surface active materials (see texts such as Refs. 5, 10, 198, 295, 991 for detailed discussion). Mathematical predictions and modeling of the surface behavior and properties of lung surfactants can provide useful information to supplement experimental measurements. However, application of interfacial theory to typical lung surfactants is difficult because of their multicomponent composition and the complex dynamic phenomena that contribute to overall system behavior. A number of theoretical analyses of phospholipid and lung surfactant adsorption, spreading, and dynamic surface activity are available (e.g., [117, 118, 378, 473, 527, 811]). Also analyzed has been the action of surfactants in creating surface tension gradients in airway lining fluid that might antagonize airway closure [297, 400, 544, 681, 812], influence the clearance of materials from the deep lung [354, 723], or affect the spreading of tracheally instilled exogenous preparations [246, 355].

Because of its compositional and functional importance in lung surfactant, several theoretical studies have analyzed the adsorption of the disaturated phospholipid DPPC [118, 527]. The extremely low critical micelle concentration of DPPC [1010], together with the aggregated aqueous phase microstructure of phospholipids in general [22, 301, 985, 1007], strongly suggests that aggregates are important in lung surfactant adsorption. Johannsen et al [527] examined several one-dimensional models of adsorption and diffusion and concluded that particulates played a key role in the adsorption of DPPC in dispersed lung surfactant, either by transporting into the interface themselves or by increasing the concentration of dissolved or adsorbable DPPC. Keough [561] has also calculated that the adsorption of individual phospholipid molecules is unlikely to be a major contributor to lung surfactant adsorption, and Schürch et al [961] have reported that a porcine lung surfactant extract adsorbs with the entry of aggregates containing large numbers of lipid molecules into the air-water interface. More quantitative models of lung surfactant adsorption will require better information on rate constants and mechanisms, including whether preferential adsorption of some components occurs and if material at the interface affects the subsequent entry of additional molecules.

The surface tension–area behavior of dynamically cycled surfactant films can also be modeled theroetically. In particular, several authors have examined the surface tension lowering properties of adsorbing surfactants at the interface of a dynamically pulsated air bubble [117, 378, 473, 811]. In addition to adsorption, such models must account for complex dynamic phenomena including viscous effects as well as film collapse, spreading, and refinement during cycling. When dynamically compressed to small areas, lung surfactant films become enriched in DPPC and related rigid disaturated phospholipids through the selective ejection or "squeeze-out" of more fluid molecules to achieve more effective surface tension lowering (Chapters 2, 3, 8). Squeeze-out and monolayer collapse during compression also influence how molecules respread back into the film during expansion to enhance surface activity on subsequent cycles. Modeling these behaviors is complicated by the fact that they are not completely understood at the molecular level. Lung surfactant films with a multicomponent composition that varies during cycling depending on the balance of adsorption, desorption, squeeze-out, collapse, and spreading are very difficult to model precisely. Surface tension–area isotherms and hysteresis loops qualitatively similar to those found experimentally can be obtained from theory [473, 811], but their interpretation is ambiguous. This is also the case for theoretical models of the spreading of tracheally instilled exogenous surfactants, which disagree on whether spreading will be helped or hindered if endogenous surfactant is also present [246, 355]. As in the case of adsorption above, the utility and accuracy of theoretical models of lung surfactant dynamic film behavior may improve as additional

---

**Theoretical Analyses of Lung Surfactant
Adsorption and Film Behavior**

The adsorption and film behavior of lung surfactants can be analyzed based on the principles of interfacial phenomena.

Theory indicates that lung surfactant phospholipids adsorb as groups of molecules as well as single molecules.

Theories of dynamic surface tension–area behavior are limited by the complex composition and phenomenology of lung surfactants.

The limitations of theory make direct experimental measurements of surface activity essential for lung surfactants.

quantitative information on relevant parameters and molecular mechanisms becomes available.

VI. Theoretical Analyses of the Role of Lung Surfactant in Alveolar Stability

As indicated earlier, initial views about the importance of lung surfactant in pulmonary mechanics and alveolar stability were developed in the late 1950's and early 1960's (e.g., [136–138, 714, 829, 830, 942]). The simplified concept that lung surfactant acts alone to stabilize independent, spherical alveoli at equilibrium is clearly not exact and neglects several factors including:

1. The interdependence of the alveolar network, where tissue septa border alveoli on two sides and are connected structurally to the entire network [293, 294, 890, 1187]
2. The existence of specialized fibrous tissue support for the alveoli *in vivo* [1166, 1167, 1169]
3. The dynamic nature of breathing and the complex morphology of alveoli [1165, 1167, 1169], which limit the applicability of the Laplace equation for spherical air-sacs at equilibrium

However, while differences in detail emerge when these and other factors are taken into account in more sophisticated theories, the ultimate conclusion that lung surfactant has important benefits in stabilizing alveoli against collapse and overdistension remains unchanged.

The many factors affecting alveolar stability have been addressed in theoretical and experimental research since the 1960's.[5] A number of studies have suggested that because of alveolar network interdependence and/or specialized support architecture, the stabilizing role of lung surfactant is less important than originally thought [293, 294, 890, 1166]. Alveolar interdependence and specialized connective tissue fibers clearly help to stabilize alveoli against collapse during expiration and to prevent overdistension during inspiration. However, the vast majority of biomechanical theories indicate that the moderation and variation of surface tension forces by lung surfactant are important in enhancing alveolar stability ([320, 340, 468, 1025] for review). This conclusion is also well documented by direct experimental observation. If lungs are rinsed with detergents or other substances to impair or remove lung surfactant and increase surface tension, alveolar collapse, uneven inflation, or decreased surface to volume ratio is observed (e.g., [39, 42, 44, 739, 1009]). Perhaps the most dramatic demonstration

---

5. For example, Refs. 39, 41–45, 293, 294, 304, 307, 320, 340, 468, 544, 797, 812, 890, 937, 941, 1009, 1025–1028, 1166, 1167, 1169, 1187–1190.

Role of Lung Surfactant in Alveolar Stability

Alveolar stability is enhanced by multiple factors including:
Pulmonary surfactant
Network structural interdependence
Specialized fibrous architecture

The consensus of biomechanical theory and experiment is that lung surfactant helps to stabilize alveoli against collapse and overdistension.

The stabilizing role of lung surfactant is shown unequivocally by the alveolar collapse and uneven inflation found in diseases where it is deficient or dysfunctional.

---

that lung surfactant is crucial for alveolar stability is given by the prominent pathological findings of alveolar collapse and overdistension demonstrated in premature infants with RDS (Chapter 10).

VII. The Alveolar Hypophase and Theories of Lung Surfactant Action

Views about the physiological roles and importance of lung surfactant have also been affected by uncertainties about the extent of the alveolar hypophase. The presumptive importance of surface tension forces in pulmonary mechanics depends on an extensive air-water interface to generate them. Due to the technical difficulty of visualizing a thin liquid lining in microscopic studies, as well as the imprecision of tracer experiments and related methods, the alveolar hypophase *in vivo* is nontrivial to view and analyze. However, the reality of significant pulmonary surface tension forces has been demonstrated repeatedly in P-V mechanical studies as discussed earlier. Although questions about the existence of the hypophase have been raised [433, 434], a thin liquid lining layer covering a substantial area in alveoli is well documented (e.g., [303, 304, 306, 307, 788, 1167–1169]). The airways also have a thin liquid lining layer [1206]. In order for lung surfactant to affect alveolar stability and pulmonary mechanics, it is not necessary for the hypophase to be of uniform thickness or fully continuous throughout the lung. The argument that pulmonary surfactant reduces pressure differences within and between alveoli as a function of radius can be made just as readily if the lining layer is variable in thickness and not continuous in extent. In fact, a fully continuous alveolar hypophase could have both positive and negative ramifications (e.g., [941]). Surface tension gradients on a continuous hypophase could lead to interfacial transport of surfactant or other material between alveoli as well as from the alveoli to the airways (net transport would be constrained by surface viscosity and by the fact that surface tension at any location fluctuates during breathing). Interalveolar surfactant transport could be detrimental if it smoothed out surface tension differences between large and small alveoli that aid stability, while surface tension–driven transport toward the airways could be beneficial in enhancing alveolar clearance.

VIII. Theoretical Roles of Lung Surfactant in the Airways

Although the major physiological actions of lung surfactant are at the alveolar level, it has also been proposed that surface active materials could have physiologic benefits if they were present in the liquid lining of small airways. Phospholipids are known to be present in the airway lining and also to be synthesized in airway cells. In addition, surfactant proteins with the exception of SP-C have been shown to be produced in airway cells, particularly nonciliated bronchiolar cells (Clara cells) as discussed in Chapter 6. While it is very unlikely that surfactant endogenously present in the airways is identical to alveolar surfactant secreted by type II cells, it could still be sufficiently surface active to influence physical behavior. Exogenous surfactants deposited in the airways during replacement therapy could also affect the properties of small airways as well as alveoli. Two actions that have been proposed for endogenous and exogenous surfactants in the airways are that they could enhance the stability of small airways against closure [237, 241, 245, 297, 400, 446, 544, 660, 681, 812] as well as contribute to airway clearance mechanisms [196, 299, 313, 354, 723, 933, 957].

At the mechanistic level, surfactants could in principle increase airway stability against closure by affecting surface tension and hydrostatic pressure in the liquid lining layer. The law of Young and Laplace for a cylindrical airway is given by $\Delta P = \sigma/R$, where R is airway radius (see Equation 6-2). Based on this equation, a subatmospheric pressure of about -3 cm H_2O would exist in the liquid lining of an airway of radius 0.25 mm (e.g., a terminal bronchiole) if its surface tension were equal to that of water. This subatmospheric pressure would tend to promote local closure in thin-walled small airways during exhalation when interstitial pressures are greatest. By lowering surface tension, surfactant in the airway lining layer would make pressures less subatmospheric so as to resist closure and would also antagonize the formation of liquid bridges that plug airways [544, 681, 812]. Experiments have verified that surfactants can help maintain an opening for airflow in tubes with a liquid lining [400, 660], and that this stabilizing effect is lost if surface tension lowering ability is compromised by inhibitors [241]. It has also been shown experimentally that lung surfactant acts to decrease airway resistance, implying an effect in maintaining airway patency [237, 245]. In addition to stabilizing airways against closure, surface active material in the airways could contribute to clearance mechanisms. In particular, surfactants could generate surface tension–driven flows in the airway lining layer to transport particulates and other material toward the central airways [299, 354, 723] or epithelium [957]. They could also interact with mucus and enhance mucociliary clearance mechanisms [196, 299, 313, 745, 933].

IX. Chapter Summary

Theories based on thermodynamics, interfacial phenomena, and biomechanics show the direct relation between alveolar surface tension–area (σ-A) behavior and pulmonary P-V mechanics. Among other applications, such theories allow the prediction of alveolar σ-A behavior from P-V data in air-filled and saline-filled lungs. The adsorption and film behavior of phospholipids and lung surfactants can also be analyzed and predicted based on interfacial theory. Although theory is not sufficiently precise to replace direct surface activity measurements, it is indispensable in understanding the basis of lung surfactant activity and function. Theory indicates that σ-A hysteresis in lung surfactant films is a major contributor to the difference in P-V mechanics found when lungs are inflated and deflated with air. Theory also demonstrates that alveolar surfactant lowers and varies surface tension during compression and expansion in qualitative agreement with surface balance studies. The linkage between σ-A and P-V behavior can also be exploited experimentally to demonstrate the importance of specific surface properties in the physiological function of lung surfactants. As an example, measurements in surfactant-deficient rat lungs are used in this chapter to show the functional relevance of rapid adsorption and low minimum surface tension in the effects of several exogenous surfactants on P-V mechanics.

The role of pulmonary surfactant in alveolar stability has also received extensive theoretical and experimental study. The quantitative contribution of lung surfactant to alveolar mechanics varies to some extent among studies, but the consistent consensus of biomechanical theory and experiment shows its importance in helping to stabilize alveolar inflation and deflation behavior. Tissue phenomena including the interdependence of the alveolar network and specialized support fibers act along with lung surfactant to enhance alveolar stability. Although the primary actions of surfactants in the lungs are at the alveolar level, surface active material in the airway lining may also enhance the stability of small airways against closure and contribute to airway clearance mechanisms. The crucial physiological importance of lung surfactant to pulmonary mechanics and to alveolar stability is shown unequivocally by the increased work of breathing, alveolar collapse and overdistension, and diffuse edema present in surfactant-deficient lung disease. The lipid and protein components of endogenous lung surfactant that generate its functional surface active properties are discussed in detail in the next chapter.

8

Functional Composition and Component Biophysics of Endogenous Lung Surfactant

I. Overview

This chapter describes the functional composition and component biophysics of endogenous lung surfactant. A functional view of composition is essential for all lung surfactants. Endogenous lung surfactant is a complex mixture of lipids and proteins, chemical classes present in all cells and tissues. Functional components in lung surfactant are those that generate the surface properties required for physiological activity. This chapter discusses the chemical components of lung surfactant in the context of their biophysical actions and importance in adsorption and dynamic surface activity. Primary functional constituents in endogenous surfactant include DPPC, a mix of additional saturated and unsaturated phosphatidylcholines and other phospholipid classes, and surfactant proteins SP-A, SP-B, and SP-C. Neutral lipids also influence surface activity, while a fourth surfactant protein SP-D does not affect biophysical function.

II. Overview of Materials and Methods Used in Studying Functional Composition

Assessing the functional composition of lung surfactants requires correlated biophysical and biochemical measurements on a variety of surface active materials. Chapters 4 and 5 detail materials and methods in lung surfactant research, and only a brief summary is given here. Endogenous lung surfactant, also called whole or natural surfactant, can be obtained from the lungs by several methods ([250, 568, 569, 829, 918, 919, 942] for review). The most direct of these is to wash surfactant out of the alveolar spaces by bronchoalveolar lavage with normal saline (Chapter 5). Surfactant can also be washed from the lungs by saline instilled through the vasculature, or it can be processed from minced or homogenized lung tissue by conventional and density gradient centrifugation, column chromatography, or related biochemical methodology. Lung surfactant obtained by different methods and from animals of different species and ages exhibits some compositional variation (e.g., [148, 486, 568, 571, 788, 936]). However, certain major features are consistently present in the composition of active preparations of pulmonary surfactant as described in subsequent sections. Studies of component activity and functional composition in endogenous lung surfactant utilize a range of materials in addition to whole surfactant itself. This includes phospholipids, neutral lipids and apoproteins purified from endogenous surfactant, as well as a variety of synthetic phospholipids, synthetic peptides, and other surface active compounds as detailed in Chapter 5.

In addition to examining a range of materials, research on component-specific surface activity in lung surfactants uses multiple experimental methods to assess biophysical properties and behavior. Surface properties of interest include how lung surfactant and its components adsorb to the air-water interface and how the interfacial film lowers and varies surface tension during dynamic compression and expansion. Common biophysical methods used in studying lung surfactants include the Wilhelmy surface balance, pulsating bubble surfactometer, and captive bubble apparatus (Chapter 4). Experiments on the Wilhelmy balance can isolate surface tension–area properties directly within spread interfacial films, and this instrument can also examine the properties of adsorbed surface films. Wilhelmy balance studies of film behavior are complemented by the pulsating bubble and captive bubble techniques, which define overall surface tension lowering from adsorption and dynamic film compression at cycling rates, area compressions, temperature, and humidity relevant for the lung *in vivo*. The latter two methods can also be used to study adsorption in the absence of dynamic film compression. Adsorption is also commonly measured in a simple dish-type apparatus with the subphase stirred to minimize diffusion. Lung surfactant research also uses spectroscopic and other sophisticated physical chemical methods to obtain additional information on molecular structure, orientation, aggregation, and interactions in films, bilayers, and multilayers. Imaging techniques including Brewster angle microscopy, epifluorescence, scanning tunneling and atomic force microscopy, transmission and scanning electron microscopy, and a number of others also help define large- and small-scale structure and aggregation in lung surfactant films and dispersions as described in Chapter 4.

III. Concept of Functional Composition

Lavaged alveolar surfactant has a composition by weight of approximately 85–90% phospholipid, 6–8% biophysically active protein, and 4–7% neutral lipid

(Chapter 6). However, compositions outside these representative ranges can be found depending on the source and methods used to obtain the surfactant. Rather than viewing lung surfactant composition in isolation, it is most useful to consider it in the context of functionality. Functional assessments of composition are an important example of the biochemical-biophysical-physiological correlates necessary in lung surfactant research. It is very difficult to sample surfactant from alveoli or type II cells without some contamination by nonsurfactant lipids and proteins. This applies even to surfactant obtained by bronchoalveolar lavage, and contaminating material is significantly increased in surfactant processed from lung tissue. Determining the relevance of specific components thus requires knowing how they affect surface activity, i.e., a functional assessment. The *functional components* of lung surfactant are those that contribute to its required surface active properties. *Functional composition* is then defined by the appropriate mole or mass ratio of these components. The physiological activity of any lung surfactant material is best understood in the context of its functional composition and component-specific surface active properties. As detailed below, phospholipids and apoproteins are the major biophysically functional components in endogenous surfactant, although neutral lipids also affect surface behavior.

**Functional Composition and Its Importance
for Lung Surfactants**

Compositional assessments for alveolar surfactant are complicated by several factors:
  Difficulty of sampling the alveolar interface and type II cell
  Contaminating cell and tissue components
  Complex multicomponent mix of surfactant lipids and proteins

The composition of endogenous surfactant requires a functional evaluation based on component contributions to surface activity.

The *functional components* of endogenous surfactant are those that contribute physiologically important adsorption or dynamic surface tension lowering properties.

The activity of exogenous surfactants can also be better understood when composition is viewed in the context of function.

IV. Lung Surfactant Phospholipids and Their Contributions to Surface Activity

A broad distribution of phospholipid classes is present in lung surfactant obtained by lavage or lamellar body isolation, and each of these phospholipid classes contains multiple individual compounds. The content and biophysical contributions of different phospholipids in lung surfactant activity are described below. The general nomenclature, molecular structure, and film and phase behavior of phospholipids have been discussed in Chapter 3. Phospholipids in lung surfactant are also detailed in a variety of reviews (e.g., [61, 148, 163, 365, 568, 798, 936, 1105]).

A. Phospholipid Classes in Endogenous Surfactant

Phosphatidylcholine (PC) is by far the most prevalent class of phospholipids in lung surfactant, accounting for about 80% of total phospholipid (Table 8-1). DPPC is the single most abundant compound, but multiple additional saturated and unsaturated PC species are also present. Endogenous surfactant also contains smaller amounts of other phospholipid classes including both anionic phospholipids (PG, PI, PS) and zwitterionic phospholipids (PE, SPH). Anionic phospholipids as a group account for about 12–15% of total phospholipid by weight (Table 8-1). The distribution of phospholipid classes in lung surfactant has several important distinctions from that found in typical cell membranes. Lung surfactant has a much higher content of PC, and a lower content of PE and SPH, than most cell membranes. The presence of PG in lung surfactant is also an identifying feature, since this class of phospholipids is atypical in mammalian cells although it is common in bacteria. In addition to distinctions based on phospholipid class, the presence of specific compounds, particularly disaturated molecular species, also differentiates surfactant phospholipids from cell membrane phospholipids.

B. Subdivisions of Lung Surfactant Phospholipids

Phospholipids in endogenous lung surfactant can be subdivided conceptually in several ways. These subdivisions are to some degree arbitrary but can be helpful in functional understanding. One simple conceptual grouping of lung surfactant phospholipids is as DPPC and remaining phospholipids. This view of lung surfactant phospholipids was very common in the early literature, where DPPC was emphasized as the major (and sometimes the only) functional component of lung surfactant. Alternative subdivisions of lung surfactant phospholipids focus on their molecular characteristics and phase behavior. For example, lung surfactant phospholipids can be grouped as disaturated phospholipids and unsaturated phospholipids. The disaturated phospholipids are sometimes approximated by disaturated phosphatidylcholine (DSPC), although some saturated phospholipid

Table 8-1 Average Content of Phospholipid Classes in Lung
Surfactant from Lavage and Lamellar Body Isolates

Phospholipid class	Cell-free lavage	Lamellar body isolates
PC	80.0 ± 0.9	77.4 ± 0.2
PG	6.8 ± 1.4	8.9 ± 1.3
PI + PS	5.4 ± 1.3	4.7 ± 0.6
PE	3.7 ± 0.4	5.0 ± 0.9
SPH	2.1 ± 0.4	1.8 ± 0.7
Other	2.0 ± 0.3	2.0 ± 0.5

Data are averages from the literature for six lavage studies [91, 390, 923, 924,
993, 1214] and four lamellar body studies [391, 397, 863, 1020] involving
multiple animal species and ages.
Source: Ref. 788.

compounds with other headgroups are also present. Another useful conceptual
classification of lung surfactant phospholipids in a biophysical sense is as
gel-phase phospholipids and liquid crystal phase phospholipids based on the
bilayer states of individual compounds at body temperature. The relevance of
various subdivisions of lung surfactant phospholipids for biophysical behavior is
apparent in the following sections.

Conceptual Subdivisions of Lung Surfactant Phospholipids

Conceptual subdivisions of lung surfactant phospholipids, while
arbitrary to some degree, can be helpful in functional understanding.

Lung surfactant phospholipids can be subdivided conceptually in
several ways:
As DPPC plus remaining phospholipids
As DSPC plus remaining phospholipids
As saturated phospholipids and unsaturated phospholipids
*As gel-phase saturated phospholipids and liquid crystal phase
saturated and unsaturated phospholipids*

These conceptual groupings reflect the differing biophysical behavior of subsets of phospholipids in the surface activity of endogenous lung surfactant as described in subsequent sections.

C. DPPC and Its Content in Lung Surfactant

The potential functional importance of DPPC in lung surfactant was apparent when the high content of this disaturated molecule in pulmonary washings was demonstrated in the early 1960's [102, 103]. With two saturated C16:0 chains, DPPC is very different from typical cell membrane phospholipids. Its rigid chains impart crucial biophysical characteristics, enabling DPPC to be highly effective in lowering and varying surface tension in a compressed surface film. The percentage of DPPC in endogenous surfactant has been variably reported to be 40–70% of PC [61, 148, 163, 365, 419, 568, 936, 1105]. DPPC contents at or exceeding the upper end of this range are present in many exogenous lung surfactant preparations. However, several careful studies indicate that the actual DPPC content of endogenous surfactant is near the low end of the reported range. Measurements both by HPLC [486, 541, 949] and by GC following phospholipase A_2 treatment [460, 1214] place the probable content of DPPC in alveolar surfactant at 40–50% of PC. If PC is assumed to account for 80% of lung surfactant phospholipids, this is equivalent to a DPPC content near one third of total phospholipid. Remaining non-DPPC phospholipids as a group thus constitute about two thirds of total lung surfactant phospholipid and are actually in the majority compared to DPPC. Not surprisingly, these "secondary" phospholipids are found to make important contributions to lung surfactant biophysics.

D. Content of Saturated versus Unsaturated Phospholipids in Lung Surfactant

As indicated earlier, the quantitative distribution of phospholipids in lung surfactant varies with source and preparation methodology. However, active isolates of lung surfactant all contain a mix of saturated and unsaturated phospholipids with PC as the predominate class. The distribution of saturated and unsaturated PC compounds in lavaged calf lung surfactant is given in Table 8-2. Approximately 55% of calf lung surfactant PC compounds are disaturated. In addition to DPPC, these compounds include 1-myristoyl-2-palmitoyl phosphatidylcholine (MPPC), 1-palmitoyl-2-myristoyl phosphatidylcholine (PMPC), and palmitoyl-stearoyl phosphatidylcholine (PSPC). Disaturated phospholipids with headgroups other than PC are not shown in Table 8-2 but are also known to be present in lung surfactant [61, 148, 163, 568, 936, 1105, 1214]. The remaining lung surfactant phospholipids are unsaturated; i.e., they contain at least one chain having one or more double bonds. Two major unsaturated PC compounds in lung surfactant are palmitoyl-palmitoleoyl-PC (PPoPC) and palmitoyl-oleoyl-PC (POPC); small amounts of other unsaturated PC compounds are also present (Table 8-2). These unsaturated PC compounds are all in the fluid liquid crystal phase at 37°C, while disaturated PC compounds with chains at least 16 carbon atoms in length are in

Table 8-2 Distribution of Phosphatidylcholines in Calf Lung Surfactant as Found by HPLC and Gas Chromatography

Phosphatidylcholine compound	Abbreviation	Fatty acids	(mole % of PC)
Dipalmitoyl-PC	DPPC	C16:0-C16:0	41.1 ± 1.5
Palmitoyl-palmitoleoyl-PC	PPoPC	C16:0-C16:1	18.7 ± 0.3
Palmitoyl-oleoyl-PC	POPC	C16:0-C18:1	14.3 ± 0.6
Palmitoyl-myristoyl-PC +	PMPC	C16:0-C14:0⎱	12.5 ± 0.1
Myristoyl-palmitoyl-PC	MPPC	C14:0-C16:0⎰	
Palmitoyl-stearoyl-PC	PSPC	C16:0-C18:0	1.1 ± 0.1
Myristoyl-palmitoleoyl-PC	MPoPC	C14:0-C16:1	2.3 ± 0.5
Dioleoyl-PC	DOPC	C18:1-C18:1	3.5 ± 0.2
Palmitoleoyl-oleoyl-PC	PoOPC	C16:1-C18:1	2.3 ± 0.1
Unknown	—	—	4.3 ± 0.2

Data are mole % of diacyl-PC (>97% total PC) from GLC and HPLC analysis of calf lung surfactant by Kahn et al [541]. PMPC and MPPC were found in a distribution ratio of approximately 65%/35%. Isomeric determinations were not done for other tabulated mixed chain PCs. *Unknown* also includes identified species present at <0.5%. Total content of disaturated PC species (DPPC, PSPC, MPPC, PMPC) is ~55%.

the rigid gel phase. Saturated PC compounds with mixed C16:0-C14:0 chains, or containing shorter chains, are in the liquid crystal phase at 37°C [301, 985, 1007]. For example, PMPC and MPPC have gel to liquid crystal T_c values of 27°C and 35°C, respectively [123, 979] (Chapter 3). These two compounds are more fluid than DPPC at body temperature, but retain a capacity for tighter packing and more condensed film behavior than unsaturated phospholipids. Lung surfactant phospholipids with headgroups other than PC also contain a mix of saturated and unsaturated molecules that influences surface active behavior.

Measurements of DSPC Content in Endogenous Surfactant. Values reported in the literature for the DSPC content of endogenous lung surfactant vary over a broad range of about 55–80% of PC (45–65% of total phospholipid). One factor potentially affecting many of the higher reported values for surfactant DSPC is that an osmium tetroxide chromatographic assay [697] commonly used in such determinations is now known to read falsely high in the presence of mono-unsaturated PC molecules [460]. Since mono-unsaturated compounds are prevalent in lung surfactant PC (Table 8-2), this assay can significantly overestimate DSPC content. Values of DSPC content in the range of 55–65% of PC (~45–50% of total surfactant phospholipid) appear reasonable based on available GC and HPLC data [460, 486, 541, 1214]. DSPC contents of this magnitude are consistent with a balance of rigid and fluid phospholipids to contribute to surface behavior. As noted above, the DSPC content of lung surfactant includes not only rigid gel-phase PC species but also compounds like PMPC and MPPC that are in the liquid crystal phase at 37°C. The saturated C14:0 and C16:0 chains in MPPC and PMPC, however, still facilitate closer and more rigid molecular packing than in films of fluid unsaturated phospholipids. DSPC content is sometimes used as equivalent to the total content of disaturated phospholipids in lung surfactant, but this is not accurate because some disaturated phospholipids with non-PC headgroups are also present.

DPPC and DSPC Pool Sizes in the Lung. Pool sizes, synthesis rates, and recycling of lung surfactant components appear to be relatively tightly regulated in the mammalian lungs (see Chapter 6 for more detailed coverage of surfactant synthesis and recycling). The fact that almost all components in lung surfactant are either recycled intact or otherwise reutilized in synthesis by type II cells is consistent with the essential physiologic importance of this material. Reuptake and recycling of alveolar surfactant by type II cells allow physiologically adequate surfactant pool sizes to be maintained despite relatively slow rates of *de novo* synthesis ([61, 515, 1199, 1201] for review). In order to cover the alveolar surface in a tight-packed monoloyer for maximal surface tension lowering, lung surfactant phospholipids must be compressed to surface concentrations of 40 $Å^2$/molecule based on fatty chain limiting areas. For a gm-molecular weight of 750 and an alveolar surface area of 1 m^2/kg body weight, this translates to 3.1 mg phospholipid/kg body weight (actual physiological requirements are less than this because surface area is smaller at end-expiration). Since the alveolar surface film is thought to become significantly enriched in DPPC and other disaturated molecules at high degrees of compression to promote surface tension lowering (see below), a nontrivial portion of this phospholipid needs to be DSPC. Substantial additional amounts of phospholipid including DPPC and DSPC are also needed to supply surfactant aggegates in the alveolar hypophase. Approximate calculations of alveolar PC pools indicate that sufficient amounts of DPPC and DSPC are available to meet these requirements. Jobe [515] has calculated estimated alveolar PC pools of 77 μmole/kg (58 mg/kg) in term newborn rabbits and 11 μmole/kg (8.3 mg/kg) in adult rabbits, and alveolar pools of a similar order of magnitude have been estimated in other animal species. If DPPC and DSPC make up 45% and 55–60% of total PC, respectively, these alveolar pools are consistent with sufficient phospholipid to supply an active surface film and to occupy subphase aggregates. However, the safety factor is not extreme, particularly in adult animals.

E. Biophysical Importance of a Balance of Rigid and Fluid Phospholipids in Lung Surfactant

Long-chain disaturated phospholipids have very different surface properties compared to fluid unsaturated phospholipids (see Chapter 3 for the general film behavior of phospholipids). To achieve low minimum surface tensions and to vary surface tension effectively during cycling, the surface film at the alveolar interface must have a significant content of rigid saturated phospholipids, particularly DPPC. Rigid PC molecules like DPPC are known to form tightly packed, solid surface films able to reach extremely low minimum surface tensions <1 mN/m under dynamic compression at body temperature [415, 416, 796, 1057, 1058,

1093, 1157]. Disaturated phospholipids in lung surfactant are thought to become enriched in the surface film at the expense of fluid phospholipids during compression to low interfacial area (high surface concentration). This phenomenon, referred to as "selective squeeze-out",[1] allows formation of a condensed film that reduces surface tension with great effectiveness. However, the same molecular characteristics that allow DPPC and related disaturated phospholipids to achieve very low surface tensions during dynamic compression constrain them from other behaviors. In particular, DPPC adsorbs very slowly to the air-water interface [569, 783, 794] and also respreads poorly in cycled surface films [270, 657, 658, 416, 785, 796, 1093, 1145]. The complete mix of phospholipids in lung surfactant has dramatically better film respreading than DPPC (Figure 8-1, Table 8-3) [1145]. Fluid phospholipids in lung surfactant do not lower surface tension as effectively as DPPC, but they have a major impact in improving respreading. In addition, surfactant apoproteins and neutral lipids contribute to increased film respreading in lung surfactant (Table 8-3). Fluid phospholipids and neutral lipids also increase adsorption relative to DPPC [783, 1146], although apoproteins are primarily responsible for maximizing lung surfactant adsorption as detailed later.

A variety of studies have shown that fluid PCs and other liquid crystal phospholipids form films having higher minimum surface tensions and/or better respreading than DPPC (e.g., [73, 415, 416, 432, 557, 785, 789, 796, 1157]). Addition of one or more fluid phospholipids to DPPC has also been shown to improve respreading in cycled mixed films at the air–water interface (e.g., [270, 560, 785, 796]) and to increase adsorption after dispersion in the aqueous phase [783, 794]. However, fluid unsaturated phospholipids decrease stability and increase surface pressure relaxation in dynamically compressed films with DPPC [432, 795], and at high concentrations can decrease surface tension lowering ability. Multiple rigid and fluid phospholipids clearly contribute to surface behavior in lung surfactant. This includes the saturated and unsaturated PC compounds in Table 8-2, as well as compounds in other phospholipid classes. In addition to chain-mediated interactions, the structure, charge, and steric bulk of the headgroup region in phospholipids significantly influence adsorption and film behavior. PC and PE compounds with identical acyl moieties, for example, have very different film and phase properties due to the differing size and hydrogen bonding of their headgroup regions (e.g., [657, 658, 1057]). Synthetic phospholipid analogs that have equivalent C16:0 saturated acyl chains but vary in headgroup structure and charge also display substantial differences in film and adsorption behavior (e.g., [657, 658, 1092, 1093]). Based on their differing headgroup and chain structure and fluid vs rigid phase behavior, lung surfactant phospholipids comprise a complex mixture of biophysically interactive components.

---

1. For studies on squeeze-out in phospholipid and lung surfactant films see Refs. 206, 415, 432, 602, 759, 761, 796, 828, 884, 954, 961, 1157 plus reviews such as Refs. 320, 560, 785, 788, 790, 798. Squeeze-out in surfactant films is also described in Chapter 2.

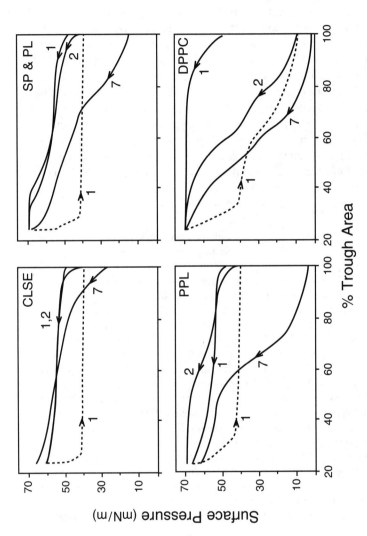

Figure 8-1 Surface pressure–area isotherms for films of the hydrophobic components of calf lung surfactant. The isotherms of cycled films containing various subfractions of lung surfactant components display substantial differences in respreading and other surface behaviors. Calculations in Table 8-3 show the greatly improved respreading of lung surfactant phospholipids as a group relative to DPPC. Films were spread to 15 Å^2/molecule in a Wilhelmy balance and cycled at 1.5 minutes/cycle at 37°C. CLSE, chloroform:methanol extract of calf lung surfactant; PPL, purified phospholipids from CLSE; SP&PL, hydrophobic apoproteins and phospholipids; DPPC, dipalmitoyl phosphatidylcholine.

Table 8-3 Respreading of Purified Lung Surfactant Phospholipids and Other Component Subfractions of Calf Lung Surfactant Extract

Film	Initial concentration ($Å^2$/molecule)	Respreading based on isotherm areas[a] between compressions	
		2/1	7/1
DPPC	15	29.6 ± 0.2	47.4 ± 0.2
PPL		0.6 ± 0.0	27.4 ± 0.5
N&PL		0.1 ± 0.0	17.6 ± 0.3
SP&PL		0.2 ± 0.0	11.0 ± 0.7
CLSE		0.2 ± 0.0	5.2 ± 0.1
DPPC	30	29.6 ± 0.1	48.0 ± 0.2
PPL		0.8 ± 0.1	38.5 ± 0.5
N&PL		0.6 ± 0.1	29.3 ± 0.4
SP&PL		0.7 ± 0.0	27.5 ± 0.4
CLSE		0.3 ± 0.0	16.0 ± 0.4

[a]Area units are arbitrary; smaller values imply better respreading (zero is perfect respreading). The greatly improved respreading exhibited by the complete mix of lung surfactant phospholipids (PPL) relative to DPPC is apparent. Complete surface pressure–area isotherms for films spread to 15 $Å^2$/molecule are given in Figure 8-1.
Abbreviations: CLSE, calf lung surfactant extract; PPL, purified phospholipids; N&PL, purified neutral lipids and phospholipids; SP&PL, purified hydrophobic proteins and phospholipids.
Source: Data from Ref. 1145.

F. Contributions of Zwitterionic versus Anionic Phospholipids to Surface Activity

The substantial content and high surface activity of PC compounds in lung surfactant necessitate that molecular interactions involving zwitterionic phospholipids are important in surface behavior. The quantitative contributions of anionic phospholipids (PG, PI, PS) to the surface active function of lung surfactant, however, are less clear. The presence of anionic phospholipids in lung surfactant is well established, as is the utility of PG and PI as developmental markers of surfactant in fetal lung fluid [318, 390–392, 517, 923, 1061, 1075]. The potential for specific molecular biophysical associations between anionic phospholipids and one or more of the surfactant apoproteins has also been widely suggested.[2] Specific associations between anionic phospholipids and surfactant proteins would be consistent, for example, with the positively charged amino acids known to be present in these materials (see later section for surfactant protein structure). Despite these considerations, it has not been shown that anionic phospholipids are essential for generating specific adsorption or dynamic surface tension lowering properties necessary for lung surfactant function. If a required role in surface activity does exist, it is likely to relate to anionic phospholipids as a group rather than to one class such as PG in particular. Active surfactant harvested from different animals has a variable content of different anionic phospholipid classes [148, 568, 936, 1105]. Lung surfactant also retains high biophysical and physiological activity when PI is enriched relative to PG by dietary manipulation in animals [68, 388]. Depletion of anionic components in lung surfactant phospholipids leads to comparatively minor decreases in surface and physiological activity when the remaining phospholipids are combined with SP-B,C [1144], indicative of a predominant importance of zwitterionic PC compounds in surface active function.

V. Lung Surfactant Neutral Lipids and Their Contributions to Surface Active Biophysics

Endogenous lung surfactant contains a small percentage of neutral lipids in addition to phospholipids. Neutral lipids constitute less than 10% of typical lung surfactant isolates by weight (e.g., [148, 568, 571, 936, 1105, 1214]). Cholesterol is the most prevalent neutral lipid, although small amounts of cholesterol esters, diglycerides, triglycerides, and free fatty acids are also found in lung surfactant compositional studies. The importance of neutral lipids to the biophysical function of lung surfactant is certainly not as great as that of phospholipids and surfactant proteins. However, neutral lipids do influence overall system behavior. Cholesterol, in particular, is well known to interact with phospholipids in bilayers and surface films [108, 120, 301, 985, 1007] (see Chapter 3). Cholesterol tends to decrease order and disrupt packing in rigid gel-phase phospholipid bilayers, while having the opposite effect on fluid liquid crystal phase phospholipids. Cholesterol thus contributes to the balance of fluid and rigid lipids in lung surfactant. The compact shape of the cholesterol molecule allows it to interdigitate effectively with phospholipids in expanded surface films, resulting in apparent area condensation effects. Cholesterol exhibits better adsorption and film respreading than DPPC, but it can also disrupt packing and impair the ability of DPPC films to reach low minimum surface tensions during dynamic compression (e.g., [796]). Cholesterol in large amounts (e.g., 10–30 wt %) also raises minimum

---

2. For example, Refs. 35, 119, 528, 560, 575, 793, 805, 807, 1069, 1111, 1217.

surface tension when added to endogenous lung surfactant or organic solvent extracts of whole surfactant (Chapters 7, 9). Studies with column-purified sub-fractions of lung surfactant components indicate that neutral lipids contribute some improvement in adsorption and film respreading when mixed with surfactant phospholipids [1145, 1146, 1215]. However, neutral lipids also lead to a small decrease in surface tension lowering during dynamic cycling when combined with purified hydrophobic apoproteins and surfactant phospholipids [1145, 1215].

VI. Lung Surfactant Proteins and Their Molecular Biophysical Characteristics

The biology, structure, and biophysics of the lung surfactant proteins are detailed in a variety of comprehensive reviews.[3] Endogenous surfactant contains about an order of magnitude less protein than phospholipid on a weight basis. However, despite their relatively low content, the proteins in lung surfactant are highly important in biophysical function. As described earlier in Chapter 6, lung surfactant contains three biophysically important proteins designated SP-A, SP-B, and SP-C based on nomenclature defined by Possmayer in 1988 [861]. A fourth surfactant protein not implicated in biophysical function was called CP4 when first discovered [846], but was subsequently named SP-D according to this system [845, 1159]. Prior to the late 1980's, diverse nomenclature was used for surfactant proteins in the literature, complicating data interpretations and comparisons between studies. Some of the molecular features of SP-A, SP-B, SP-C, and SP-D are summarized in Table 8-4, and further details on their discovery and molecular behavior are given below. The molecular genetics, metabolism, and cellular origin of SP-A, SP-B, SP-C, and SP-D are summarized in Chapter 6.

A. Discovery of Lung Surfactant Proteins

The existence and functional importance of proteins in endogenous surfactant were controversial for years after the initial discovery of the surfactant system. The early literature is replete with studies investigating whether functional lung surfactant contained proteins other than those derived from contaminating plasma and cells ([319, 569, 790, 942] for review). Surfactant proteins not only had to be distinguished from plasma- and cell-derived components but also required specialized purification and characterization methods (Chapter 5). The low concentration, small size, and extreme hydrophobicity of SP-B and SP-C made their

---

3. For example, reviews of lung surfactant proteins are given in Refs. 163, 362, 417–420, 423, 426, 529, 610, 614, 696, 722, 798, 861, 1162, 1174, 1176. Apoprotein synthesis, metabolism, and nonbiophysical roles are discussed in Chapter 6.

Table 8-4 Molecular Characterisitics of Lung Surfactant Proteins

Surfactant protein	Selected molecular structural characteristics
SP-A	MW 26–38 kDa (monomer)
	Most abundant surfactant apoprotein, relatively hydrophilic
	Acidic glycoprotein with multiple post-translational isoforms
	Member of C-type lectin family and collectin family
	Oligomerizes to an active octadecamer (six triplet monomers)
SP-B	MW 8.5–9 kDa (monomer), 78–79 AA in length
	Hydrophobic with at least two amphipathic helical regions
	Human form has a significant content of disulfide-linked dimer
	Contains both α-helical and non-helical structural domains
	Contains multiple charged amino acids at neutral pH
SP-C	MW 3.5–4.2 kDa (monomer), 33–35 AA in length
	Most hydrophobic surfactant protein
	Forms oligomers of possible biophysical significance
	Human form has two palmitoylated cysteine residues
	Contains two charged basic amino acids at neutral pH
	Primarily α-helical in structure
SP-D	MW 39–46 kDa (monomer)
	Has significant structural similarity to SP-A
	Member of C-type lectin family and collectin family
	Not implicated in lung surfactant biophysics
	Oligomerizes to a docamer (four triplet monomers)

The sequences, structures, molecular biophysics, and surface activity of surfactant apoproteins are detailed in subsequent sections of this chapter. Surfactant proteins, particularly SP-A and SP-D, have additional nonbiophysical roles in pulmonary biology (Chapter 6).

identification and characterization particularly difficult. The existence of an amphipathic lung surfactant protein that eventually became identified as SP-A was first demonstrated in the early 1970's (e.g., [569, 574]). The possibility that lung surfactant contained additional functional proteins that were sufficiently hydrophobic to extract into solvents like chloroform:methanol was not recognized until later [853]. Hydrophobic organic solvent extracts of whole surfactant containing small amounts of protein material were widely used in early research to demonstrate that the major surface activity of lung surfactant resided in its phospholipid fraction (e.g., [286, 287, 571, 572, 581, 744, 1032]). The fact that this hydrophobic protein was made up of specific, active entities was not recognized until the 1980's. The existence of two distinct hydrophobic proteins, now known as SP-B and SP-C, was established definitively in 1987 [180, 317, 422].

B. Structure and Molecular Biophysics of SP-A
([163, 172, 362, 417, 529, 614, 696, 722, 1162] for review)

SP-A is the most abundant apoprotein in endogenous lung surfactant and has an amino acid sequence that is highly conserved among animal species [66, 92, 272, 1130, 1173]. The N-terminal portion of the molecule is collagen-like, while its C-terminal carbohydrate binding sequence places it in the family of mammalian C-type lectins and the collectin family. The multiple isoforms of monomeric SP-A have a molecular weights of 28–36 kDa, and a representative value of 35 kDa is often used for convenience. In addition to a 20 AA signal peptide, the human SP-A proprotein contains 228 amino acids with a mix of hydrophilic and hydrophobic residues (Table 8-5). This monomer has a short N-terminal domain of 7 amino acids followed by a collagen-like region with 23 proline-rich Gly-X-Y repeats broken between the 13th and 14th repeats with a Pro-Cys-Pro-Pro sequence (Table 8-5). SP-A also contains an intermediate amphipatic helical domain thought to contribute to lipid binding and a calcium-dependent C-terminal lectin-like carbohydrate binding region [363, 364, 366, 927, 1130] (Figure 8-2). This latter region contains invariant amino acids including four cysteine residues that form intrachain disulfide bonds essential for binding carbohydrates. After glycosylation and other post-translational modifications, SP-A oligomerizes to an active octadecamer made up of 6 sets of trimers held together with covalent and noncovalent bonding similar to complement factor C1q [366, 577, 1130] (Figure 8-2). Hydroxylation of proline residues in SP-A is essential for oligomerization and for thermal stability in interacting with lipids [709]. The trimers in the active octadecamer result from the organization of collagen-like regions in three SP-A monomers into a triple helix. The break in the Gly-X-Y repeat in the monomers introduces a flexible kink so that beyond the 13th repeat the six trimers fan out from a more tightly associated stem (Figure 8-2).

The major molecular biophysical action of SP-A is to increase the aggregation and order of phospholipids in a calcium- and pH-dependent fashion.[4] One molecule of SP-A binds two or three molecules of Ca^{++} [364], which may in part act to neutralize carboxylate ions on the protein near neutral pH [226, 529]. SP-A can bind phospholipid vesicles and influence the exchange of phospholipids between them [109], and it interacts with both zwitterionic and anionic phospholipids. The large size of each SP-A octadecamer (molecular weight 630 kDa based on a 35 kDa monomer) influences the organization and properties of a substantial number of phospholipid molecules. SP-A promotes the association of

---

4. For example, Refs. 67, 109, 226, 280, 298, 364, 421, 685, 872, 930, 932, 1053, 1115, 1125, 1184, 1200.

Table 8-5   cDNA-Derived Amino Acid Sequence of Human SP-A[a]

Glu - Val - Lys - Asp - Lys - Cys - Val - Gly - Ser - Pro - Gly - Ile - Pro - Gly - Thr - Pro - Gly - Ser - His - Gly - 20
Leu - Pro - Gly - Arg - Asp - Gly - Arg - Asp - Gly - Leu - Lys -Gly - Asp - Pro - Gly - Pro - Pro - Gly - Pro - Met - 40
Gly - Pro - Pro - Gly - Glu - Met - Pro - Cys - Pro - Pro - Gly - Asn - Asp - Gly - Leu - Pro - Gly - Ala - Pro - Gly - 60
Ile - Pro - Gly - Glu - Cys - Gly - Glu - Lys - Gly - Glu - Pro - Gly - Glu - Arg - Gly - Pro - Pro - Gly - Leu - Pro - 80
Ala - His - Leu - Asp - Glu - Glu - Leu - Gln - Ala - Thr - Leu - His - Asp - Phe - Arg - His - Gln - Ile - Leu - Gln - 100
Thr - Arg - Gly - Ala - Leu - Ser - Leu - Gln - Gly - Ser - Ile - Met - Thr - Val - Gly - Glu - Lys - Val - Phe - Ser - 120
Ser - Asn - Gly - Gln - Ser - Ile - Thr - Phe - Asp - Ala - Ile - Gln - Glu - Ala - Cys - Ala - Arg - Ala - Gly - Gly - 140
Arg - Ile - Ala - Val - Pro - Arg - Asn - Pro - Glu - Glu - Asn - Glu - Ala - Ile - Ala - Ser - Phe - Val - Lys - Lys - 160
Tyr - Asn - Thr - Tyr - Ala - Tyr - Val - Gly - Leu - Thr - Glu - Gly - Pro - Ser - Pro - Gly - Asp - Phe - Arg - Tyr - 180
Ser - Asp - Gly - Thr - Pro - Val - Asn - Tyr - Thr - Asn - Trp - Tyr - Arg - Gly - Glu - Pro - Ala - Gly - Arg - Gly - 200
Lys - Glu - Gln - Cys - Val - Glu - Met - Tyr - Thr - Asp - Gly - Gln - Trp - Asn - Asp - Arg - Asn - Cys - Leu - Tyr - 220
Ser - Arg - Leu - Thr - Ile - Cys - Glu - Phe -   228

[a]The 228 amino acid sequence of human SP-A derived from cDNA by Floros et al [272] and the genomic clone of White et al [1173] is shown from Hawgood and Benson [420]. Minor differences in amino acid sequence are found in other cDNA-derived sequences for human SP-A (e.g., [420, 943].) A signal peptide 20 AAs in length (not shown) precedes the tabulated sequence. Extensive post-translational processing of SP-A occurs; the final active form of SP-A is an oligomer containing six sets of monomer triplets (Figure 8-2). Collagenous domain: residues 8–80; break in collagenous Gly-X-X repeat: residues 47–50; potential glycosylation site (Asn - X - Thr): residues 187–189.

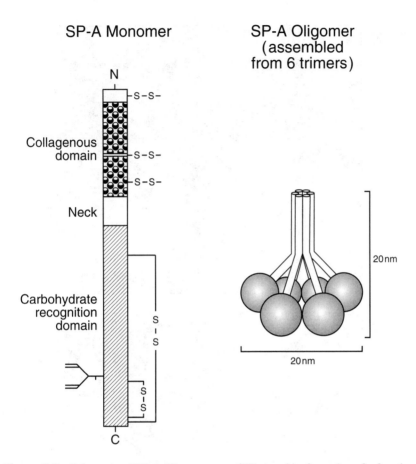

Figure 8-2 Schematic of SP-A. The structure of SP-A and its formation of a functional octadecamer containing six sets of monomer triplets is illustrated. SP-A is the most abundant and hydrophilic of the three biophysically active surfactant proteins. (Adapted from Ref. 362.)

phospholipids into a variety of large aggregate forms in water (Figure 8-3). The most striking of these is tubular myelin, a three-dimensional lattice of intersecting phospholipid bilayers with a repeat spacing of about 50 nm whose formation also requires SP-B and calcium (Chapters 3, 6). Oligomeric SP-A is apparent as particles aligned at regular intervals along these bilayers and projecting from their intersections (e.g., [413, 1125, 1182]). Tubular myelin and other large phospholipid aggregates whose formation is facilitated by SP-A are associated with rapid lung surfactant adsorption as described in earlier chapters and in a later section here.

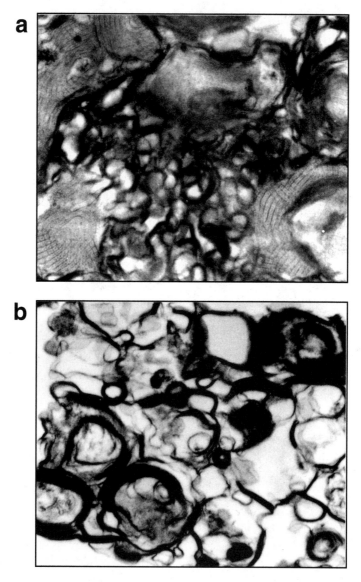

Figure 8-3 Calcium-dependent phospholipid aggregate formation facilitated by SP-A. (a) Lavaged lung surfactant in 0.15 M NaCl plus 1.4 mM Ca^{++}; (b) lavaged lung surfactant in 0.15 M NaCl with 5 mM EDTA and no added calcium. SP-A facilitates the aggregation of phospholipids into large aggregates, with its most pronounced effects including tubular myelin formation found in the presence of calcium (a). See text for details. Approximate magnification ×30,000. (From Ref. 791.)

C. Structure and Molecular Characteristics of SP-D

([163, 172, 362, 426, 529, 610, 614, 696, 722] for review)

SP-D is distinct from the other named surfactant apoproteins in that it is not a functional biophysical constituent of pulmonary surfactant. This hydrophilic, calcium-dependent C-type lectin has significant structural analogy to SP-A. The molecular structure of monomeric SP-D contains an N-terminal collagenous domain, a linking region, and a C-terminal carbohydrate-binding domain [844–846] (Figure 8-4). The N-terminus has a short initial noncollagen region, followed by a collagen-like domain containing 59 Gly-X-Y repeats that is significantly longer than in SP-A. This is followed by a short linking sequence ("neck")

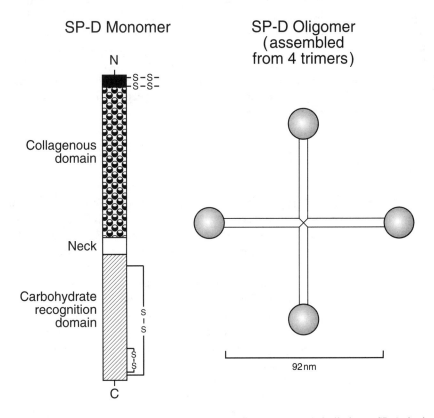

Figure 8-4 Schematic of SP-D. SP-D has significant structural similarity to SP-A; both are C-type lectins and members of the collectin family. The oligomeric form of SP-D is a dodecamer with four sets of triplet monomers. Although classed as a surfactant apoprotein, SP-D is not directly involved in the biophysical function of lung surfactant. (Adapted from Ref. 362.)

connected to the C-terminal carbohydrate recognition domain containing invariant amino acids including four cysteine residues to allow intramolecular disulfide bridging as in SP-A. The length of the human SP-D monomer based on cDNA cloning is about 355 amino acids [671], significantly larger than SP-A. SP-D has multiple isoforms and like SP-A undergoes extensive post-translational glycosylation and other processing. The molecular weight of monomeric SP-D is reported as 39–46 kDa depending on its state of processing, with a value of about 43 kDa for the glycoprotein form [845, 846]. SP-D oligomerizes into a dodecamer composed of four sets of triplet chains organized into 46 nm rods terminating in carbohydrate binding heads [166, 670] (Figure 8-4). This structure is well suited for binding and agglutinating carbohydrates but does not affect the surface activity of phospholipids [844]. In contrast to SP-A, no role has been demonstrated for SP-D in the surface activity of lung surfactant. SP-D is not found either in lamellar bodies or in tubular myelin [167, 1129]. The majority of SP-D in lavage also does not sediment with the other surfactant phospholipids and proteins [610, 612, 613]. Although not participatory in lung surfactant biophysics, SP-D, like SP-A, is a member of the collectin family of host defense proteins and it may also participate in surfactant metabolism and other aspects of pulmonary biology [172, 362, 696] (Chapter 6).

D. Structure and Molecular Biophysics of SP-B and SP-C ([163, 426, 529, 560, 614, 722, 1162, 1174, 1176] for review)

Despite their common hydrophobicity, SP-B and SP-C differ markedly in their amino acid sequences and secondary structure. Both proteins undergo post-translational processing from larger precursors to generate final small, active peptides that are highly conserved among animal species (Chapter 6). The final human SP-B monomer contains 79 amino acids and has a molecular weight of 8.7 kDa [317, 422] (Table 8-6). SP-B also forms dimers and higher oligomers through intermolecular cysteine-linked sulfhydryl bridges [528, 1064]. The dimer form of SP-B has been proposed as being particularly active in biophysical function, although this has not been demonstrated definitively (see below). In addition to its significant content of hydrophobic amino acids, SP-B contains multiple polar residues that can interact with phospholipid headgroups (Table 8-6). SP-B has several predicted amphipathic helical domains [317, 1155, 1174], and it contains 25–45% α-helical structures by spectroscopic analysis [749, 839, 996, 1111, 1174]. The polar residues and amphipathic helices of SP-B are thought to constrain a significant portion of this protein to the vicinity of lipid headgroups, leading to a relatively peripheral location in phospholipid bilayers [174, 749, 1162, 1174] (Figure 8-5). The final active monomeric form of SP-C is much smaller than SP-B, with a sequence length of only 35 amino acids in humans

Table 8-6 Amino Acid Sequences of Active Human SP-B and SP-C Peptides[a]

SP-B

PHE - PRO - ILE - PRO - LEU - PRO - TYR - CYS - TRP - LEU - CYS - ARG - ALA - LEU - ILE -	15
LYS - ARG - ILE - GLN - ALA - MET - ILE - PRO - LYS - GLY - ALA - LEU - ARG - VAL - ALA -	30
VAL - ALA - GLN - VAL - CYS - ARG - VAL - VAL - PRO - LEU - VAL - ALA - GLY - GLY - ILE -	45
CYS - GLN - CYS - LEU - ALA - GLU - ARG - TYR - SER - VAL - ILE - LEU - LEU - ASP - THR -	60
LEU - LEU - GLY - ARG - MET - LEU - PRO - GLN - LEU - VAL - CYS - ARG - LEU - VAL - LEU -	75
ARG - CYS - SER - MET - ASP 80	

SP-C

PHE - GLY - ILE - PRO - CYS - CYS - PRO - VAL - HIS - LEU - LYS - ARG - LEU - LEU - ILE -	15
VAL - VAL - VAL - VAL - VAL - LEU - ILE - VAL - LEU - VAL - ILE - VAL - GLY - ALA -	30
LEU - LEU - MET - GLY - LEU 35	

[a]These final active SP-B and SP-C peptides are processed from larger precursors (see Chapter 6). ARG and LYS basic residues are positively charged at pH 7, while GLU and ASP acidic residues are negatively charged. In SP-B, intrachain disulfide bridges are thought to link cysteines 8–77, 11–71, and 35–46, while cysteine 48 can link with another SP-B monomer to form a dimer [528, 529, 1174]. See text for details. Sequences from Refs. 316, 317, 534, 1156 as reported in Refs. 420, 943.

Model of SP-B in the Bilayer

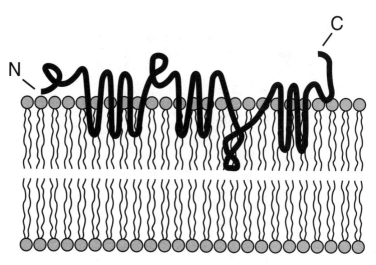

Figure 8-5 Schematic of SP-B and its hypothetical bilayer association. SP-B is the larger of the two hydrophobic surfactant apoproteins. It has multiple charged residues to interact with phospholipid headgroups plus amphipathic helices to interact with phospholipid chains. SP-B presumptively occupies a peripheral bilayer position, although its actual location in bilayers and surface films is not certain. (Adapted from Ref. 1162.)

(Table 8-6). SP-C contains a high content of valine, leucine, and isoleucine and is the most hydrophobic surfactant apoprotein. In fact, it is one of the most hydrophobic proteins known to exist in any biological system. Human SP-C has a monomer molecular weight of 3.7 kDa (4.2 kDa including cysteine-linked palmitoyl groups at positions 5 and 6 in humans) [179, 534]. SP-C also forms oligomers [37, 161, 1064], but their biophysical significance is uncertain. Due to its hydrophobicity, SP-C is thought to locate largely in the interior of phospholipid bilayers with its helical axis aligned with lipid acyl chains [827, 1112, 1162, 1174] (Figure 8-6). A range of 45–90% α-helical structure in SP-C has been reported, with a probable value near 70% [37, 536, 537, 839, 1112, 1142, 1174].

A significant body of research has addressed the molecular biophysical behavior of SP-B and SP-C in phospholipid bilayers and surface films.[5] Both

---

5. For example, Refs. 35–37, 119, 143, 160, 164, 174, 298, 474–476, 535–537, 749, 762, 804, 805, 807, 827, 838–842, 891, 996, 1002, 1111, 1112, 1122.

Model of SP-C in the Bilayer

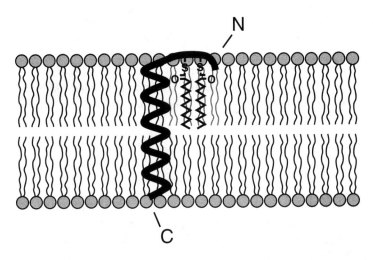

Figure 8-6 Schematic of SP-C and its hypothetical bilayer association. SP-C, the most hydrophobic surfactant protein, is thought to form a transmembrane α-helix that interacts primarily with phospholipid fatty chains. The cysteine-linked acyl groups at positions 5 and 6 in humans can extend into the bilayer interior as shown or alternatively could extend out to interact with adjacent bilayers. (Adapted from Ref. 1162.)

hydrophobic apoproteins have extensive molecular interactions with phospholipids. SP-B has been shown to increase phospholipid aggregation, as well as to disrupt and fuse phospholipid bilayers and vesicles and promote phospholipid insertion into surface films [119, 164, 804, 805, 807]. SP-B affects phospholipids not only through hydrophobic interactions but also through interactions between its polar residues and the headgroups of both zwitterionic and anionic phospholipid classes [35, 36, 143, 804, 807, 838, 1111]. Calcium is not required for the biophysical activity of SP-B, although some of its effects on anionic phospholipids may be calcium dependent. Spectroscopic studies show that SP-B can order the headgroup region of lipid bilayers [35, 36, 749, 1111, 1122], while its amphipathic helices also interact with phospholipid acyl chains (Figure 8-5). SP-B can increase the gel to liquid crystal transition temperature of phospholipids but also broadens transition width, indicating a mixed effect on overall bilayer order [35, 36, 474, 476, 749, 838, 996]. SP-C primarily interacts with phospholipids through hydrophobic forces, although its two positively charged amino acid residues can also interact with phospholipid headgroups. SP-C appears to have less selectivity than SP-B for particular phospholipid classes [474, 476, 838].

SP-C can promote the fusion of some phospholipid vesicles [160, 807] and also increases lipid vesicle binding and insertion into interfacial films [161, 804, 805]. Surface balance studies indicate that SP-C disorders the acyl chain region of desaturated phospholipids and/or expands their film behavior [476, 842, 1068, 1072]. SP-C also lowers the gel to liquid crystal transition temperatures of desaturated phospholipids [474, 476, 1002] and broadens the transition width [474, 476, 838, 1002], consistent with a similar disordering of acyl chains. An increase in chain disorder is also typically observed when SP-C is present together with SP-B in mixtures with phospholipids [298, 891].

Theoretical Modeling of the Molecular Structure of SP-B and SP-C. One valuable source of information about the molecular structural features of surfactant apoproteins comes from theoretical modeling ([529, 530, 532, 560, 1174] for review). This approach uses known information on amino acid charge, polarity, relative hydrophobicity/hydrophilicity, potential for disulfide bridging, and structural effects in other proteins to predict the molecular characteristics and higher structure of surfactant proteins based on their primary sequences. Theoretical modeling has identified three potential amphipathic helical regions in SP-B and predicts one extended hydrophobic helical region spanning all but the N-terminal third of SP-C. Modeling suggests that the predicted amphipathic helices of SP-B have their nonpolar residues grouped to form hydrophobic faces to interact with lipid fatty chains, while polar residues localized to other parts of each helix take part in interacting with phospholipid headgroups [1174] (Figure 8-7). Molecular models also indicate that intrachain disulfide bridges in human SP-B link cysteines 8–77, 11–71, and 35–46, while cysteine 48 can link with another SP-B monomer to form a dimer (e.g., [528, 529, 1174]). Modeling also predicts that the major α-helical portion of SP-C has dimensions that neatly span a phospholipid bilayer, maximizing interactions between this protein and hydrophobic chains in the bilayer interior. SP-C presumptively orients so as to allow the N-terminal portion of the molecule to expose its two charged residues (Lys and Arg at positions 11 and 12) to the polar headgroup region of the bilayer. The two cysteine-linked palmitic acid moieties of SP-C also presumptively locate near the bilayer periphery, where they can either internalize and interact with lipid fatty chains as in Figure 8-6 or alternatively extend out from the surface to interact with adjacent bilayers.

Functional Oligomerization in SP-B and SP-C. Whereas the importance of oligomerization for the surface active function of SP-A in lung surfactant is well established, this is not true for the two hydrophobic apoproteins. Both SP-B and SP-C are known to oligomerize to dimers and larger aggregates, but the importance of these different forms for surface active interactions with phospholipids is still speculative ([426, 529, 560, 614, 1174] for review). Perhaps the most attention has been directed at dimeric SP-B, which is the predominant biochemical form when human SP-B is isolated on nonreducing gels (e.g., [528, 600]). Although this Cys 48–linked dimer has been suggested as specifically important in surface active function, its quantitative impact on surface activity compared to monomeric or other oligomeric forms of SP-B has not been examined in detail. A number of different forms of SP-B may be highly active in improving the adsorption and dynamic film behavior of phospholipids.

Hydrophobic faces of SP-B helices

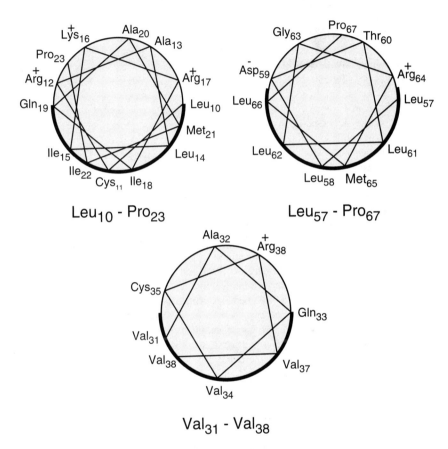

Figure 8-7 Helical wheel projections of three potential amphipathic helical domains in human SP-B. Three predicted amphipathic helices in SP-B have hydrophobic faces (thick lines) formed by nonpolar amino acids to interact with lipid fatty acyl chains in bilayers and films. The more polar amino acids of the helix orient so as to interact with lipid headgroups. In addition, the N-terminal helix (Lue$_{10}$-Pro$_{23}$) and the C-terminal helix (Leu$_{57}$-Pro$_{57}$) are near mirror images and may align in antiparallel fashion to facilitate intramolecular disulfide bridging between Cys$_8$ and Cys$_{77}$ and between Cys$_{11}$ and Cys$_{71}$. Intermolecular disulfide bridging at Cys48 (not shown) is thought to be responsible for dimer formation in human SP-B. (Redrawn from Ref. 1174.)

Similar considerations also apply to oligomeric forms of SP-C. A dimer form of SP-C has been identified and examined for activity [37, 161], but comprehensive activity assessments relative to SP-C monomer or higher oligomers remain to be done.

Functional Regions of Amino Acid Sequence in SP-B and SP-C. Identifying regions of primary sequence associated with functionally important molecular characteristics in SP-B and SP-C has obvious relevance for understanding endogenous and exogenous lung surfactants. Both theoretical modeling noted above as well as experimental research using spectroscopic and related techniques have examined the secondary and higher structures of SP-B and SP-C and the regions of sequence that contribute to them (e.g., [426, 530, 560, 614, 840, 1174] for review). While research of this kind is highly useful, the precision of theory and the complexity of apoprotein structure, purification, and interactions with lipids hamper quantitative analysis of sequence-specific biophysics. One important approach for studying regional behavior in SP-B and SP-C has utilized synthetic hydrophobic peptides containing defined segments of apoprotein sequence with or without specific amino acid substitutions (see sections on synthetic peptides in Chapters 5, 14 for detailed literature citations). Peptide studies indicate, for example, that the N- and C-terminal regimes of SP-B containing predicted amphipathic helices contribute to its ordering of phospholipid headgroups [36], while the membrane-spanning α-helical region of SP-C is important in its interactions with phospholipid acyl chains (e.g., [139, 535]). Although sequence-specific activity information can be obtained from synthetic peptide research, interpretations can also be affected by differences in higher structure or other complex molecular characteristics between these materials and endogenous SP-B and SP-C, which are processed from larger proproteins and packaged with lipids in the type II cell. One reflection of this is that even full-length SP-B and SP-C peptides synthesized by solid-state techniques do not generate surface or physiological activity equivalent to that of purified native hydrophobic proteins in mixtures with phospholipids [529, 530, 532]. Another approach to defining active regions in SP-B and SP-C is to make focused modifications in the native apoproteins and determine the functional consequences. Analyses of this kind indicate that the two positively charged amino acid residues in SP-C (Lys and Arg at positions 11 and 12) contribute to its ability to promote interactions between subphase phospholipid vesicles and interfacial films [162]. The functional importance of acylation in SP-C has also been addressed as described below.

Functional Importance of Cysteine-Linked Acylation in SP-C. Acylation may be functionally important in SP-C for several reasons. A number of studies have demonstrated that palmitoylation in endogenous SP-C influences its secondary structure including α-helix formation [37, 532, 535, 536, 841, 1112, 1142]. Deacylation of endogenous SP-C leads to a significant decrease in α-helical structure [535, 536, 1112, 1142]. NMR data indicate that removal of the two palmitoyl groups from SP-C causes the helix to start at or near residue 15 rather than at residue 9 as in the native protein [536]. The α-helical structure of proteins in general is known to affect their incorporation in lipid bilayers, and mismatches between helix length and bilayer thickness can lead to phase separation and reduced interactions. In addition, the molecular region in SP-C that is affected by acylation includes the functionally implicated charged Lys and Arg at residues 11 and 12. The presence of C16:0 palmitic chains in SP-C could also be important in surface active function through direct interactions with similar acyl moieties in DPPC and other saturated phospholipids. However, despite the known and potential molecular effects of acylation in SP-C, the quantitative importance of this modification in surface activity is not certain. Wang et al [1142] found that acylated SP-C was much more active than deacylated SP-C in increasing the adsorption and dynamic surface tension lowering of lung surfactant phospholipids. However, other studies have reported less substantial surface activity differences between mixtures of phospholipids combined with acylated vs deacylated native SP-C or nonacylated recombinant SP-C [160, 878, 968]. Several forms of recombinant SP-C used as components in active exogenous surfactants are either not acylated or have been reported not to be greatly reduced in activity by deacylation [184, 374, 375, 425, 638, 967, 968] (Chapter 14). Even if acylation in native SP-C is important in surface activity, this may not be the case in some forms of recombinant SP-C particularly when amino acid substitutions relative to endogenous sequence are present. A mixture of dipalmitoy-lated (15%) and nonpalmitoylated forms of recombinant human SP-C has recently been produced in the baculovirus expression system [1113], and surface activity comparisons utilizing these materials may help to further define the impact of acylation on the function of recombinant SP-C.

VII. Effects of Lung Surfactant Apoproteins on Phospholipid Surface Activity

SP-A, SP-B, and SP-C all have substantial, functionally important actions in increasing surface activity in endogenous lung surfactant. As detailed below, these apoproteins significantly increase the adsorption of surfactant lipids into the air-water interface. Apoproteins also enhance surface activity in the interfacial film itself, improving respreading and participating in film refining to optimize surface tension lowering during dynamic cycling.

A. Effects of SP-A on Phospholipid Surface Activity

SP-A facilitates the adsorption of lung surfactant phospholipids largely through molecular biophysical interactions that promote the formation of tubular myelin and other large aggregates. Large aggregates isolated by differential centrifuga-tion from endogenous surfactant have maximal adsorption and dynamic surface tension lowering ability (e.g., [67, 350, 379, 685, 872, 873, 1200]) (Chapter 6). These large aggregate surfactant subtypes contain a higher content of SP-A and hydrophobic apoproteins than small aggregate subtypes [67, 685, 872, 1115, 1200]. They also have a calcium-dependent microstructure rich in tubular myelin, although many other less specific large aggregate forms are also present [67, 421, 685, 872, 1200]. SP-A is cooperative with SP-B in the formation of tubular myelin, and the greatest effects on surface activity occur when both proteins are present together. Lung surfactant has maximal adsorption under conditions where tubular myelin is formed (Figure 8-8) [791]. If EDTA is added to lung surfactant to chelate endogenous calcium and inhibit tubular myelin formation, adsorption is reduced at low surfactant concentration (Figure 8-8). However, large aggregates other than tubular myelin also have significant adsorption facility, and the absence of this microstructure can be overcome by increasing surfactant concentration (Figure 8-8, curve 3). SP-A has also been shown to influence the molecular organization and refining of phospholipid surface films during cycling [761, 931], and it can enhance adsorption and/or dynamic surface activity when added exogenously to phospholipids [226, 421, 576, 894, 927, 1070]. Whether or not SP-A is absolutely necessary for the surface active function of alveolar surfactant

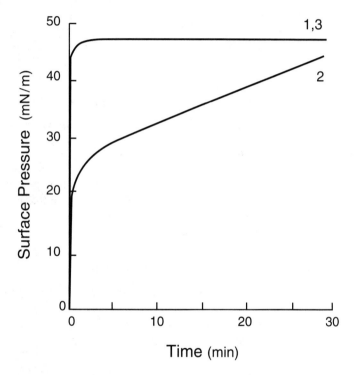

Figure 8-8 Calcium-dependent adsorption of lung surfactant. Curve 1: 0.25 mg/ml calf lung surfactant (LS) plus 1.4 mM CaCl$_2$; curve 2: 0.25 mg/ml LS + 5 mM EDTA; curve 3: 0.63 mg/ml LS + 5 mM EDTA. Lung surfactant adsorbs rapidly at the air-water interface without calcium, but it has maximal adsorption at the lowest concentration when this divalent cation is present. This behavior correlates with the formation of tubular myelin and other large aggregates mediated in part through the actions of SP-A. The maximum adsorption surface pressure of 47–48 mN/m at 37°C is equivalent to an equilibrium surface tension of 22–23 mN/m. See text for details. (From Ref. 791.)

has been the subject of some debate. SP-A gene knockout in mice is not lethal [597]. Moreover, hydrophobic organic solvent extracts of lung surfactant that lack SP-A but contain surfactant lipids and SP-B/C can be formulated *in vitro* with very high surface activity. The ability of these extracts to resist inactivation by plasma proteins and other inhibitory substances, however, is improved by addition of SP-A [149, 1116, 1220] (Chapters 9, 14). The significant molecular interactions and effects of SP-A on phospholipid surface activity are consistent with its designation as a biophysically functional component of endogenous lung surfactant.

B. Effects of the Hydrophobic Apoproteins on Phospholipid Surface Activity

The hydrophobic surfactant proteins have a striking impact in improving the adsorption and dynamic surface activity of phospholipids. This is illustrated, for example, by comparing the adsorption of subfractions of phospholipids, neutral lipids, and hydrophobic proteins separated from calf lung surfactant by gel-permeation chromatography [1146] (Figure 8-9). The complete mix of surfactant phospholipids adsorbs more rapidly than DPPC, but adsorption is still poor compared to lung surfactant itself. Adsorption is improved only slightly if surfactant neutral lipids and phospholipids are combined. However, when hydro-phobic SP-B/C are present with surfactant phospholipids, adsorption becomes almost equivalent to that of solvent-extracted lung surfactant (Figure 8-9). Similar substantial improvements in adsorption and/or dynamic surface tension lowering are found when SP-B and SP-C are added alone or together to mixtures of synthetic phospholipids.[6] The activity of the hydrophobic surfactant proteins in facilitating adsorption correlates with their ability to disrupt and fuse phos-pholipid bilayers and to promote the insertion and transfer of phospholipids into vesicles and surface films [119, 160,161, 164, 804, 805, 807]. The hydrophobic proteins have also been shown to affect the aggregation and sqeeze-out of phospholipids in dynamically compressed surface films [164, 206, 602, 653, 656, 761, 762, 1072] and to improve film respreading during dynamic cycling [1067–1069, 1072, 1142–1145]. In order to directly affect surface behavior in this fashion, SP-B and SP-C must remain either within the interfacial film or in close association with it during cycling.

---

6. For example, Refs. 180, 422, 460, 793, 804, 805, 877, 893, 914, 967, 968, 1008, 1052, 1064, 1066, 1116, 1117, 1142, 1143, 1145, 1215, 1216, 1218.

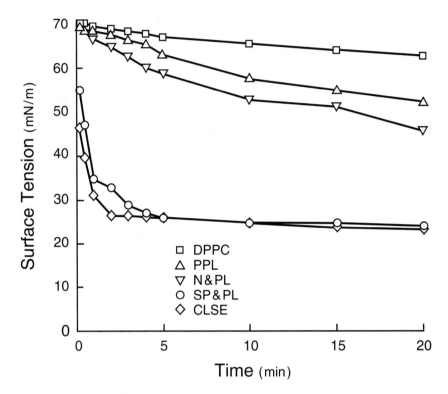

Figure 8-9 Adsorption of component subfractions of calf lung surfactant extract (CLSE). The significant ability of hydrophobic SP-B/C to enhance the adsorption of surfactant phospholipids is illustrated by comparing the adsorption of component subfractions of calf lung surfactant. Column-purified surfactant phospholipids (PPL) or purified neutral lipids plus phospholipids (N&PL) adsorb better than DPPC but much worse than CLSE. Mixed hydrophobic SP-B/C plus phospholipids (SP&PL) have adsorption almost equal to that of CLSE. Phospholipid concentration was 0.25 mM in 0.15 M NaCl. Data were measured at 37°C in a bubble surfactometer in the static mode and include a diffusion resistance [1146].

C. Relative Effectiveness of SP-B versus SP-C in Improving Surface Activity

Although SP-B and SP-C both enhance the adsorption and dynamic surface activity of phospholipids, a variety of studies indicate that SP-B is more effective in doing so (e.g., [180, 532, 804, 805, 807, 893, 939, 968, 1143, 1216]). Activity differences between SP-B and SP-C are found on a weight basis and are even more pronounced on a molar basis due to the smaller molecular weight of SP-C. An example of the greater activity of SP-B vs SP-C in enhancing the adsorption of phospholipids is shown in Figure 8-10 [1143]. The increased adsorption of mixtures of phospholipids with SP-B correlates with the fact that this apoprotein has a capacity for binding lipid vesicles to interfacial films that is about four times greater than that of SP-C on a weight basis [804]. In terms of effects on film behavior, SP-B and SP-C both participate in refining phospholipid films during compression and have relatively similar effects in improving film respreading during dynamic cycling [1067–1069, 1072, 1143]. However, aqueous dispersions containing phospholipids mixed with SP-B vs SP-C exhibit significantly better

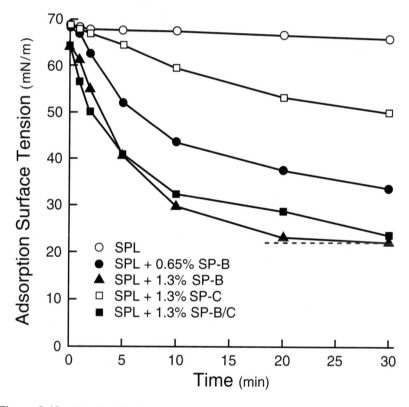

Figure 8-10 Relative effectiveness of SP-B vs SP-C in enhancing phospholipid adsorption. Surface tension–time adsorption is significantly better in mixtures of synthetic phospholipids (SPL) with SP-B vs SP-C. A low surfactant concentration of 0.031 mg phospholipid/ml was used to highlight apoprotein activity differences. The dashed line is the equilibrium surface tension reached by whole calf lung surfactant. Similar results were found in mixtures of purified lung surfacant phospholipids with SP-B vs SP-C (not shown). SPL contained DPPC:egg PC:egg PG (50:35:15, molar ratio). Apoprotein contents are in weight percent; mixed SP-B/C is 1:1 by weight. (Data are means from Ref. 1143 at 37°C.)

overall surface tension lowering during dynamic cycling at physiologic rates and area compressions (Figure 8-11) [1143]. Mixtures of phospholipids with SP-B vs SP-C also have an improved ability to resist inhibition by serum albumin [968, 1143]. These biophysical findings are generally consistent with physiological activity studies. Instilled exogenous surfactants containing lipids combined with SP-B have been shown to give more substantial improvements in pulmonary mechanics and function in animal lungs than corresponding mixtures of lipids

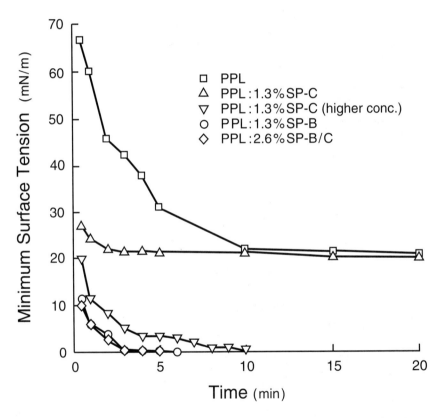

Figure 8-11 Dynamic surface tension lowering in mixtures of lung surfactant phospholipids with SP-B vs SP-C. Minimum surface tensions are plotted as a function of time for suspensions of purified calf lung surfactant phospholipids (PPL) combined with SP-B vs SP-C vs mixed 1:1 SP-B/C examined on a pulsating bubble apparatus (37°C, 20 cycles/min, 50% area compression). Curves for 1.3% SP-B and 2.6% SP-B/C are equivalent; other curves are statistically different. Apoprotein contents are in weight percent. Phospholipid concentration is uniform at 0.5 mM except high concentration SP-C results are at 4 mM. (Data are means from Ref. 1143.)

with SP-C [898]. In addition, supplementation with SP-B is found to improve the surface and physiological activity in animals of several clinical exogenous surfactants lacking or deficient in this apoprotein [382, 740] (Chapter 14). The high relative activity of SP-B also correlates with the known lethal nature of congenital SP-B deficiency in humans [201, 403, 778, 779] (Chapter 10). Mice genetically deficient in the SP-B gene (SP-B "knockouts") are similarly subject to lethal respiratory failure [132].

D. Degree of Synergy of SP-B and SP-C in Increasing Phospholipid Surface Activity

Whether or not SP-B and SP-C are synergistic in interacting with phospholipids has obvious relevance for lung surfactant function. Although SP-B exhibits cooperativity with SP-A in forming tubular myelin and facilitating adsorption [422, 1184], this does not appear to be the case for many of the biophysical actions of the hydrophobic proteins. The results shown in the preceding section for delipidated isolates of purified SP-B and SP-C suggest that these two proteins are not significantly synergistic in enhancing phospholipid adsorption and dynamic surface activity (Figures 8-10 and 8-11) [1143]. This is consistent with fluorescence energy transfer data indicating that SP-B and SP-C do not interact directly in lipid bilayers [474] and with additional studies showing that these apoproteins have independent effects on phospholipid adsorption [805] and film behavior [1069]. The surface properties affected by SP-B and SP-C are complex and depend on multiple molecular interactions. Some of these molecular interactions may have an element of apoprotein cooperativity, but currrent data suggest that SP-B and SP-C act independently in generating a significant portion of their enhancement of phospholipid surface activity.

VIII. Summary of Functional Lung Surfactant Components and Their Actions

The multiple chemical constituents of endogenous lung surfactant interact biophysically to determine overall surface active behavior. Phospholipids and apoproteins are the major functional components of lung surfactant, although neutral lipids can also influence surface active function. In terms of component-specific roles, the extreme surface tension lowering power of endogenous lung surfactant is generated largely by DPPC and related rigid disaturated phospholipids. These disaturated phospholipids are also primarily responsible for the ability of lung surfactant films to vary surface tension with area during cycling. SP-A, SP-B, and SP-C are the major components responsible for facilitating the adsorption of surfactant phospholipids into the air-water interface, and these apoproteins also affect film refining during compression and contribute to increased film respread-

---

**Biophysical Actions and Effects of Lung Surfactant
Apoprotein on Surface Activity**

SP-A
   Enhances phospholipid aggregation and order
   Necessary for tubular myelin formation (with SP-B, calcium)
   Enhances phospholipid adsorption
   May participate in film refining and in enhancing respreading
   during cycling

SP-B
   Necessary for tubular myelin formation (with SP-A, calcium)
   Most active hydrophobic protein in improving overall surface
   activity
   Interacts with both phospholipid headgroups and chains
   Orders lipid headgroups and has mixed overall effects on
   bilayer order
   Can disrupt and fuse phospholipid bilayers and vesicles
   Greatly enhances phospholipid adsorption
   Increases respreading and contributes to film refining during
   cycling

SP-C
   Decreases order and packing density in phospholipid bilayers
   Interacts largely with phospholipid hydrophobic chains
   Can disrupt and fuse phospholipid bilayers and vesicles
   Greatly enhances phospholipid adsorption
   Increases respreading and contributes to film refining during
   cycling

---

ing during cycling. Fluid lung surfactant phospholipids have a pronounced effect in improving film respreading, and neutral lipids also enhance this surface property. Fluid phospholipids and neutral lipids also provide some facilitation of adsorption relative to DPPC alone. A conceptualized overview of system behavior is as follows:

1. During expiration, alveolar area decreases and the surface film is compressed. Fluid lipids with less ability to sustain high surface pressures are preferentially ejected ("squeezed out") to enrich the film in DPPC and other disaturated phosphatidylcholines and rigid

---

**Major Functional Components in
Endogenous Lung Surfactant**

DPPC and related rigid, disaturated phospholipids
Give low minimum surface tension on dynamic compression
Vary surface tension with area during cycling

Surfactant proteins SP-A, SP-B, and SP-C
Greatly increase phospholipid adsorption to the air-water interface
Improve film respreading on successive cycles
Participate in refining the surface film during compression

Fluid phospholipids (and neutral lipids)
Greatly increase film respreading during cycling
Increase adsorption relative to DPPC

---

phospholipids. Surfactant proteins participate in this selective refining process, influencing the structure and organization of the surface film and aggregates of ejected material in the interfacial region.

2. Near end-expiration, the enrichment of the surface film in DPPC and related disaturated phospholipids generates very low surface tensions as film compression proceeds into the collapse regime.[7] At high degrees of compression, rigid as well as fluid components leave the film and enter collapse structures above or below the interface.

3. As inspiration begins, there is an immediate rise in surface tension as the solid-like condensed film expands and breaks apart. Components ejected from the surface during compression respread and reintegrate back into the film. Fluid phospholipids and surfactant proteins, as well as neutral lipids, facilitate the reentry and respreading of saturated phospholipids back into the interfacial film.

4. Continuing adsorption of surfactant components from tubular myelin and other aggregates in the alveolar hypophase also replenishes film material in addition to respreading from interfacial region structures.

---

7. Reaching low minimum surface tensions <1 mN/m in films of lung surfactant requires rapid dynamic compression and is associated with the formation of metastable surface states by rigid phosphatidylcholines like DPPC (see Chapters 3 and 7).

The sum of these complementary processes results in a regenerating film of alveolar surfactant able to lower and vary surface effectively during cycling so that normal pulmonary mechanics and gas exchange are maintained.

IX. Chapter Summary

Lung surfactant is a functional mixture whose overall behavior reflects contributions from multiple chemical components. No single constituent has all the surface behaviors required for physiological activity. The major biophysically functional components of endogenous surfactant are phospholipids and apoproteins, although neutral lipids also influence surface behavior. About 80% of surfactant phospholipids are phosphatidylcholines, with smaller amounts of other zwitterionic and anionic phospholipid classes. Lung surfactant contains a balance of fluid and rigid phospholipids based on film and bilayer behavior. DPPC accounts for about one third of total surfactant phospholipid, and disaturated-PC (DSPC) compounds as a group make up about 45–50% of total phospholipid. DPPC and related rigid disaturated phospholipids give lung surfactant the ability to lower and vary surface tension effectively during dynamic cycling. Fluid phospholipids greatly improve film respreading during cycling and also increase adsorption relative to DPPC.

The structure, molecular biophysics, and effects of surfactant apoproteins on surface activity are complex as detailed in this chapter. SP-A, SP-B, and SP-C have crucial roles in surfactant activity, but SP-D is not involved in biophysical function. SP-A, SP-B, and SP-C are the major components responsible for facilitating adsorption in endogenous surfactant. These apoproteins, particularly SP-B, and SP-C, also help refine film composition and improve dynamic surface tension lowering and respreading during cycling. At the molecular biophysical level, SP-A facilitates the formation of active large phospholipid aggregates. Tubular myelin, formed by phospholipids in the presence of SP-A, SP-B, and calcium, is just one example of such aggregation. SP-B and SP-C disrupt and fuse phospholipid bilayers and facilitate phospholipid insertion into surface films. SP-B interacts with both phospholipid headgroups and hydrophobic chains and is thought to locate more peripherally in bilayers than SP-C, which primarily interacts with acyl chains in bilayer interior. SP-B orders phospholipid headgroups and has a mixed overall effect on bilayer order and packing, while SP-C decreases bilayer order. Both SP-B and SP-C significantly increase the adsorption and dynamic surface activity of phospholipids, but SP-B has greater relative effectiveness in doing so, making it a particularly important functional component of lung surfactant.

9

Lung Surfactant Dysfunction

I. Overview

Lung surfactant dysfunction occurs when surface activity is compromised, a phenomenon also called lung surfactant inactivation or inhibition. One important cause of lung surfactant inactivation is direct interactions with endogenous inhibitors that act biophysically to reduce the adsorption and dynamic film activity of alveolar surfactant. The term lung surfactant dysfunction is also used in a broader sense to encompass losses of surface activity following a number of other detrimental processes. These include, for example, chemical degradation of functional surfactant components by lytic enzymes or reactive oxidants present in the lungs during inflammation. Physicochemical or metabolic alterations in lung injury that compromise or selectively deplete the most active large surfactant aggregates can also lead to decreased surface activity. This chapter summarizes characteristic surface activity changes found in lung surfactant dysfunction and gives examples of relevant inhibitors and the mechanisms by which they act. The consequences of decreased surface activity for the physiological effects of lung surfactants on P-V mechanics are also illustrated.

II. Concept of Lung Surfactant Inhibition or Inactivation

For many biological phenomena, inhibition refers to biochemical or cellular events. An inhibitory compound binds to an enzyme, ligand, or membrane receptor and generates its actions through effects on chemical reactions or other cellular processes. For lung surfactants, surface active function rather than cellular function is of primary importance. Consequently, lung surfactant inhibition or inactivation commonly refers to a decrease in surface activity. As described in more detail in subsequent sections, many inhibitors of lung surfactant activity act through direct biophysical interactions to impair adsorption or dynamic surface tension lowering, while others interact chemically with lung surfactant to degrade its functional components. The actions of these latter inhibitors can also produce reaction products like lysophospholipids and free fatty acids that futher reduce surface activity. Inhibitory substances or inducers of lung injury can also decrease lung surfactant activity by depleting or compromising the large, most active surfactant aggregate subtypes. Decreases in lung surfactant activity may also be caused by injury-induced alterations to type II cells that reduce surfactant synthesis or recycling and lead to an acquired surfactant deficiency. If the surface active function of lung surfactant is sufficiently compromised, impaired respiration results. Lung surfactant dysfunction is an important contributor to the pathophysiology of acute lung injury and clinical ARDS (Chapter 10).

III. Endogenous Compounds Able to Directly Inhibit Lung Surfactant Activity

A variety of compounds have been identified that can interact physically or chemically with lung surfactant and reduce its surface activity.[1] Some of these inhibitors are present endogenously in the lungs, while others such as caustic or toxic chemicals are inhaled, aspirated, or ingested exogenously. Discussion here focuses primarily on endogenous inhibitor substances that interact either physically or chemically with lung surfactant to impair surface activity.

A. Biophysical Inhibitors of Lung Surfactant Activity

Probably the most widely recognized biophysical inhibitors of lung surfactant are plasma proteins such as albumin, fibrinogen, and fibrin monomer.[2] Hemoglobin, which can be present in the lungs through lysis of red blood cells in hemorrhagic injury, is a related protein inhibitor [448, 453, 1116, 1117]. Lung surfactant can also be inactivated biophysically by interactions with cellular lipids including cholesterol [796, 1083], fluid membrane phospholipids and lysophospholipids [147, 149, 453, 461, 1147], and glycolipids [887]. Lysophospholipids not only are able to inhibit surface activity through direct biophysical interactions but also can

---

1. For example, Refs. 20, 131, 147, 149, 244, 282, 310, 371, 380, 381, 448, 449, 453, 456, 459, 461, 495, 562, 590, 592, 750, 776, 796, 859, 887, 909, 970, 972, 1083, 1100, 1116, 1117, 1147 plus reviews in Refs. 455, 643, 784, 786, 798, 969.
2. For example, Refs. 147, 149, 282, 380, 448, 453, 456, 459, 461, 495, 562, 592, 776, 972, 1083, 1116, 1117, 1147.

damage the integrity of the pulmonary endothelial-epithelial barrier and increase the concentration of plasma-derived inhibitors [774]. Fluid free fatty acids such as oleic acid also generate direct biophysical detriments to surface activity [380, 381, 970, 1083] as well as causing increased capillary endothelial permeability [687, 988, 999, 1097, 1222, 1227]. Another relevant inhibitor of lung surfactant activity is meconium, a complex fetal product containing multiple inhibitory cell membrane lipids, tissue proteins, and fluid fatty acids that cause severe respiratory failure in infants if aspirated at birth [131, 750, 1079]. The biophysical mechanisms of action of several of these inhibitory substances are detailed in a later section.

B. Chemically Acting Inhibitors of Lung Surfactant

Reductions in the surface tension lowering ability of lung surfactant also result if the concentration of functional surface active components is reduced. Examples of inhibitors released in the lungs during inflammation or injury that can react with and degrade functional surfactant lipids or proteins are proteases [859], phospholipases [244, 449, 1100], and reactive oxidant species [20, 310, 371, 372, 909, 970]. Antibodies to surfactant apoproteins can also combine chemically with essential components of lung surfactant to impair their activity and lead to decreased surface activity [590, 917, 1039], although such antibodies are not commonly present endogenously. As noted earlier, chemically acting inhibitors not only decrease lung surfactant activity by degrading or reacting with func-

---

**Examples of Endogenous Inhibitors
of Lung Surfactant Activity**

Biophysically acting inhibitors
　　Plasma and blood proteins (e.g., albumin, hemoglobin, fibrinogen)
　　Unsaturated cell membrane phospholipids
　　Lysophospholipids
　　Cholesterol
　　Fluid free fatty acids
　　Meconium
Chemically interacting inhibitors
　　Lytic enzymes (proteases, phospholipases)
　　Reactive oxidants
　　Antibodies to surfactant proteins

tional constituents but also can generate reaction products that cause additional inhibition. For example, degradation of phospholipids by phospholipases leads to the production of inhibitory unsaturated free fatty acids and lysophospholipids.

IV. Characteristic Surface Activity Changes Found During Inactivation of Endogenous Lung Surfactant

Inhibition of lung surfactant can involve reductions in adsorption or dynamic surface tension lowering or both. In most cases, the detriments to surface activity depend strongly on surfactant and inhibitor concentrations. The inhibitory effects of the plasma protein albumin on the adsorption of endogenous lung surfactant are illustrated in Table 9-1. When albumin is added to bovine lung surfactant (LS), the rate of adsorption is reduced and the equilibrium surface tension is raised in a concentration-dependent manner. The inhibitory effects of albumin on lung surfactant adsorption are most apparent when the concentration of surfactant is low (Cases 3 and 4, Table 9-1). When surfactant concentration is increased, inhibition from albumin decreases substantially (Case 6, Table 9-1). Similar concentration-dependent patterns of adsorption inhibition are found for endogenous lung surfactant in the presence of other blood proteins such as hemoglobin and fibrinogen (e.g., [282, 453, 1117]) (Figure 9-1).

Table 9-1 Inhibitory Effects of Albumin on the Adsorption of Endogenous Lung Surfactant

Surfactant and/or inhibitor	Lung surfactant concentration (mg PL/ml)	Plasma protein concentration (mg/ml)	Adsorption surface tension (mN/m) at time (min):				
			0.25	5	10	15	20
1. Albumin	—	0.03–1.9	56	53	53	53	53
2. LS	0.063	—	53	24	23	23	23
3. LS + Alb	0.063	1.1	52	48	45	32	25
4. LS + Alb	0.063	1.9	52	49	47	46	43
5. LS	0.125	—	25	23	23	23	23
6. LS + Alb	0.125	2.5	51	29	25	24	23

Lavaged bovine lung surfactant (LS) or albumin or both were added at time zero to a stirred subphase containing 0.15 M NaCl + 1.4 mM $CaCl_2$, and surface tension was followed as a function of time during adsorption. Albumin inhibited LS adsorption most significantly at low surfactant concentration. Complete surface pressure–time adsorption isotherms showing a similar pattern of inhibition for hemoglobin are given in Figure 9-1.
Source: Data from Ref. 456.

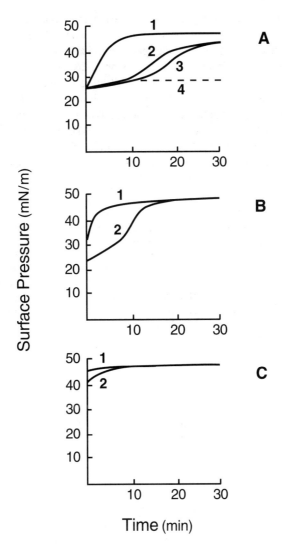

Figure 9-1 Inhibition of the adsorption of endogenous lung surfactant by hemoglobin. Hemoglobin (Hb) inhibits the adsorption of lavaged bovine lung surfactant (LS) when the two are added simultaneously to a stirred subphase (0.15 M NaCl + 1.4 mM CaCl$_2$) at 37°C. The surface pressure–time adsorption isotherms depend on the concentration of Hb and LS in analogy with the results for albumin in Table 9-1. Corresponding surface tensions are found by subtracting surface pressures from 70 mN/m. (A) 1: LS, 0.065 mg/ml; 2: LS, 0.065 mg/ml + Hb, 0.31 mg/ml; 3: LS, 0.065 mg/ml + Hb, 0.63 mg/ml; 4: Hb, 0.31–2.5 mg/ml. (B) 1: LS, 0.13 mg/ml; 2: LS, 0.13 mg/ml + Hb, 0.63-2.5 mg/ml. (C) 1: LS, 0.28 mg/ml; 2: LS, 0.28 mg/ml + Hb, 2.5–3.75 mg/ml. (Data from Ref. 453.)

In addition to inhibiting adsorption, plasma and blood proteins can compromise the dynamic surface activity of lung surfactant (Table 9-2). At a low surfactant concentration of 0.4 mg phospholipid/ml (0.5 mM), albumin and hemoglobin raise the minimum surface tension of lavaged bovine surfactant from <1 mN/m to more than 20 mN/m on the pulsating bubble surfactometer. These inhibitory effects on dynamic surface tension lowering again exhibit a strong dependence on lung surfactant concentration. When the concentration of surfactant phospholipid is raised to 0.75 mg/ml (1 mM), the detrimental effects of albumin and hemoglobin in elevating minimum surface tension during dynamic cycling are reduced even if blood protein concentration is increased (Table 9-2).

Table 9-2 Effects of Blood Proteins and Cell Membrane Lipids on the Dynamic Surface Activity of Endogenous Lung Surfactant

Surfactant and/or inhibitor	Lung surfactant concentration (mg PL/ml)	Plasma protein concentration (mg/ml)	Minimum surface tension (mN/m) at time (min):		
			0	5	10
Albumin	—	2-200	45	45	45
Hemoglobin	—	2-200	36	35	35
LS	0.4	—	20	6	<1
LS + albumin	0.4	5	45	40	13
LS + albumin	0.4	10	45	44	30
LS + hemoglobin	0.4	25	34	30	27
LS + membrane lipids	0.4	0.4	24	22	21
LS	0.75	—	19	3	<1
LS + albumin	0.75	100	44	3	<1
LS + hemoglobin	0.75	100	36	25	<1
LS + membrane lipids	0.75	0.4	22	20	19
LS	1.5	—	12	<1	<1
LS + hemoglobin	1.5	200	34	—	<1
LS + membrane lipids	1.5	0.4	20	—	<1

Lavaged bovine lung surfactant (LS) suspended in 0.15 M NaCl + 1.4 mM $CaCl_2$ was studied in a pulsating bubble surfactometer (37°C, 20 cycles/min, 50% area compression) in the presence and absence of inhibitors. Time 0 was <15 sec from the start of pulsation. Blood proteins and red blood cell membrane lipids raised minimum surface tension at low surfactant concentration, but inhibition was overcome at high surfactant concentration. See text for details.
Source: Data from Refs. 453, 456.

At a higher lung surfactant concentration of 1.5 mg phospholipid/ml (2 mM), even large amounts of hemoglobin (200 mg/ml) have very little inhibitory effect on minimum surface tension. The ability of lung surfactant to resist inactivation as its concentration is raised is widely observed, although the quantitative details of this process vary depending on the specific inhibitor substance involved. For example, inhibition by red blood cell membrane lipids is more difficult to overcome by increasing surfactant concentration than inhibition by plasma proteins (Table 9-2). This is also true for inhibition by oleic acid as shown in Figure 9-2. When oleic acid is present at a high molar ratio of 0.67 or 0.75 relative to

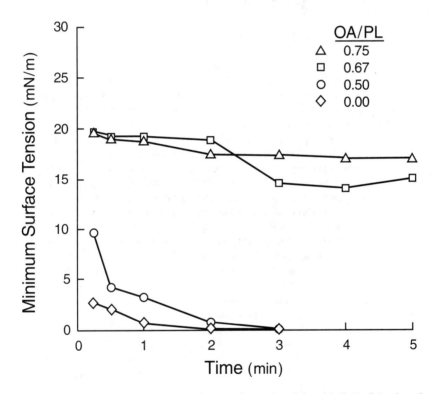

Figure 9-2 Inhibition of endogenous lung surfactant by oleic acid (OA). OA raises the minimum surface tension of lavaged bovine lung surfactant in a pulsating bubble apparatus (37°C, 20 cycles/min, 50% area compression). Inhibition is present at a molar ratio of OA to surfactant phospholipid (PL) of 0.67 or 0.75 even though surfactant concentration is very high (12 mM or 9 mg/ml for the data shown). A lower OA/PL molar ratio of 0.5 is not inhibitory at this high surfactant concentration, but was found to elevate minimum surface tension when surfactant concentration was reduced to 0.5 mM (not shown). (Data are means from Ref. 380.)

lung surfactant phospholipid, it elevates minimum surface tension substantially even at a very high surfactant concentration of 12 mM (Figure 9-2) [380]. If the molar ratio of oleic acid to phospholipid is lowered to 0.5, minimum surface tension is no longer elevated at the high surfactant concentration in the figure. However, inhibition does occur at this lower compositional ratio of oleic acid when lung surfactant concentration is only 0.5 mM phospholipid [380]. To fully overcome inhibition by oleic acid, lung surfactant concentration needs to be high enough so that the molar ratio of oleic acid to surfactant phospholipid is less than 0.5. The modified concentration dependence of inhibition for oleic acid and red cell membrane lipids relative to blood proteins relates to the fact that these materials act through different biophysical mechanisms as described in a later section.

Additivity of Lung Surfactant Inhibitors. The possibility of additive or synergistic interactions between inhibitors of lung surfactant activity has received relatively little detailed study. In principle, such interactions could increase the severity of surfactant dysfunction in situations where multiple inhibitory substances are present in alveoli. One important factor affecting the additivity of inhibitors is whether they act by complementary mechanisms. For example, since surface activity is typically impaired most easily at low surfactant concentration, inhibitors that degrade lung surfactant and reduce its functional composition might be expected to exhibit additivity with those acting biophysically to reduce adsorption or dynamic surface tension lowering. Current research suggests that although inhibitor additivity can occur under some conditions, it may not be a major factor in surfactant dysfunction in many forms of lung injury [1147]. Hemoglobin and red blood cell membrane lipids have been shown to cause slightly more severe inhibition of lung surfactant surface activity when present together than when either is present alone at the same concentration (Figure 9-3). However, the magnitude of additivity is not pronounced, and inhibition is still overcome by raising lung surfactant concentration (Figure 9-3). A similar small degree of additivity, also reversible by increasing surfactant concentration, has been reported for mixtures of hemoglobin or albumin with lysophosphatidylcholine or cell membrane lipids [1147]. In some instances, mixtures of inhibitors can have reduced rather than increased effects on surface activity. This occurs in mixtures of albumin and free fatty acids, where albumin binding of fatty acid reduces the effective inhibitor concentration and leads to smaller detriments in surface activity than when the same amount of free fatty acid is present alone [1147]. The potential additivity of other combinations of lung surfactant inhibitors is still under investigation.

V. Inactivation of Exogenous Lung Surfactants

The concentration-dependent inactivation processes described in the preceding sections for endogenous lung surfactant also occur for exogenous lung surfac-

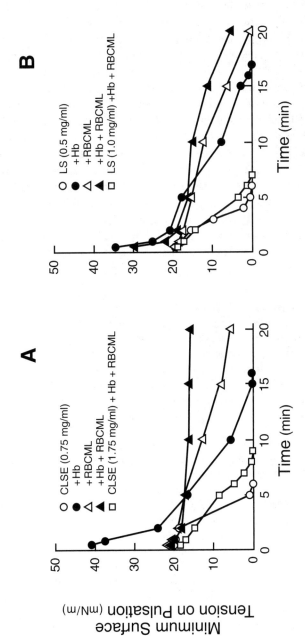

Figure 9-3 Additivity of lung surfactant inhibitors. Red blood cell membrane lipids (RBCML) or hemoglobin (Hb) or both were added to lavaged calf lung surfactant (LS) or its chloroform:methanol extract (CLSE) and studied in a pulsating bubble apparatus at 37°C and a rate of 20 cycles/min. These inhibitors exhibit some additivity in elevating minimum surface tension in mixtures, but additivity is not pronounced and inhibition is overcome by increasing surfactant concentration. (A) CLSE; (B) whole LS. Inhibitor concentrations: Hb: 2.5 mg/ml; RBCML: 0.1 mM. (Data are means from Ref. 1147.)

tants. A number of exogenous surfactants have been shown to be able to resist inactivation as their concentration is increased, although current preparations at best approach the inhibition resistance of lavaged endogenous surfactant. Organic solvent extracts of lavaged alveolar surfactant containing its lipid and hydrophobic protein constituents have a significant ability to overcome inactivation by plasma proteins, membrane lipids, and other substances (e.g., [147, 382, 448, 453, 456, 461, 967, 968, 1116, 1147]). Such extracts form the basis of several current clinical exogenous surfactants including Alveofact, bLES, and Infasurf (Chapter 14). Even these extracts, however, can be improved toward whole surfactant in inhibition resistance by adding SP-A [149, 1116, 1220]. Most exogenous surfactants containing one or more of the hydrophobic surfactant apoproteins or related synthetic peptides have at least some ability to resist inactivation as their concentration is raised. The inhibition resistance of Surfactant-TA, an organic solvent extract of processed bovine lung tissue supplemented with synthetic DPPC, tripalmitin, and palmitic acid, is illustrated in Figure 9-4 [282]. The inhibition resistance of clinical exogenous surfactants is discussed further in Chapter 14.

Importance of Apoproteins in the Inhibition Resistance of Endogenous and Exogenous Lung Surfactants.   The ability of endogenous and exogenous lung surfactants to resist inactivation at high surfactant concentrations has been linked at least in part to their apoprotein constituents. As noted in the preceding section, organic solvent extracts of lung surfactant or processed lung tissue containing hydrophobic surfactant proteins have a substantial ability to resist inactivation as surfactant concentration is raised. Mixtures of synthetic lipids with hydrophobic SP-B and/or SP-C also display this pattern of inhibition resistance (e.g., [968, 1116, 1117, 1143]). In contrast, several synthetic lipid mixtures that exhibit high surface activity in the absence of inhibitors have been shown to have compromised surface tension lowering ability in the presence of relatively small amounts of plasma proteins even at high surfactant concentrations [459]. SP-A has also been found to increase the ability of surfactant mixtures to resist inactivation by plasma proteins and other inhibitors as noted above [149, 1116, 1220] and has been shown to improve their effects on pulmonary mechanics in surfactant-deficient animals in the presence of plasma proteins [1220]. Conversely, antibodies to SP-A or to SP-B can increase the sensitivity of endogenous surfactants to inactivation by plasma proteins [590, 917, 1039]. The specific mechanisms by which surfactant apoproteins improve inhibition resistance in mixtures with phospholipids are not fully understood but are almost certainly related to the strong molecular biophysical interactions that exist between these substances (Chapter 8). The addition of supplemental apoproteins to improve the surface activity and inhibition resistance of clinical exogenous surfactants is discussed further in Chapter 14.

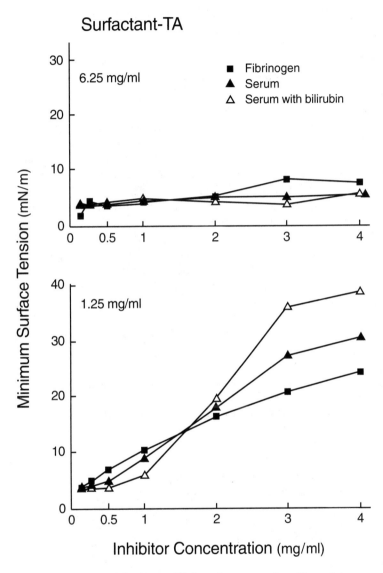

Figure 9-4 Inhibition of Surfactant-TA by plasma proteins. The minimum surface tension reached by the clinical exogenous surfactant Surfactant-TA on a pulsating bubble apparatus is raised by serum and serum components. Inhibition is most pronounced at low surfactant concentration (top) and is reduced or overcome as concentration is raised (bottom). The minimum surface tension shown is after 10 minutes of pulsation (37°C, 20 cycles/min, 50% area compression). (Data are means redrawn from Ref. 282.)

VI. Clinical Relevance of Concentration-Dependent Surfactant Inactivation

The ability to mitigate or overcome inactivation by raising surfactant concentration in the continued presence of inhibitors has important therapeutic ramifications. In particular, it provides a direct rationale for the use of exogenous surfactant supplementation in acute lung injury and ARDS, where dysfunction of endogenous surfactant is known to occur. Although inactivation from some inhibitors is easier to overcome than from others, biophysical studies clearly suggest the potential benefits of alveolar delivery of active exogenous surfactant material. The ability to reverse surfactant inactivation in patients with ARDS depends on a number of factors. One obvious requirement is that exogenous surfactants used in therapy must have the ability to resist inhibition, and current clinical surfactants vary in how effectively they accomplish this (Chapter 14). Even if surfactants with maximal activity and inhibition resistance are utilized, edema in the alveoli and terminal airways could compromise their delivery or make it impossible for a stable air-hypophase interface to be maintained. Exogenous surfactants could also be degraded by inflammatory enzymes or reactive oxidants, reducing their effective concentration. This could be particularly important in overcoming inhibition by fluid fatty acids or membrane lipids that require higher surfactant concentrations to reverse inactivation. Exogenous sur-

---

Summary of Lung Surfactant Dysfunction and Its Reversal

Lung surfactant dysfunction (inactivation, inhibition) can be caused by multiple substances and mechanisms. Many forms of inactivation can be reversed by raising surfactant concentration, a phenomenon called inhibition resistance.

Surfactant apoproteins enhance the ability to resist inactivation. Endogenous surfactant containing phospholipids plus SP-A, SP-B, and SP-C appears to have optimal inhibition resistance.

Many exogenous surfactants including hydrophobic organic solvent extracts of endogenous surfactant have been shown to have significant inhibition resistance.

The ability to overcome inhibition by raising surfactant concentration gives a rationale for using exogenous surfactant therapy in ARDS-related lung injury in addition to RDS.

factant therapy in ARDS may also be limited because it targets only one aspect of a multifaceted lung injury pathology that requires additional complementary therapeutic interventions. Surfactant therapy in ARDS is discussed further in Chapters 11 and 13.

VII. Mechanisms of Lung Surfactant Inactivation

Lung surfactant inhibitors act by a number of different mechanisms. These mechanisms include competive adsorption and interfacial shielding, as well as penetration into the surface film itself to fluidize it or otherwise alter its behavior during dynamic compression. The surface tension lowering ability of lung surfactant can also be compromised because key functional components of the surfactant system are degraded chemically by lytic proteases and phospholipases or by a variety of reactive oxygen and nitrogen species. Activity reductions can also occur through physicochemical interactions between functional components of lung surfactant and specific antibodies directed against them. In addition, the surface activity of lung surfactant can be decreased as a result of alterations in the most active large surfactant aggregate subtypes, a process that is found to occur in several forms of lung injury. The list of inhibitors and mechanisms involved in lung surfactant inactivation will undoubtedly continue to be expanded by future research. The remainder of this section summarizes several relevant mechanisms of lung surfactant inactivation in more detail and gives representative examples of the inhibitors involved.

A. Inactivation by Competitive Adsorption and Interfacial Shielding (e.g., Plasma and Blood Proteins)

Essentially all plasma and blood proteins have some degree of polar/nonpolar structure and hence have intrinsic surface activity. When these large proteins adsorb at the air-water interface, they hinder the entry of lung surfactant compo-

---

Examples of Lung Surfactant Inactivation Mechanisms

Competitive adsorption and interfacial shielding
Surface film fluidization
Chemical degradation of surfactant components
Selective depletion or alteration of large surfactant aggregates

nents into the surface film. Since plasma and blood proteins are not as effective as lung surfactant components in lowering surface tension, this reduces overall surface activity. The ability of plasma proteins to competitively adsorb and block the air-water interface to inhibit lung surfactant activity has been documented experimentally [448, 461, 562]. In particular, the adsorption of albumin and extracted lung surfactant has been studied under conditions designed to determine the influence of competitive adsorption and film penetration (Table 9-3) [448]. Albumin alone adsorbs to an equilibrium surface tension of 49 mN/m at 37°C, while CLSE reaches a much lower value of 23 mN/m equivalent to endogenous surfactant. When albumin at a concentration of 1.25 mg/ml adsorbs simultaneously with CLSE at a low concentration of 0.063 mg phospholipid/ml, inhibition is severe and the final surface tension is equal to that of albumin alone (Case 3, Table 9-3). When CLSE concentration is raised to 0.25 mg/ml so that surfactant adsorption is more rapid, simultaneous addition of albumin at 1.25 mg/ml does not inhibit equilibrium surface tension at all (Case 5, Table 9-3). This behavior is modified, however, if either albumin or CLSE is initially added alone and allowed to preform a surface film. In this case, subsequent addition of the second material even at high concentration has little impact, indicating that film penetration does not occur. Thus, if CLSE at 0.063 mg/ml is allowed to preform a surface film, albumin added subsequently at 2.5 mg/ml does not raise surface tension (Case 4, Table 9-3). Conversely, if albumin at 1.25 mg/ml is allowed to preform a film, injection of CLSE at 0.25 mg/ml beneath the film does not lower surface tension below the albumin equilibrium of 49 mN/m (Case 6).

Table 9-3 Inhibition of Lung Surfactant Adsorption by Albumin Under Different Conditions

Surfactant and/or inhibitor	Concentration (mg/ml)	Experimental condition	Equilibrium surface tension (mN/m)
1. CLSE	0.063 or 0.25	Adsorbing alone	23 ± 1
2. Albumin	1.25 or 2.5	Adsorbing alone	49 ± 2
3. CLSE + albumin	0.063 + 1.25	Simultaneous addition	49 ± 2
4. CLSE + albumin	0.063 + 1.25	CLSE added first	23 ± 1
5. CLSE + albumin	0.25 + 1.25	Simultaneous addition	23 ± 1
6. CLSE + albumin	0.25 + 1.25	Albumin added first	49 ± 2

Calf lung surfactant extract (CLSE) and albumin were studied during adsorption at 37°C in a Teflon dish with a stirred 0.15 M NaCl subphase at the indicated concentrations. If one substance was added first, it was allowed to reach its equilibrium surface pressure, followed by addition of the second substance and determination of the final tabulated equilibrium pressures. The data are consistent with albumin inhibition by competitive adsorption as described in the text.
Source: Data from Ref. 448.

The ability of albumin to inhibit CLSE by competitive adsorption, but not by film penetration, is also indicated by the results in Figure 9-5. In this experiment, a surface film of CLSE was initially formed by adsorption at an equilibrium surface pressure of 47 mN/m (surface tension of 23 mN/m at 37°C). Albumin injected to a final subphase concentration of 2.5 mg/ml beneath this film did not alter its surface pressure (arrow, Figure 9-5). However, when the CLSE film was subsequently aspirated so that competitive adsorption could occur, inhibition was severe and the final surface pressure reached was equivalent to that of albumin alone. Results similar to these were also demonstrated for the blood proteins fibrinogen and hemoglobin in inhibiting lung surfactant activity [448]. In additional experiments, the inhibitory actions of plasma proteins on adsorption and dynamic surface activity were found to be abolished if simple

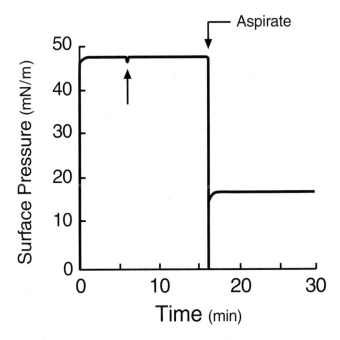

Figure 9-5 Albumin inhibition of CLSE in a preformed film and during competitive adsorption. A CLSE film was preformed by adsorption at a low concentration of 0.063 mg phospholipid/ml at 37°C to give an equilibrium surface pressure of 47 mN/m (surface tension of 23 mN/m). A concentration of albumin (2.5 mg/ml) that could inhibit CLSE during simultaneous adsorption was then injected beneath the surface but had no effect on surface pressure (First arrow). When the CLSE film was subsequently aspirated so that CLSE and albumin could adsorb competitively, severe inhibition was apparent. (From Ref. 448.)

centrifugation was used to separate them from CLSE [448]. This lack of strong binding interactions is again consistent with the mechanism of competitive adsorption being the major factor in plasma protein–induced inhibition of lung surfactant activity.

B. Lung Surfactant Inactivation by Film Penetration and Fluidization (e.g., Fluid Fatty Acids, Lysophospholipids, Membrane Lipids)

Rather than inhibiting lung surfactant by competitive adsorption, fluid fatty acids and lysophospholipids act largely through a different mechanism. These compounds primarily mix into the interfacial film itself, compromising surface tension lowering ability during dynamic compression [380, 461]. Cholesterol and fluid cell membrane phospholipids may also inhibit lung surfactant activity through this mechanism. Oleic acid and C16:0 lysophosphatidylcholine (LPC) have been shown to be able to penetrate and interact in spread films of DPPC or lung surfactant [380, 461], while plasma proteins have relatively little effect on dynamic surface tension lowering if added directly to DPPC surface films [1057]. LPC [461] and oleic acid [380] also affect lung surfactant adsorption, but not as prominently as found with plasma proteins.

An example of the differing inhibition behavior exhibited by LPC and albumin is shown in Figure 9-6 from the work of Holm et al [461]. Films of solvent-extracted calf lung surfactant were initially formed by adsorption in a bubble apparatus and cycled to generate a stable film with a minimum surface tension <1 mN/m at 37°C (Pre-exchange, Figure 9-6). This CLSE film was then isolated on a hypophase of buffered saline and cycled briefly to verify its ability to reach minimum surface tensions <1 mN/m (1st Exchange, Figure 9-6). The film was then exposed to a second hypophase containing C16:0 LPC or albumin and studied for surface tension lowering ability during dynamic cycling (2nd Exchange, Figure 9-6). The results show that LPC was able to penetrate the lung surfactant film and raise minimum surface tension while albumin could not. Minimum surface tension was increased to about 10 mN/m after 5 minutes of cycling on a hypophase containing LPC but remained at the control value of <1 mN/m on a hypophase containing albumin (Figure 9-6). Separate experiments verified that albumin as noted earlier could compete with lung surfactant for the air-water interface and inhibit its adsorption [461]. LPC also was found to interact with lung surfactant during adsorption [461], consistent with its high critical micelle concentration [1024] and ability to affect phospholipid vesicle structure and permeability [83, 633]. However, film penetration and reduction of dynamic surface tension lowering ability was a major factor in the inhibitory actions of LPC in contrast to those of albumin [461].

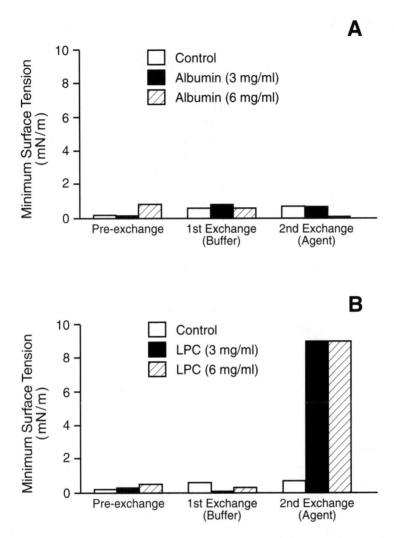

Figure 9-6 Inhibitory effect of LPC and albumin on minimum surface tension in preformed CLSE films. A specialized hypophase exchange system for the pulsating bubble surfactometer allowed isolation of a CLSE film with stable $\sigma_{min} < 1$ mN/m (1st Exchange). The ability of LPC or albumin at concentrations of 3 and 6 mg/ml to penetrate this film and raise σ_{min} was then studied (2nd Exchange). LPC could penetrate the CLSE film and inhibit surface tension lowering but albumin could not. Minimum surface tension is after 5 minutes of cycling (37°C, 20 cycles/min, 50% area compression). (Data are means from Ref. 461.)

The differing mechanisms by which LPC and albumin reduce lung surfactant activity lead to different patterns of concentration dependence for the inhibition process. Mixtures of CLSE and LPC exhibit increased minimum surface tensions as a function of LPC content: 14 wt % LPC raises σ_{min} to 5 mN/m, 25–50 wt % LPC raises σ_{min} to 15 mN/m, and 67 wt % LPC raises σ_{min} to >20 mN/m even at high CLSE concentrations of 3 and 6 mg/ml (Figure 9-7) [461]. In contrast, albumin inhibition of lung surfactant activity is not stoichiometric and is more easily reversed at high surfactant concentrations (Figure 9-7). Albumin (3 mg/ml) raises σ_{min} to >20 mN/m at a CLSE concentration of 1 mg/ml, but inhibition is abolished at a surfactant concentration of 3 mg/ml even when much larger amounts of albumin are present (Figure 9-7). As described earlier (Figure 9-2), oleic acid inhibition of lung surfactant activity exhibits a concentration dependence related to that found for LPC. In particular, oleic acid inhibition of lung surfactant depends on the molar ratio of the two and can persist even at high surfactant concentrations if the relative content of oleic acid is large. However, inhibition by either oleic acid or LPC can still be resisted if surfactant concentration is raised so that the content of inhibitors is decreased sufficiently.

C. Lung Surfactant Inactivation by Chemical Degradation (e.g., Lytic Enzymes, Reactive Oxidants)

A number of enzymes released during inflammatory lung injury are able to degrade functional components in lung surfactant. Two of the most prominent classes of such enzymes are proteases and phospholipases. Neutrophil elastase is one example of an inflammatory protease that has been shown to interact chemically with lung surfactant to reduce surface activity [859]. Phospholipases A_1, A_2, C, and D are all capable of cleaving both lung surfactant and tissue phospholipids. Phospholipase A_1 and A_2 act at the junctional linkage of the fatty chains at the *sn*-1 and *sn*-2 positions, respectively [1043]. Phospholipase C cleaves phospholipids at the junction between the phosphate group and glycerol backbone at the *sn*-3 carbon, and phospholipase D acts at the base of the phosphate group proximal to the N-headgroup [1043]. Aside from lytic phospholipases and proteases, reactive oxidants are another major group of materials able to react with lung surfactant components (Figure 9-8) [116]. Some oxidants such as superoxide anion ($\cdot O_2^-$) and hydroxyl radical ($\cdot OH$) are free radicals that contain unpaired electrons, while others are reactive nonradicals such as hydrogen peroxide (H_2O_2), organic hydroperoxides (ROOH), and peroxynitrite ($ONOO^-$). These reactive oxidants are antagonized by a variety of antioxidant enzymes in the normal lung, three of the most of important of which are catalase, superoxide dismutase, and glutathione peroxidase (Figure 9-8). The use of antioxidant enzymes to mitigate the effects of reactive oxidant species in acute lung injury and ARDS is covered in Chapter 13.

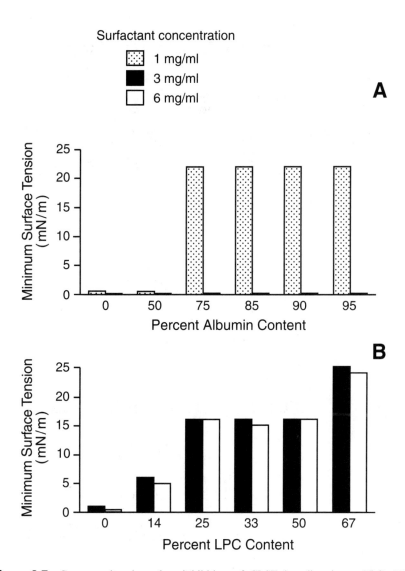

Figure 9-7 Concentration-dependent inhibition of CLSE by albumin vs LPC. (A). Minimum surface tension as a function of albumin content; (B). Minimum surface tension as a function of LPC content. The different mechanisms of action of albumin vs LPC lead to different concentration-dependent patterns of inhibition. LPC induces inhibition at higher surfactant concentrations due to its film penetrating ability. Minimum surface tension is after 5 minutes of cycling on the pulsating bubble surfactometer; % content is % by weight of inhibitors in total mixture with CLSE. (Data are means from Ref. 461.)

A   Formation of Reactive Oxidants

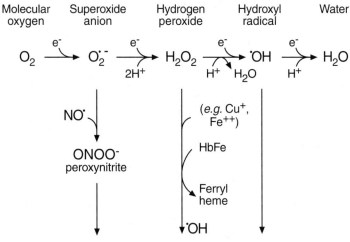

oxidation of surfactant lipids or proteins

B   Reduction of Reactive Oxidants

$$2O_2^{\cdot -} + 2H^+ \xrightarrow{\text{SOD}} H_2O_2 + O_2$$

$$2H_2O_2 \xrightarrow{\text{CAT}} 2H_2O + O_2$$

$$ROOH + 2GSH \xrightarrow{\text{GSH-Px}} GSSG + ROH + H_2O$$

Figure 9-8 Production and removal of selected reactive oxidant species. (A) Chemical pathways producing reactive oxidant species that cause lung injury and can potentially degrade or compromise functional components in lung surfactant. (B) The actions of antioxidant enzymes in removing selected reactive oxidants. NO˙ nitric oxide; Hb, hemoglobin; Cu^+, copper; Fe^{++}, ferrous iron; SOD, superoxide dismutase; CAT, catalase; GSH, reduced glutathione; GSSG, oxidized glutathione; GSH-Px, glutathione peroxidase. (Adapted in part from Ref. 116.)

D. Lung Surfactant Inactivation by Alterations in Surfactant Aggregate Subtypes

Another pathway of lung surfactant inactivation known to occur in some forms of acute lung injury involves alterations in subphase surfactant aggregates. As described in Chapter 6, phospholipid-rich aggregates of different sizes and densities exist in the aqueous phase microstructure of endogenous lung surfactant. These aggregate subtypes vary in surface activity, with larger aggregates having the highest apoprotein content and displaying the best surface tension lowering ability (e.g., [67, 350, 379, 685, 872, 873, 1200]). A net decrease in overall surface tension lowering ability thus results if large surfactant aggregates are selectively depleted or are altered so as to have reduced activity. This effect can be present even if the total amount of lavageable surfactant phospholipid remains normal or is increased. Abnormalities in the concentration or surface activity of large surfactant aggregates have been demonstrated in a number of animal models of acute lung injury [64, 348, 349, 379, 640, 871]. The percentage of large aggregates and their content of SP-A have also been shown to be reduced in broncho-alveolar lavage from patients with ARDS [361, 1114]. The mechanisms by which lung surfactant aggregates are altered in lung injury are not fully understood. Inhibitors may interact directly with subphase aggregates to change their structure and activity. Large aggregate changes could also occur secondary to alterations in surfactant metabolism in type II cells or in intra-alveolar processing by serine proteases or other mediators involved in large to small aggregate conversion [348, 352, 871].

VIII. Physiological Correlates of Lung Surfactant Inhibition

The direct connection between lung surfactant inactivation and decreased physiological activity can be demonstrated experimentally. Two examples showing physiological correlates of surfactant inactivation in excised rat lungs are given below. This mechanical model was used in Chapter 7 to illustrate the physiological importance of lung surfactant surface properties such as rapid adsorption and the generation of low minimum surface tension. Additional physiological correlates of lung surfactant activity and inhibition in animal models of RDS and ARDS are detailed in Chapter 11.

A. Example of Concentration-Dependent Pulmonary Mechanical Changes from Surfactant Inhibitors

The concentration-dependent behavior with which inhibitors decrease the surface activity of lung surfactant *in vitro* can also be replicated in physiological studies in intact lungs. As detailed earlier, the effects of inhibitors on surface tension lowering are generally most pronounced at low surfactant concentration and are mitigated as surfactant concentration is raised. The physiological correlate expected from this behavior is that detriments to P-V mechanics from inhibitors should be largest when levels of endogenous surfactant are reduced and should be reversed if sufficient amounts of active exogenous surfactant are delivered to overcome inactivation. These behaviors can be observed in experiments with excised rat lungs (Figure 9-9) [380, 453]. When albumin, hemoglobin, or cell membrane lipids are instilled into the freshly excised lungs, P-V deflation mechanics are altered in the direction of decreased compliance (top panel, curves B, C vs curve A). Mechanical detriments become even more pronounced if these inhibitors are instilled into lungs made partially surfactant deficient by a single lavage with normal saline (bottom panel, curves C, D vs curve B). These mechanical changes can be reversed if surfactant levels in the lungs are augmented by supplementation with exogenous surfactants able to resist inactivation at high concentration. Installation of 25 mg of CLSE into normal or partially surfactant-deficient rat lungs almost completely reverses these impaired P-V mechanics despite the continued presence of inhibitors (top, curve D, and bottom, curve E, Figure 9-9). Similar results are obtained when a fluid fatty acid (oleic acid) rather than blood proteins or cell membrane lipids is instilled into excised rat lungs and supplementation is done with whole surfactant [380].

B. Example of P-V Mechanical Correlates Related to Inhibition Resistance

As noted earlier, exogenous surfactants vary significantly in their ability to resist inhibition by plasma proteins and other substances. Although some exogenous surfactants can resist inactivation effectively, this is by no means true for all preparations. If such surfactants are delivered to the lungs, they can be inactivated and exhibit low physiological activity. As an example showing this behavior, Holm et al [459] studied the physiological activity of mixtures of synthetic DPPC with 10% palmitic acid (PA), dipalmitin (DP), or diolein (DO) in comparison to calf lung surfactant extract (CLSE). The synthetic surfactants all reached low minimum surface tensions <1 mN/m at a concentration of 5 mg/ml on a pulsating bubble apparatus [459]. However, they were sensitive to inhibition by albumin and achieved minimum surface tensions of only 30–35 mN/m in the presence of 2 mg/ml of this protein even at high surfactant concentrations of 20 mg/ml. In contrast, CLSE at a concentration of 2 mg/ml reached minimum surface tensions <1 mN/m in the presence of 200 mg/ml concentrations of albumin. When instilled into lavaged, surfactant-deficient rat lungs, 9:1 DPPC:PA, 9:1 DPPC:DP, or 9:1 DPPC:DO had very little effect in improving P-V mechanics relative to CLSE (Figure 9-10). When the synthetic surfactants were recovered from the lungs by

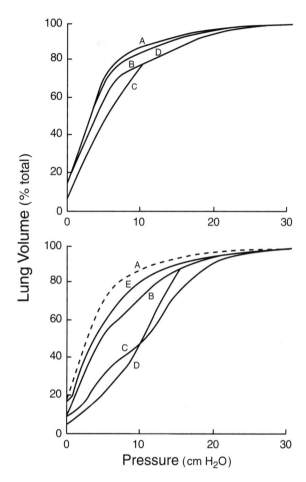

Figure 9-9 Effects of endogenous inhibitors on P-V mechanics in normal and partially surfactant-depleted excised lungs. Albumin, hemoglobin, or red blood cell membrane lipids were instilled into freshly excised lungs (top) or excised lungs partially depleted in endogenous surfactant by a single lavage (bottom). P-V mechanical changes are consistent with inhibition of endogenous surfactant that is most pronounced at low surfactant concentrations. Installation of 25 mg CLSE in 2.5 ml saline restored air-filled mechanics to normal, correlating with reversal of inhibition by increased surfactant concentration. (Top) A: Normal lungs, no inhibitors; B: 5 mg red blood cell membrane lipids instilled; C: 400 mg albumin or hemoglobin instilled; D: 400 mg albumin/hemoglobin or 5 mg cell membrane lipids instilled, followed by 25 mg CLSE. (Bottom) A: Normal lung; B: partially surfactant-deficient lung (one lavage); C: partially deficient lung with 400 mg albumin or hemoglobin instilled; D: partially deficient lung with 5 mg cell membrane lipids instilled; E: partially deficient lung with 400 mg albumin/hemoglobin or 5 mg cell membrane lipids instilled, followed by 25 mg CLSE. (Redrawn from Ref. 453.)

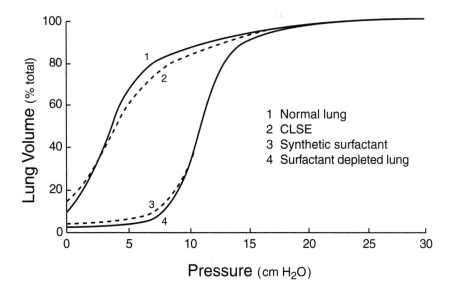

Figure 9-10 Lack of response of P-V mechanics after instillation of inhibitable exogenous surfactants into surfactant-deficient lungs. Curve 1: Mechanics of freshly excised, surfactant-sufficient rat lungs; curve 4: mechanics of surfactant-deficient lungs after multiple lavage; curve 2: lavaged lungs instilled with 30 mg of CLSE in 2.5 ml saline; curve 3: lavaged lungs instilled with 30 mg of 9:1 DPPC:palmitic acid, 9:1 DPPC:diolein, or 9:1 DPPC:dipalmitin in 2.5 ml saline. CLSE with high inhibition resistance fully restores normal mechanics but inhibitable synthetic surfactants do not. Biophysical data indicate these latter surfactants were inactivated by small amounts of plasma proteins in the lungs. See text for details. (Data at 37°C from Ref. 459.)

lavage, plasma proteins were shown to be present and minimum surface tensions were elevated to >28 mN/m. Separation of the plasma proteins from the synthetic lipid mixtures by centrifugation or extraction restored surface activity to the high preinstillation levels exhibited in the absence of inhibitors [459]. These results demonstrate the importance of inhibition resistance in the activity of exogenous surfactants. They also illustrate that the inactivation of exogenous surfactants in the lungs can give an apparent noncorrespondence between preinstillation biophysical properties and physiological activity. Exogenous surfactants and their effects on pulmonary mechanics and function are detailed further in subsequent chapters.

IX. Chapter Summary

Lung surfactant dysfunction (inhibition, inactivation) can be caused by a variety of substances and can occur by multiple mechanisms. One important cause of surfactant dysfunction involves inhibitors that interact biophysically with lung surfactant to reduce its adsorption or dynamic surface activity. Lung surfactant dysfunction can also occur from chemically acting inhibitors that degrade or alter essential phospholipid or apoprotein components. Examples of compounds that can inhibit lung surfactant activity are plasma and blood proteins, cell membrane lipids, fluid free fatty acids, lysophospholipids, cholesterol, reactive oxidants, and lytic proteases and phospholipases. Plasma and blood proteins inhibit surface activity primarily by competitive adsorption, while fluid fatty acids, lysophospholipids, and cell membrane lipids mix into the surface film to fluidize it and raise minimum surface tension during dynamic cycling. Reactive oxidants and lytic enzymes degrade lung surfactant chemically, removing functional components and also generating detrimental reaction products. Reductions in lung surfactant surface activity from the selective depletion or alteration of active, large aggregate subtypes have also been demonstrated in several kinds of acute lung injury.

Many forms of lung surfactant inactivation can be mitigated or overcome by raising surfactant concentration. Endogenous lung surfactant containing all the biophysically active apoproteins appears to have a maximal ability to resist many inhibitors. Organic solvent extracts of endogenous surfactant containing hydrophobic surfactant proteins also have high inhibition resistance, although there is significant variation among preparations. The ability to reverse inactivation by raising lung surfactant concentration gives a rationale for exogenous surfactant therapy in acute lung injury and ARDS in addition to RDS. Inhibition of lung surfactant surface activity can be correlated experimentally with detriments to P-V mechanics in intact lungs. Instillation of inhibitory substances alters P-V mechanics and compliance consistent with lung surfactant inactivation. P-V detriments are greatest when endogenous surfactant is reduced in concentration, consistent with the known concentration dependence of surfactant inactivation in biophysical studies. Instillation of exogenous surfactants able to resist inhibition can restore mechanics to normal despite the continued presence of inhibitors, while instillation of surfactants unable to resist inactivation is ineffective. The consequences of lung surfactant deficiency and dysfunction, and their reversal by exogenous surfactant therapy in animals and humans, are detailed further in Chapters 10–14.

10

Diseases of Lung Surfactant Deficiency or Dysfunction

I. Overview

Lung surfactant deficiency and dysfunction contribute to the pathophysiology of several important respiratory diseases. The major disease of lung surfactant deficiency worldwide is the respiratory distress syndrome (RDS) of premature infants, also called hyaline membrane disease (HMD). Developing clinically effective exogenous surfactant therapy for RDS has been a major driving force in lung surfactant research. A second, much more rare form of surfactant-deficient lung disease occurs when a functional component of lung surfactant is not produced. This is the case in hereditary SP-B deficiency, a genetic disease associated with lethal respiratory failure in infants. The most important conditions associated with surfactant dysfunction are acute lung injury (ALI) and the acute respiratory distress syndrome (ARDS). Surfactant deficiency can also occur in ALI and ARDS, but dysfunction from biophysical or chemical interactions with inhibitors is typically more prominent. The complex pathophysiology of acute lung injury and ARDS described in this chapter presents a significant challenge for surfactant replacement interventions.

II. The Neonatal Respiratory Distress Syndrome (RDS, HMD)

RDS is a disease of prematurity. The nomenclature of HMD derives from the distinctive protein-rich shiny "hyaline membranes" found in the alveoli and small airways of affected infants at autopsy. The possibility that elevated surface tension contributed to atelectasis and hyaline membrane formation in the lungs of premature infants was first proposed in 1947 by Gruenwald [357]. The linkage

233

of HMD with abnormal pulmonary surface tension was further emphasized by Gruenwald in 1955 [356], and the likely importance of surfactant deficiency in this condition was suggested several years later by Pattle [830]. A direct connection between HMD and decreased surfactant function was provided in 1959 by Avery and Mead [34], who showed that pulmonary washings from premature infants dying from this disease had impaired surface tension lowering ability. However, the surfactant-deficient etiology of HMD was not uniformly accepted by physicians, and respiratory distress in premature infants was widely described as "idiopathic" (of unknown cause) in the early 1960's [1001]. Indeed, the use of idiopathic in association with HMD and RDS was not uncommon into the 1970's. The underlying cause of RDS is now understood to be a primary deficiency in alveolar surfactant due to a lack of functional type II pneumocytes in the premature lung.

A. Incidence of RDS in Premature Infants

As many as 50,000–60,000 premature infants in the United States alone each year are at risk for RDS. In the absence of prophylactic surfactant therapy, the incidence of RDS is >50% in infants less than 29 weeks gestation (e.g., [369, 370, 484, 606, 765]). The incidence of RDS decreases as gestational age and birth weight increase. Premature infants of 32–34 weeks gestation have an incidence of about 10–20%, and infants ≥35 weeks gestation have a very low incidence of 5% or less. Maternal diabetes, male sex, Caucasian vs black ethnicity, and perinatal asphyxia are among the factors associated with an increased risk for RDS [52, 484, 606, 1061]. Multiple gestations also have a significantly increased risk of RDS, at least partly in association with the lower birth weight of these infants. Prolonged fetal stress accompanying maternal hypertension, drug abuse, or chronic infection can decrease the incidence of RDS, possibly by inducing a higher level of endogenous steroid hormones. The incidence of RDS is decreased by maternal administration of glucocorticoid steroids for 24–72 hours prior to birth to enhance fetal type II cell development [154, 173, 648]. Amniotic fluid testing for fetal lung maturity based on PG/PI or lecithin/sphingomyelin (L/S) ratios is generally done to assess fetal lung maturity and the need for maternal steroid therapy [318, 392, 517, 1061, 1075], which has the potential to be synergistic with postnatal surfactant therapy. RDS was the major cause of neonatal mortality in the United States and other developed nations for decades prior to the 1990's [369, 370, 765]. It still remains a clinically significant entity, although effective surfactant therapy has greatly improved the prognosis and outcome of small premature infants (Chapter 12).

B. Pathophysiology of RDS

The lung is the limiting organ of viability in humans from near the end of the second trimester through about 32 weeks gestation. This largely reflects the timing of type II pneumocyte development. Type II cells with characteristic osmiophilic lamellar bodies are visible in the human lungs as early as 22–24 weeks gestation. However, type II cell numbers and surfactant stores in most infants are insufficient for normal respiration until much later in gestation. This surfactant deficiency is the underlying cause of RDS, although other pathology including lung injury and multiple complications of prematurity and intensive care can enter the course of disease (Figure 10-1). Infants with RDS exhibit alveolar collapse (atelectasis), alveolar overdistension, increased work of breathing (decreased compliance), and pulmonary edema. Ventilation and perfusion in the lung are mismatched (intrapulmonary shunting), and arterial oxygenation is reduced (hypoxemia). Increased arterial carbon dioxide concentrations (hypercapnia) cause respiratory acidosis, and a systemic metabolic acidosis can also be present. Superimposed acute injury from mechanical ventilation and high inspired oxygen concentrations is one of many factors that can affect the course of RDS, giving rise to pulmonary inflammation and cellular damage as well as to surfactant inactivation. Chronic lung injury including bronchopulmonary dysplasia also often occurs. Patent ductus arteriosus, intraventricular hemorrhage, necrotizing enterocolitis, infection, and pneumothorax and other forms of pulmonary air leak are among the many complications of prematurity and intensive care that can also occur in premature infants with RDS. The pathophysiology and clinical features

---

The Neonatal Respiratory Distress Syndrome (RDS)

The major surfactant-deficient lung disease worldwide, also known as hyaline membrane disease.

Up to 50,000–60,000 premature infants annually in the United States alone are at risk for RDS.

Infants <32 weeks gestation and <1,500 gm birth weight are primarily affected, although disease can occur in larger infants.

The underlying cause of RDS is surfactant deficiency due to a lack of functional type II pneumocytes.

Lung injury and many other complications of prematurity and intensive care can enter the course of this disease.

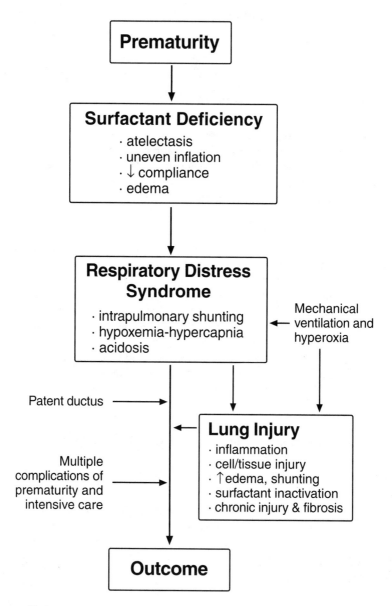

Figure 10-1 Pathophysiology of RDS in premature infants. Surfactant deficiency is the initiating cause of RDS, but acute and chronic lung injury and a variety of complications of prematurity and neonatal intensive care can enter the course of disease. See text for details.

of RDS are covered in further detail elsewhere (e.g., [175, 250, 606, 918, 919, 989, 1061, 1075]).

C. Clinical Aspects of RDS in Premature Infants

Infants with RDS typically show signs of impaired respiration in the first few hours of life. Many of these infants display a classic diagnostic pentad of cyanosis in room air, tachypnea, intercostal and subcostal retractions, expiratory grunting, and nasal flaring. Prior to the introduction of continuous positive airway pressure (CPAP) and positive end-expiratory pressure (PEEP) in mechanical ventilation at the beginning of the 1970's, most infants <28 weeks gestation or <1,250 gm birth weight survived only a brief time after birth [95, 765, 1001]. The smallest infants often died before the development of classic symptoms associated with HMD and were sometimes classified as having lethal anatomical immaturity of the lung. The clinical diagnosis of HMD used to be most common in premature infants of 1,000–2,000 grams birth weight and 28–34 weeks gestation [33]. As discussed above, it is now known that surfactant deficiency as opposed to anatomical immaturity is the primary factor causing RDS in infants from about 24–25 weeks gestation on.

Several clinical features commonly associated with RDS are noted in Table 10-1. A variety of scoring and classification systems based on these and other clinical signs and symptoms have been developed over the years but there is no clear consensus on a single optimal description (e.g., [33, 52, 95, 1050, 1061]). In some infants, the course of RDS is relatively mild and responds to supplemental oxygen and small amounts of positive pressure such as from nasal CPAP. In its most severe form, RDS is characterized by progressive hypoxemia despite oxygen therapy and intensive mechanical ventilation. Generalized atelectasis is apparent on chest radiography, with low inflation volumes and hazy "ground glass" lung fields. So-called air bronchograms are present due to the outlining of airways in contrast to bordering atelectatic parenchyma. At autopsy, infants have characteristic massive atelectasis with overdilatation of adjacent lung

Table 10-1 Selected Clinical Features Associated with RDS

Premature birth with respiratory distress in the first few hours of life
Classic diagnostic features of cyanosis, tachypnea, chest retractions, expiratory grunting, and nasal flaring
Decreased breath sounds
Chest radiograph with granular air bronchograms showing evidence of atelectasis
Hypoxemia and hypercapnia responsive in many cases to positive pressure ventilation and supplemental oxygen

Source: Adapted from Ref. 52.

areas, along with hyaline membranes in the lumen of airspaces [606]. Uncompli-
cated RDS typically resolves in 2–5 days as production of endogenous surfactant
increases. In the absence of injury, type II cells mature rapidly even in very
premature infants due to mechanical stimuli and hormones such as steroids
released at or near the time of birth. However, the course of RDS in many infants
is prolonged by acute and chronic lung injury and by other complications of
prematurity and intensive care as noted earlier (Figure 10-1). These infants
require multiple medical and pharmacological interventions in addition to exogenous surfactant, and their clinical outcome is not simply a reflection of this latter
therapy alone. Therapies for premature infants are detailed in standard pediatric
texts (e.g., [1061, 1075]).

III. Congenital Deficiency of Functional Surfactant Components (SP-B Deficiency)

Even if the overall amount of lung surfactant is normal, an effective state of
surfactant deficiency exists if one or more functional components is depleted or
absent. This occurs in hereditary SP-B deficiency [201, 403, 404, 778, 779, 1128,
1176]. This condition is typified by severe, persistent respiratory distress in
full-term infants within 24–48 hr of birth, typically in association with alveolar
proteinosis in the absence of cardiac disease or infection. The existence of a
familial form of neonatal alveolar proteinosis was noted by several investigators
in the 1980's (e.g., [151, 583]). It is now clear that this familial alveolar
proteinosis reflects a congenital deficiency in SP-B. In addition to being deficient
in SP-B, affected infants have a reduced amount of some surfactant phospholipids
(e.g., phosphatidylglycerol), plus abnormalities in SP-C processing and accumulation of proSP-C [403, 404, 1128, 1176]. The majority of infants with congenital
SP-B deficiency have a homozygous mutation in exon 4 of the SP-B gene at
codon 121, referred to as 121ins2 [778, 779]. This form of SP-B deficiency is
autosomal recessive, and heterozygous individuals do not typically have significant pulmonary pathology. There is, however, some genetic heterogeneity and
symptomatic variability in congenital SP-B deficiency. Although the pathophysiology of congenital SP-B deficiency is still being studied, impaired surface
activity from the absence of SP-B is clearly a major factor. SP-B gene knockout
in mice also leads to lethal respiratory failure with compromised lung surfactant
activity [132]. Congenital SP-B deficiency is not effectively treated with exogenous surfactant therapy [403], which requires eventual normalization of endogenous surfactant production. Extracorporeal membrane oxygenation is also not
effective in treating SP-B deficient alveolar proteinosis [752]. Lung transplantation is the only current option for affected infants [404, 405], who otherwise die
within the first months of life despite intensive care and mechanical respiratory

---

Hereditary SP-B Deficiency

A rare, lethal genetic condition associated with neonatal alveolar proteinosis and respiratory failure.

Most common form is autosomal recessive, with heterozygous individuals relatively unaffected.

SP-B is absent, and abnormalities in SP-C processing also exist.

Not effectively treated with surfactant replacement therapy, and requires lung transplantation or gene therapy.

---

support. Gene-based therapy may in future be effective in treating hereditary SP-B deficiency. Fortunately, SP-B deficiency is very rare, and this is also likely to be true for congenital deficiencies in other lung surfactant components.

IV. Acute Lung Injury and ARDS

ARDS was originally described in adult patients [28, 848] and was generally termed the "adult" respiratory distress syndrome during the 1970's and 1980's. ARDS is now known as the *acute* respiratory distress syndrome, reflecting its rapid onset and pathophysiological origin in association with acute lung injury. ARDS affects patients of all ages including term infants, children, and adults. An overview of the incidence, risk factors, pathophysiology, and clinical features of ARDS is given below.

A. Definition, Incidence, and Risk Factors for ARDS

ARDS is a reflection of inflammatory acute lung injury from multiple pulmonary and nonpulmonary causes (Table 10-2). Patients initially exhibit symptoms associated with the underlying disorder (Table 10-2) but then develop clinical evidence of edema that rapidly progresses to respiratory failure. Patients with ARDS can have extensive pathology outside the lungs, including sepsis and multiorgan disease. However, pulmonary criteria are used in defining ARDS. A once common definition was based on a qualitative set of respiratory signs and symptoms in the absence of left heart failure: bilateral pulmonary edema, decreased lung compliance and volumes, and intrapulmonary shunting apparent as arterial hypoxemia resistant to high levels of inspired oxygen [467, 848]. The American-European Consensus Committee in 1994 defined clinical ARDS based

Table 10-2 Etiologies Associated with Clinical Acute Lung Injury and ARDS-Related Respiratory Failure

Aspiration	**Inhaled toxic gases**
Caustic acids, bases, other chemicals	Oxygen
Gastric contents	Other oxidants (NO_x, ozone)
Meconium	Smoke
Blood transfusions (multiple)	**Near-drowning**
Cardiopulmonary bypass	**Pancreatitis**
Disseminated intravascular coagulation	**Physical injuries/trauma**
Drug overdose or exposure	Burns
Barbiturate overdose	Chest trauma
Heroin, methadone overdose	Head injury
Paraquat	Multiple fractures with fat emboli
Immunosuppression	Pulmonary contusion
Infection	**Radiation pneumonitis**
Sepsis and sepsis syndrome	**Shock of any cause**
Pulmonary bacterial infection	**Uremia**
Pulmonary viral infection	
Fungal or mycoplasmal infection	

The most common causes of ARDS in adults include sepsis and sepsis syndrome, gastric aspiration, shock with multiple transfusion, diffuse pulmonary infection, and mechanical trauma including head injuries [377, 606]. The most common causes of ARDS in children include sepsis, near-drowning, hypovolemic shock, and closed space burn injury [1075]. Infectious agents associated with ARDS include a variety of gram-negative and gram-positive bacteria, respiratory syncytial virus, herpesvirus, mycoplasma, pneumocystis, and tuberculosis, among others.
Source: Compiled from Refs. 252, 377, 467, 606, 1075.

on more quantitative respiratory criteria (Table 10-3) [76]. The Consensus Committee definition for clinical acute lung injury (ALI) is identical to that of ARDS except that a higher value of 300 mmHg is used for PaO_2/FiO_2 ratio (Table 10-3) [76]. The Consensus Committee definitions of ARDS and ALI are now in widespread use, supplemented by a variety of lung injury scores such as that of Murray et al [756] The incidence of ARDS has been variably reported to be 50,000–150,000 cases per year in the United States depending on the specific criteria used to define the syndrome [76, 191, 483, 488, 599, 735, 756, 1084, 1121]. By any definition, ARDS affects substantial numbers of patients and has a poor prognosis. Survival rates in ARDS were very low when this syndrome was first identified in the late 1960's, and mortality still approaches 50% in severely affected patients despite significant advances in medical intensive care [76, 191, 220, 428, 488, 599, 735, 1084].

The Acute Respiratory Distress Syndrome (ARDS)

A clinical syndrome involving severe respiratory failure due to acute lung injury from a variety of causes. Chronic lung injury can also be present in the later phase of disease.

Originally called the "adult" respiratory distress syndrome, but can affect patients of all ages including term infants, children, and adults.

Affects 50,000–150,000 patients in the United States alone each year depending on the clinical definition used.

Surfactant dysfunction contributes to a complex lung injury pathophysiology. Surfactant deficiency can be present but is typically less prominent.

B. Pathophysiology of Lung Injury and ARDS

The intimate association between acute lung injury and ARDS has been recognized for several decades (e.g., [467, 906]). All patients with ARDS have clinical ALI [76]. Lung injury not only arises from the etiologies in Table 10-2 but also can be exacerbated by ventilator-induced or hyperoxia-induced injury during intensive care. The pathology of lung injury is complex and includes damage to the alveolocapillary membrane, edema, ventilation-perfusion abnormalities, inflammation, oxidant injury, and surfactant dysfunction among its many aspects. Multiple cell types are involved in lung injury, as are an almost bewildering number of inflammatory mediators, products, and transduction and regulatory

Table 10-3 Clinical Diagnosis of ARDS as Defined by the American-European Consensus Committee [76]

Acute onset
$PaO_2/FiO_2 \leq 200$ mmHg (regardless of PEEP)
Bilateral infiltrates on frontal chest radiograph
Pulmonary capillary wedge pressure ≤ 18 mmHg (if measured)
    or no evidence of left atrial hypertension

Clinical ALI has an identical definition except for a less severe ratio of $PaO_2/FiO_2 \leq 300$ mmHg [76]. See text for details.

factors.[1] Examples of inflammatory mediators and factors involved in the pathophysiology of ALI and the acute phase of ARDS are shown in Table 10-4. Additional factors are implicated in chronic injury and late phase ARDS, which involves remodeling, repair, and fibrosis [80, 1153, 1224]. The complexity of inflammatory lung injury and its clinical manifestations are among the most challenging aspects of current pulmonary research.

There are several pathways by which detriments to the pulmonary surfactant system can enter the pathophysiology of ALI and ARDS ([447, 455, 643, 786, 798, 969] for review) (Figure 10-2). Initiators of lung injury can directly damage type II cells and compromise the synthesis, packaging, secretion, or recycling of pulmonary surfactant. Surfactant metabolism can also be disrupted if type II cells, which are stem cells for the alveolar epithelium, are altered in response to type I cell injury. However, direct physicochemical effects on alveolar surfactant are generally more prominent in ALI. The surface activity of alveolar surfactant can be compromised by interactions with plasma proteins, cellular lipids, and a variety of other biophysical inhibitors (see Chapter 9 for details and literature citations involving lung surfactant dysfunction and its mechanisms). Components in lung surfactant can also be degraded or altered by reactive oxidants or lytic phospholipases and proteases. In addition, the most active large surfactant aggregates can be depleted or altered by interactions with inhibitors or through changes in recycling and metabolism. Biochemical and biophysical abnormalities in pulmonary surfactant lavaged from patients with ARDS are well documented (eg, [344, 346, 361, 398, 847, 858, 971, 1114]). The physiological consequences of decreased surfactant activity include atelectasis, loss of lung volume, increased compliance, and intrapulmonary shunting. These factors are additive with the detrimental effects of alveolocapillary membrane damage, edema, vascular dysfunction, and inflammation.

C. Clinical Aspects of ARDS

Pharmacologic agents, ventilation therapies, and the general intensive care management of patients with ARDS-related respiratory failure are covered in detail elsewhere.[2] ARDS is sometimes divided conceptually into early and late phases. A majority of patients exhibit respiratory signs within 24 hours of the onset of the predisposing condition (Table 10-2). Initial clinical signs include increased respi-

---

1. For example, for reviews of inflammatory mediators relevant for ALI and ARDS see Refs. 76, 80, 116, 368, 429, 435, 548, 675, 716, 906, 980, 1055, 1153, 1224.

2. Therapies for ARDS are detailed in reviews and critical care textbooks such as Refs. 27, 80, 252, 292, 377, 482, 548, 594, 636, 907, 980, 1061, 1075, 1076, 1170. Selected therapeutic agents are also noted in the context of multimodal interventions in Chapter 13.

Selected Aspects of Lung Injury Contributing to the Pathophysiology of ARDS

Capillary and epithelial cell injury
  Increased alveolocapillary membrane permeability
  Alveolar and interstitial edema
  Type I cell injury and death
  Type II cell hyperplasia and dedifferentiation
  Impaired surfactant synthesis or recycling

Pulmonary inflammation
  Multiple inflammatory mediators produced
  Reactive oxidants increased, antioxidants decreased
  Lytic enzymes released
  Vascular permeability increased
  Acute damage to alveolar, airway, interstitial cells
  Tissue repair, remodeling, and fibrosis

Lung surfactant dysfunction
  Biophysical inactivation by endogenous inhibitors
  Chemical degradation by lytic enzymes
  Chemical interactions with reactive oxidants
  Altered alveolar surfactant aggregates/subtypes

Microvascular dysfunction
  Reactive vasoconstriction and vasodilation
  Capillary permeability increased
  Mismatching of ventilation and perfusion

---

ratory rate and dyspnea that progress in severity. The early (exudative) phase of ARDS is dominated by ALI and is characterized by severe, progressive bilateral edema that leads to respiratory failure in the majority of patients by 72 hours. Histology reveals diffuse exudative alveolar damage including hyaline membranes and evidence of protein-rich edema fluid and remnants of necrotic epithelial cells. The late (fibroproliferative) phase of ARDS incorporates aspects of chronic lung injury and repair and is associated with tissue remodeling and the start of pulmonary fibrosis. Several of the prominent pulmonary aspects of clinical ARDS are noted in Table 10-5. Multiorgan involvement includes cardiac and renal failure, although cardiac pathology is not the initial causative lesion in this condition. One obvious focal point of therapy is maintenance of oxygenation

Table 10-4 Selected Inflammatory Mediators, Receptors, and Factors Associated with ALI and Early Phase ARDS

Cytokines/growth factors

EGF
G-CSF
GM-CSF
INF-γ
IL-1β, 4, 9 (pro-inflammatory)
IL-6, 10 (anti-inflammatory)
KGF
TGF$_\alpha$
TGFβ_1
TNF$_\alpha$
VEGF

Chemokines

ENA-78   MIP-1α
GRO$_\alpha$ RANTES
IL-8, MIP-2
IP-10
MCP-1

Reactive oxygen/nitrogen species

Free radicals   *Nonradicals*
Hydroxyl ($\cdot$OH)   Peroxynitrite (ONO_2^-)
Peroxyl ($RO\cdot_2$)   Alkyl peroxynitrite (ROONO)
Alkoxyl (RO$\cdot$)   Hydrogen peroxide (H_2O_2)
Hydroperoxyl ($HO\cdot_2$)   Hydroperoxide (ROOH)
Superoxide (O_2^-)
Nitric oxide (NO$\cdot$)

Antioxidants

Enzymes   *Nonenzymes*
Catalase   Ascorbate
GSH peroxidases   GSH
SODs α-Tocopherol
                                                     Uric acid

Membrane receptors/ligands/adhesion molecules

CD14 (LPS receptor)   LPS binding protein
CD40/CD40-ligand   L-Selectins (e.g., CD62-L)
Glucocorticoid receptors   sTNFR
ICAM-1   TNFR1,2
β_1-Integrins (e.g., $\alpha_v\beta_1$)   VCAM-1
β_2-Integrins (e.g., CD11a/CD18,
                              CD11b/CD18)

Transcription factor families	Other mediators/compounds	
AP-1(fos, jun)	CBG	HSPs
C/EBP (e.g., NF-IL-6)	CCSP	Lactate
HSF	Complement (and fragments)	LPS
IκB	Eicosanoids	Neuropeptides
NFκB	Leukotrienes	NOSs
	PGs (E,F,I families)	PAF
	Thromboxanes	PAF-AcH

Tabulated mediators of ALI are representative only.

Abbreviations: CBG, corticosteroid binding globulin; CCSP, clara cell secretory protein; C/EBP, cyclic AMP/enhancer binding protein; EGF, epidermal growth factor; ENA-78, epithelial cell-derived neutrophil activator 78; G-CSF, granulocyte colony-stimulating factor (CSF); GM-CSF, granulocyte macrophage CSF; GRO_α, growth-related oncogene α; GSH, glutathione; HSF, heat shock transcription factor; HSPs, heat shock proteins; ICAM-1, intracellular adhesion molecule-1; IFN, interferon; IL, interleukin; LPS, lipopolysaccharide; KGF, keratinocyte growth factor; MCP, monocyte chemoattractant protein; MIP, macrophage inflammatory chemokine; NF, nuclear factor; NOSs, nitric oxide synthetases; PAF, platelet activating factor; PAF-AcH, PAF-acetylhydrolase; PGs, prostaglandins; RANTES, regulated on activation normal T expressed and secreted; SODs, superoxide dismutases; TGF, transforming growth factor; sTNFR, soluble TNF receptor; TNF, tumor necrosis factor; TNFR1,2, TNF receptor 1,2; VCAM-1, vascular cell adhesion molecule-1; VEGF, vascular-endothelial growth factor.

Source: compiled from Refs. 11, 26, 63, 121, 126, 219, 332, 333, 427, 540, 548, 674, 707, 716, 718, 764, 825, 868, 977.

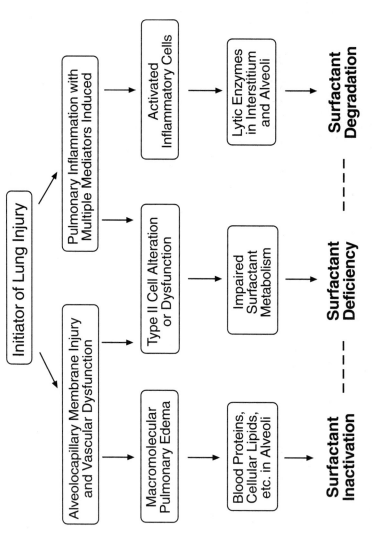

Figure 10-2 Schematic of surfactant-related events in ARDS-related lung injury. As part of the pathophysiology of ALI and ARDS, surfactant can be inactivated by biophysical interactions with inhibitory compounds present in the lungs through edema and inflammation. Active surfactant components can also be degraded or altered chemically, and surfactant can become deficient due to type II cell injury or alteration. Surfactant dysfunction/deficiency can also result from depletion or alteration of active large surfactant aggregate subtypes. Not shown are the multiple aspects of inflammatory lung injury unrelated to lung surfactant that also contribute to ARDS. See text for details.

Table 10-5 Common Pulmonary Features in Patients with ARDS-Related
Respiratory Failure

Presence of a systemic process or direct pulmonary insult associated with acute lung
    injury
Early tachypnea and dyspnea, followed by progressive hypoxemic respiratory failure
Bilateral alveolar edema on chest radiograph
Decreased lung volumes (FRC, TLC)
Decreased lung compliance (<40 ml/cm H_2O)
Hypoxemia resistant to high oxygen concentrations (PaO_2/FiO_2 <200 torr)
Normal pulmonary capillary wedge pressure (<18 mmHg) and lack of atrial hypertension
Signs of chronic lung injury and fibrosis may be apparent in later stage of disease

Abbreviations: FRC, functional residual capacity; TLC, total lung capacity; compliance, $\Delta V/\Delta P$.
Important predisposing conditons for ARDS are given in Table 10-2.
Source: Adapted from Ref. 377.

and gas exchange with supplemental oxygen and mechanical respiratory support
using a variety of techniques and strategies. Exogenous surfactant therapy is
most applicable for early phase ARDS, when lung injury is less established.
The development and current status of exogenous surfactant therapy for ARDS
and RDS are discussed in Chapters 11–13. The possible use of exogenous
surfactant therapy with other agents as part of a multimodal approach directed
against several aspects of lung injury pathophysiology in ARDS is also detailed
in Chapter 13.

V. Chapter Summary

RDS (or HMD) is the major disease of lung surfactant deficiency worldwide.
As many as 50,000–60,000 premature infants are at risk for RDS in the United
States alone each year. Although RDS is initiated by a deficiency of surfactant in
the premature lung, superimposed acute and chronic injury, surfactant dysfunction, and diverse complications of prematurity and intensive care can contribute
to patient outcomes. The association between surfactant deficiency and RDS was
defined in the 1950's, providing a significant driving force for subsequent basic
and clinical research. Along with general advances in medical and ventilation
technology, pharmacologic agents, and neonatal intensive care, the development
of effective exogenous surfactant therapy has had a significant impact in preventing and treating RDS in premature infants (Chapter 12). A much more rare form
of surfactant-deficient lung disease is the autosomal recessive disorder of hereditary SP-B deficiency. The chronic lack of SP-B in these infants cannot be treated
effectively with exogenous surfactant, and lung transplantation or the develop-

ment of effective gene therapy is needed. The major disease of lung surfactant dysfunction is ARDS, a clinical syndrome associated with acute lung injury from a variety of causes. ARDS-related respiratory failure occurs in all age groups and affects 100,000–150,000 patients in the United States each year. ARDS has a multifaceted pathophysiology that includes inflammation, alveolocapillary membrane injury, pulmonary edema, surfactant dysfunction, and vascular dysfunction. Patients with ARDS can also have chronic lung injury and fibrosis, as well as sepsis and severe multiorgan pathology. Surfactant dysfunction from substances present as a result of edema and inflammation is an important contributor to ARDS-related respiratory failure. Surfactant deficiency is typically less prominent. Surfactant therapy in ARDS is complicated by the complex patient population and multifaceted pathology of this condition. The remainder of this book focuses on the development and efficacy of exogenous surfactant therapy for RDS and ARDS in animals and humans.

11

Lung Surfactant Replacement in Animal Models

I. Overview

This chapter describes animal models of lung surfactant deficiency and dysfunction used in studying exogenous surfactant replacement therapy for RDS and ARDS. Research in animal models of surfactant deficiency has been essential for understanding the physiological actions of lung surfactant and for building a foundation for successful clinical trials of exogenous surfactant therapy for RDS in premature infants. Research in animal models has also been important in defining the physiological consequences of lung surfactant dysfunction and the potential clinical utility of exogenous surfactant therapy for ARDS-related acute lung injury. This chapter discusses a variety of animal models of RDS and acute lung injury and illustrates their use in studying the efficacy of exogenous surfactant replacement interventions.

II. Overview of Animal Models for Evaluating Surfactant Therapy in RDS

Animal studies provide a crucial link between laboratory research and clinical medicine. Research in animal models has been invaluable in defining the clinical relevance and benefits of exogenous surfactant therapy for RDS. In conjunction with basic research indicating that surfactant was deficient in the premature lungs, animal studies showed that biophysically active surfactants could be instilled exogenously to improve pulmonary mechanics and gas exchange. The majority of animal models of RDS utilize premature animals that are already deficient in endogenous surfactant, although several involve adult animals or animal lungs where endogenous surfactant has been depleted by bronchoalveolar lavage. As noted earlier in Chapter 5, surfactant-deficient animal models can be grouped as those primarily emphasizing measurements of quasi-static P-V mechanics or those providing detailed assessments of pulmonary function over an appreciable (but still acute) timescale *in vivo*. In addition to functional assessments, this latter category of animal models can also include measurements of quasi-static P-V mechanics. Examples of common animal models used in studying exogenous surfactant therapy for RDS are noted in Table 11-1. Animal models of RDS and of ARDS-related lung injury are also detailed in a variety of reviews (e.g., [114, 452, 455, 620, 643, 784, 786, 912, 913, 969, 1192]).

III. Surfactant-Deficient Animal Models Primarily Emphasizing Static P-V Mechanics

Two widely used animal models in this category are 1) excised, lavaged lungs from adult rats (e.g., [74, 380, 382, 453, 459, 492, 500, 784, 787]) and 2) surfactant-deficient premature rabbit fetuses of 27–28 days gestation (e.g., [238, 239, 242, 740, 898, 914, 916]). In the adult rat lung model, the quasi-static P-V deflation behavior of the freshly excised lungs is compared with that measured after depletion of endogenous surfactant by lavage and after subsequent delivery of exogenous surfactants. Premature rabbit fetuses at 27–28 days gestation (term = 31 days) are naturally deficient in endogenous surfactant. However, the degree of surfactant deficiency is variable and is not measured directly. Instead, exogenous surfactant activity is assessed by comparing the P-V mechanics of groups of surfactant-treated and control animals. Lung function can also be assessed in this model during a brief period of mechanical ventilation, and histological assessments of lung expansion can be done.

Table 11-1 Representative Surfactant-Deficient Animal Models for Studying Exogenous Surfactant Therapy for RDS

Surfactant-deficient model	Representative examples
Surfactant-depleted adult animals	Excised, lavaged adult rat lungs
	In vivo lavage of adult rabbits or other species[a]
Premature small animals	Premature 27–28 day gestation rabbit fetuses
Premature large animals	Premature lambs of 120–135 days gestation
	Premature baboons
	Premature monkeys

[a]In vivo lavage of adult animals is also used as a model of lung injury despite its major component of surfactant deficiency (see later section on animal models of ALI and ARDS). Examples using the other tabulated RDS animal models are in Refs. 9, 74, 115, 176, 181, 184, 224, 227, 238–240, 242, 382, 453, 459, 490–492, 494, 498, 500–502, 511, 516, 522, 683, 684, 740, 787, 799, 892, 898, 914, 916.

A. Surfactant Replacement in Lavaged, Excised Rat Lungs

Mechanical studies in lungs excised from adult animals such as rats yield highly reproducible and consistent data if standardized degassing and stress-relaxation procedures are followed. In the widely used model of Bermel et al [74], lungs are carefully excised from adult rats, degassed, and rapidly inflated with air to 30 cm H_2O (total lung capacity, TLC) at 37°C (Figure 11-1). Air leakage is ruled out by a stringent stress-relaxation requirement at this inflation pressure, and an initial P-V curve is measured under slow (quasi-static) deflation to define normal mechanics. The lungs are then made surfactant deficient by a series of lavages with normal saline (Figure 11-2). This lavage procedure is highly effective in depleting endogenous surfactant and does not alter lung tissue forces [74]. Following lavage, the lungs are degassed, reinflated, and again subjected to stress relaxation prior to a second P-V deflation curve to define surfactant-deficient behavior. An exogenous surfactant in saline is then instilled or aerosolized into the lungs, and a third P-V curve is measured with identical methods to assess the degree of restoration of mechanics toward normal. P-V behavior in this model can be returned almost completely to normal when a sufficient amount of extracted natural surfactant is instilled (Figure 11-3). Additional examples illustrating the use of the excised rat lung model in studying surfactant activity, inhibition, and replacement are given in other chapters (Chapters 7, 9, 14).

The excised rat lung model described in Figures 11-1 to 11-3 has been extensively characterized for its reproducibility, stability, and consistency with surface activity measurements [73, 74, 380, 382, 453, 459, 784, 786, 787]. The model responds with substantial mechanical improvements for surfactants with high surface activity and inhibition resistance, and exhibits a reduced or minimal response to less active surfactants or when surfactant inactivation is present. Measured P-V data with volume normalized as percent TLC are also highly reproducible in both the freshly excised normal state and the surfactant-deficient state following multiple lavage. Thermodynamic calculations demonstrate that the post-lavage P-V curve, which exhibits increased recoil pressures at all volumes below 90% TLC, is consistent with increased alveolar surface tension [74]. In contrast to this increase in surface forces, P-V curves for saline filling are not altered after lavage, indicating that tissue forces in the lungs are unchanged. Only minimal trapped lavage liquid (<2 ml or 8% TLC) remains in the lungs after lavage, and this is accounted for in subsequent measurements and calculations. The quantitative effects of exogenous surfactants in the excised rat lung model are also consistent with functional data in other animal models. For example, Figure 11-3 indicates that about 25 μmoles (~50 mg/kg animal weight) of extracted calf lung surfactant are required to fully normalize mechanics in the excised rat lung model. Improvements in lung function are found when similar doses of this surfactant extract are tracheally instilled to premature lambs [176, 227, 787] (see later section). The reproducibility, consistency, and internal control of the excised rat lung model of Bermel et al [74] have made it an FDA- approved methodology in quality control

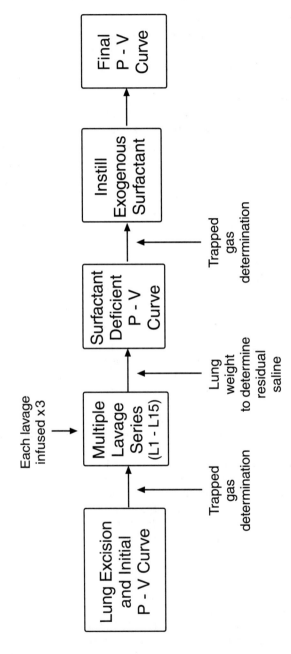

Figure 11-1 Flow diagram for a widely used excised rat lung model of surfactant deficiency. Lungs are excised from adult rats and P-V mechanics are defined in the normal surfactant-sufficient state and after a series of lavages to fully deplete endogenous surfactant. The effects of exogenous surfactants in restoring mechanics toward normal are then assessed. Standardized stress-relaxation and degassing procedures are used to give highly reproducible mechanical data as detailed in the text. (Adapted from Ref. 74.)

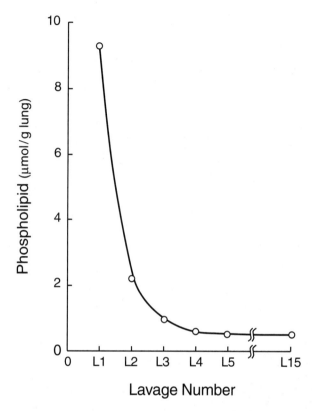

Figure 11-2 Reduction in endogenous surfactant phospholipid by multiple lavage. The mean phospholipid content in 15 successive saline lavages of excised adult rat lungs is shown. Most of the endogenous surfactant is removed in the initial four lavages. Saline P-V curves and thermodynamic analyses were used to verify that surfactant-deficient P-V mechanics were achieved without changes in lung tissue forces. See text for details. (From Ref. 74.)

assessments during the manufacture of several clinical exogenous surfactants (Survanta, Abbott Laboratories; Infasurf, ONY, Inc).

B. Surfactant Replacement in Premature Rabbit Fetuses

The lungs of fetal rabbits mature relatively late in gestation. Fetal rabbit lungs at 27 days gestation (>85% of term gestation of 31 days) have only a small percentage of type II pneumocytes and low levels of phospholipid and phosphatidylcholine (Table 11-2). Studies in premature rabbit fetuses by Enhorning, Robertson, and co-workers in the early 1970's were particularly important in

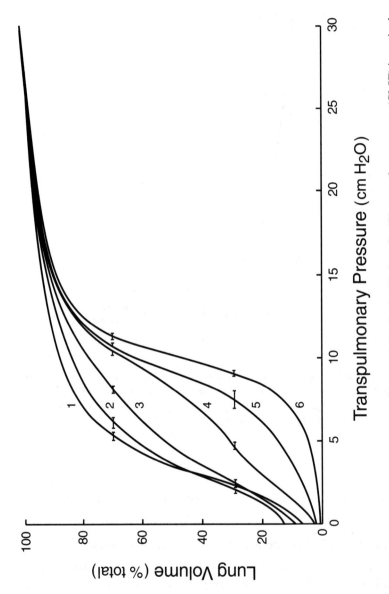

Figure 11-3 Improved pulmonary mechanics following instillation of calf lung surfactant extract (CLSE) in excised, lavaged rat lungs. The chloroform:methanol extract of lavaged calf lung surfactant restores P-V mechanics to normal in a dose-dependent fashion in this surfactant-deficient excised lung model. Curve 1: normal lungs; curve 6: surfactant-deficient lungs after standard multiple lavage; curves 2–5: lavaged lungs instilled with 24.6, 13.6, 10.6, and 7.0 μmoles CLSE in 2.5 ml of 0.15 M NaCl at 37°C. (Data from Ref. 74.)

demonstrating the ability of exogenous surfactant replacement to reverse surfactant deficiency in premature lungs [238, 239, 242, 916]. These studies were the first to show that active surfactant material could be delivered exogenously by intratracheal or pharyngeal instillation to improve pulmonary mechanics and function, increase alveolar inflation, and prolong survival in premature animals. These results demonstrated that surfactant deficiency was an important contributor to lung disease in premature animals and explicitly indicated the potential clinical utility of exogenous surfactant replacement therapy. The exogenous surfactant used in these studies was whole surfactant obtained by lavage from normal adult animals. When this material was instilled into 27–28 days gestation rabbit fetuses before the first breath, P-V mechanics were dramatically improved compared to untreated control rabbit fetuses [242] (Figure 11-4). Surfactant-treated fetuses required less pressure to open and inflate the alveoli, and more air was retained in the lungs at all deflation pressures relative to saline-treated controls. Functional residual capacity at zero transpulmonary pressure was also significantly increased, as was inflation-deflation P-V hysteresis and compliance (Figure 11-4). Histological examinations of the lungs showed that surfactant-treated rabbits had better alveolar expansion than controls, and survival times in these animals were also increased [238, 239, 916]. Enhorning, Robertson and co-workers subsequently replicated these positive findings for instilled natural surfactant in primates [181, 240]. This, along with related surfactant replacement studies with natural surfactant in animals by Adams, Ikegami, and co-workers at the end of the 1970's [9, 490, 492, 500], set the stage for effective clinical surfactant therapy.

Table 11-2 Percentage of Type II Pneumocytes and Phospholipid Content in Fetal Rabbit Lung

Rabbit gestational age	% type II pneumocytes[a]	μg phospholipid per gm lung	Percent phosphatidylcholine
27 days	1.21 ± 0.32	9.2 ± 2.1	28.6 ± 1.8
30 days	1.92 ± 0.23	16.2 ± 2.6	52.7 ± 3.4
31 days	—	37.6 ± 8.0	68.1 ± 1.3
2-day-old newborn	10.1 ± 0.9	477 ± 11	81.5 ± 1.1

[a]Type II pneumocytes were quantitated by laser flow cytometry in cell populations dissociated from fetal rabbit lungs with dilute trypsin and elastase; cells were stained with the lipophilic dye Phosphine 3R and percentages in the light scatter/fluorescence region occupied by adult rabbit type II pneumocytes are tabulated [631, 632]. Phospholipid data are from analysis of broncholaveolar lavage from fetal rabbits in a separate study [924]. All data are mean ± SEM.

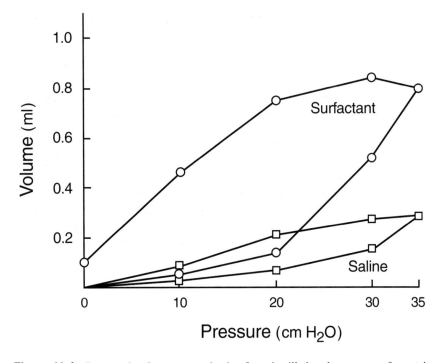

Figure 11-4   Improved pulmonary mechanics from instilled endogenous surfactant in premature rabbits. The surfactant-deficient P-V mechanics of 27–28 day gestation premature rabbit fetuses are significantly improved by tracheal instillation of endogenous surfactant lavaged from adult rabbits. Control rabbits received saline of equal volume. See text for details. (Data from Ref. 242.)

IV.   General Aspects of Animal Models of Prematurity Emphasizing Lung Functional Responses to Exogenous Surfactants

Although the effects of exogenous surfactants on respiratory function can be studied in small fetal animals, the most comprehensive functional assessments are made in larger premature animals (e.g., ovine, canine, porcine, primate species). Studies in these animals permit pulmonary function and gas exchange to be evaluated in detail over many hours so that clinically relevant aspects of surfactant replacement therapy can be fully tested and evaluated. Studies of this kind provide the final bridge between basic research and the treatment of human patients with RDS. One widely studied large animal model of prematurity is lambs between 120 and 135 days of gestation (term = 145–150 days) (see following section). Premature members of other large animal species such as baboons [683, 684] and monkeys [181, 240, 892] are also used in research on surfactant replacement therapy for RDS (Table 11-1). Premature animals used in studying surfactant activity must be of sufficient gestational age that the anatomical maturity of the lungs allows surface active function to be expressed. The lungs of human infants mature earlier in gestation than those of many animal species. The human lungs have substantial alveolation by 26 weeks of gestation (65% of term, 40 weeks). The vast majority of infants are surfactant deficient at this time, but by 32 weeks (80% of term) most have enough endogenous surfactant to make mechanical ventilation and exogenous surfactant replacement unnecessary. In contrast, fetal lambs are profoundly surfactant deficient at this stage of development, and fetal rabbits are still surfactant deficient at more than 90% of gestation (Table 11-2). Similar considerations apply to many other species of premature animals. The following section gives several examples of surfactant replacement studies in premature lambs as a representative large animal model of surfactant deficiency and RDS.

V. Studies of Surfactant Replacement in Premature Lambs with RDS

Research in premature lambs has provided a major source of information on the efficacy of exogenous surfactant replacement for RDS.[1] Particularly important in supporting the clinical use of surfactant therapy in infants were studies by Adams, Jobe, Ikegami, and co-workers in the late 1970's and early 1980's showing that natural surfactant lavaged from adult animals could significantly improve mechanics and gas exchange in premature lambs [9, 490, 494, 498, 511, 516, 521, 522]. Premature lambs were also used in the early 1980's by Egan, Notter, and co-workers [227, 787] to demonstrate that very high physiological activity was also achieved by chloroform:methanol extracts of lavaged calf lung surfactant having only a small content of hydrophobic protein. Premature lambs of several gestational age ranges are used in lung surfactant research. Lambs of 120–125 days gestation are profoundly surfactant deficient and die within hours of delivery in the absence of exogenous surfactant supplementation even with maximal respiratory support. Premature lambs at older gestational ages of 132–135 days have greater amounts of endogenous surfactant, allowing control animals to be maintained for much longer periods. Studies in lambs younger than 130 days gestation are less likely to be confounded by contributions from interactions with endogenous surfactant in the alveoli compared to studies in older lambs but are more technically demanding and can necessitate a shorter time scale of study.

---

1. For example, Refs. 9, 115, 176, 184, 224, 227, 490, 491, 494, 498, 501, 502, 511, 516, 521, 522, 526, 684, 787, 913.

**A. Example of Exogenous Surfactant Therapy with
      Extracted Calf Lung Surfactant in Premature Lambs**

Egan et al [227] and Notter et al [787] examined the effects of a chloroform:methanol extract of lavaged calf lung surfactant on pulmonary function and mechanics in premature lambs. The surfactant extract preparation was initially called CLL but was renamed CLSE when clinical findings in premature infants were reported [615, 990]. CLSE forms the basis of the clinical exogenous surfactant Infasurf (Chapter 14). Lamb experiments were designed so that biochemical and mechanical data could be obtained at several time points to complement functional assessments [227, 787]. Lambs were exteriorized from the ewe with the umbilical circulation left intact for 2 hr, followed by clamping and study over a more extended period. At the start of each experiment, lung liquid was aspirated, sampled, and then reinstilled after being mixed with the desired dose of CLSE (controls received equal volumes of lung liquid alone). Lambs were continuously ventilated with 100% oxygen at fixed rate, with ventilator pressures varied to attempt to maintain pH and arterial CO_2 in the normal range. Lung liquid samples and static P-V measurements were obtained at 30 and 120 minutes while animals were supported maternally and again at the end of study. Arterial blood gases were measured thoughout the experimental period. A ventilator efficiency index (VEI) with units of dynamic compliance was defined to facilitate comparisons between animals with different ventilator settings (VEI = alveolar ventilation/ ventilator excursion pressure × rate [787]). After final static P-V measurements, lungs were either assessed for wet-to-dry weight, lavaged to determine alveolar fluid composition and surface activity, or fixed for histology. Table 11-3 shows data on lung function, VEI, and static P-V mechanics for three groups of premature lambs: a control group receiving mechanical ventilation and oxygen only and two treatment groups receiving 15 or 100 mg/kg of CLSE at delivery. Surfactant instillation led to rapid and substantial functional and mechanical improvements. Lambs receiving even 15 mg/kg of CLSE had improved gas exchange and compliance compared to controls, and respiratory status was further improved in animals receiving 100 mg/kg of this surfactant (Table 11-3).

**B. Example of Exogenous Surfactant Therapy with
      Surfactant-TA in Premature Lambs**

Maeta et al [684] examined the effects of the clinical exogenous surfactant preparation Surfactant TA on respiratory function and mechanics in premature lambs of several gestational ages. Lambs were delivered by ceasarian section, hand-ventilated and separated from the ewe, intubated through a tracheostomy, and then supported on mechanical ventilation with 100% oxygen [684]. Respirator settings were adjusted throughout the course of study to attempt to normalize blood gas values at the lowest mean airway pressure. Surfactant-TA (100 mg/kg)

Table 11-3 Improved Oxygenation and Mechanics in Premature Lambs Given Exogenous CLSE by Tracheal Instillation at Birth

	Control lambs	CLSE (15 mg/kg)	CLSE (100 mg/kg)
N (number)	5	5	8
Gestation (days)	126 ± 1	126 ± 1	128 ± 1
Weight (kg)	2.6 ± 0.2	2.6 ± 0.3	2.9 ± 0.2
After 30 min ventilation			
pH	7.23 ± 0.03	7.35 ± 0.03	7.40 ± 0.02
PaO_2	30 ± 1	83 ± 20	199 ± 53
$PaCO_2$	65 ± 4	48 ± 5	39 ± 3
VEI	0.054 ± .004	0.087 ± .016	0.088 ± .008
$V_T/V_{10}/V_0$	9/7/2	28/19/12	33/26/17
After 2 hr ventilation			
pH	7.21 ± 0.04	7.48 ± 0.06	7.48 ± 0.03
PaO_2	39 ± 5	237 ± 63	309 ± 38
$PaCO_2$	74 ± 7	45 ± 5	33 ± 4
VEI	0.046 ± .007	0.146 ± .026	0.124 ± .023
$V_T/V_{10}/V_0$	11/9/4	38/27/12	45/35/19
At end-experiment			
pH	6.92 ± 0.05	7.52 ± 0.06	7.37 ± 0.03
PaO_2	41 ± 16	253 ± 51	336 ± 43
$PaCO_2$	138 ± 12	26 ± 5	38 ± 5
VEI	0.023 ± .001	0.227 ± .007	0.163 ± .045
$V_T/V_{10}/V_0$	8/6/1	35/25/12	46/33/20
Wet/dry	7.5 ± 0.6	5.6 ± 0.2	5.8 ± 0.3

CLSE, calf lung surfactant extract. Data are mean ± SEM from Notter et al [787] for arterial pH, PaO_2 (mmHg), and $PaCO_2$ (mmHg) for control and CLSE-treated premature lambs ventilated with 100% oxygen. VEI is recalculated from the original study to give units of ml/cm H_2O/kg rather than ml/mmHg/kg [787]. $V_T/V_{10}/V_0$ are quasi-static volumes (ml/kg) at 35, 10, and 0 (FRC) cm water. Wet-to-dry lung weight ratio is at end-experiment (death by 4 hr for controls or sacrifice at 10–12 hr for surfactant-treated animals).

was instilled into the endotracheal tube at 1.3 ± 0.5 hours of age, followed by study for 8 hours or until death, whichever was earlier [684]. Results for surfactant-treated and control premature lambs of average gestation 125 days are shown in Figure 11-5. Lambs treated with Surfactant-TA had significantly improved arterial oxygenation throughout most of the study period (Figure 11-5, top) and also had better quasi-static P-V mechanics at the end of the study (Figure 11-5, bottom). Complementary experiments in baboons of 140 days gestation also showed improved oxygenation after instillation of Surfactant-TA [683, 684], and studies in lambs of 132 days gestation did not reach statistical significance [684].

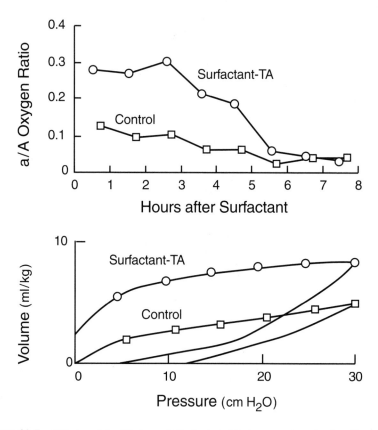

Figure 11-5 Effects of instillation of Surfactant-TA in premature lambs. Premature lambs of 125 days gestation instilled with 100 mg/kg of Surfactant-TA had significantly improved arterial oxygenation (top) and better P-V mechanics on inflation/deflation (bottom) compared to control lambs. Both groups were mechanically ventilated with 100% oxygen prior to sacrifice and P-V measurements at end-experiment. Data are means for Group I animals studied by Maeta et al [684].

C. Examples Comparing the Activity of Exogenous Surfactants in Premature Lambs

Animal models can be used not only to demonstrate the basic efficacy of exogenous surfactant therapy but also to compare the specific physiological activities of different surfactant preparations. The results of such comparisons, particularly when they are consistent across multiple animal models and supported by correlations involving composition and surface activity, can be highly useful in guiding clinical applications. A number of studies have documented physiological activity differences between exogenous surfactants in premature lambs, and several examples are summarized below.

Comparison of CLSE to 7:3 DPPC:PG in Premature Lambs

The relative effectiveness of CLSE and synthetic 7:3 DPPC:egg PG in improving respiratory function and mechanics in premature lambs was examined by Egan et al [227]. Average gestational ages for the groups of lambs studied were CLSE (128 days), control (130 days), and 7:3 DPPC:egg PG (132 days). Results showed that CLSE was significantly more active in improving arterial oxygenation and P-V mechanics than synthetic 7:3 DPPC:PG in these groups of lambs (Figures 11-6 and 11-7). Similar differences in activity were also found in experiments with individual pairs of twin lambs treated with CLSE vs 7:3 DPPC:PG [227]. Findings consistent with these were reported by Ikegami et al [494], who showed that natural sheep surfactant was beneficial to lung function in lambs of 120 days gestation while synthetic 9:1 DPPC:PG was not. The high physiological activity of CLSE and whole surfactant in premature lambs correlates with the known rapid adsorption and high dynamic surface activity of these materials as described in other chapters. Suspensions of synthetic DPPC:PG do not adsorb nearly as well as whole or extracted endogenous surfactant. Mixtures of DPPC:PG can give low minimum surface tensions in films spread directly at the air-water interface but do not have similar high surface activity after being dispersed in water [227, 783, 794].

Comparison of Exosurf, Infasurf, and Survanta in Premature Lambs

The effects of the clinical exogenous surfactants Exosurf, Infasurf, and Survanta in improving lung function and P-V mechanics in lambs of 126 days gestation were studied by Cummings et al [176]. The clinically recommended dose of each surfactant was instilled intratracheally at birth, and measurements of respiratory function were made during an extended period of mechanical ventilation with 100% oxygen. Figure 11-8 shows arterial/alveolar oxygen ratios in control and surfactant-treated lambs over the first 12 hours of ventilation (top) and static P-V deflation mechanics measured at end-experiment (bottom). Lambs given Infasurf showed the greatest improvements in arterial oxygenation and P-V mechanics, followed by lambs treated with Survanta. Lambs receiving Exosurf were not significantly improved over controls receiving oxygen and mechanical ventilation alone. Calculated values of ventilator efficiency index VEI were in the range of 0.14–0.15 ml/cm H_2O/kg for Infasurf, 0.05–0.06 ml/cm H_2O/kg for Survanta, and 0.025–0.03 ml/cm H_2O/kg for Exosurf over the time course of study [176]. This overall order of physiological activity of Infasurf > Survanta > Exosurf is consistent with results found for these preparations in other animal studies [382, 491,

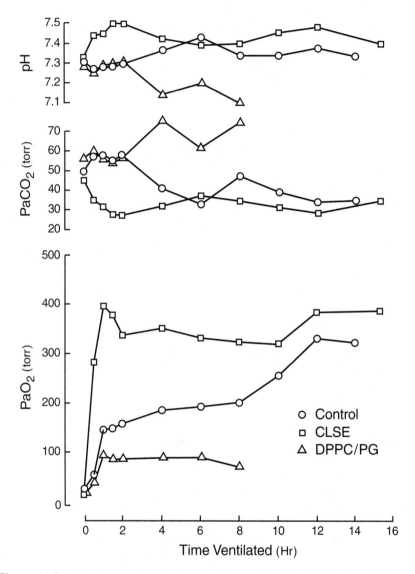

Figure 11-6 Blood gases in premature lambs treated with CLSE or synthetic 7:3 DPPC:eggPG. Premature lambs received 15–30 mg/kg CLSE (called CLL at the time of study) or 7:3 DPPC:eggPG by instillation into fetal lung liquid. CLSE-treated lambs had improved gas exchange compared to control lambs or lambs receiving DPPC:PG. The umbilical circulation was intact for the first 2 hours of study. Ventilator FiO_2 was 1.0 and rate was 60/min, while inspiratory and expiratory pressures and PEEP were varied to try to maintain normal arterial pH and CO_2. See text for details. (Data are means from Ref. 227.)

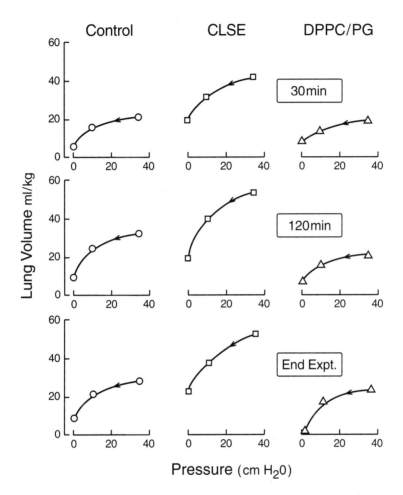

Figure 11-7 Improved P-V mechanics in premature lambs treated with CLSE vs 7:3 DPPC:egg PG. Quasi-static deflation P-V curves were measured for the lambs in Figure 11-6 at 30 minutes, 2 hours, and at end-experiment. Lambs receiving CLSE had significantly better P-V mechanics compared to control lambs or lambs receiving 7:3 DPPC:PG. See text for details. (Data are means from Ref. 227.)

740]. It also agrees with the differing adsorption and dynamic surface tension lowering ability of these surfactant preparations in biophysical studies [382, 967]. Infasurf and Survanta have also been shown to be more active than Exosurf in controlled clinical trials in premature infants [471, 480, 481, 1119]. Clinical exogenous surfactants and their activity differences are detailed more fully in Chapters 12 and 14.

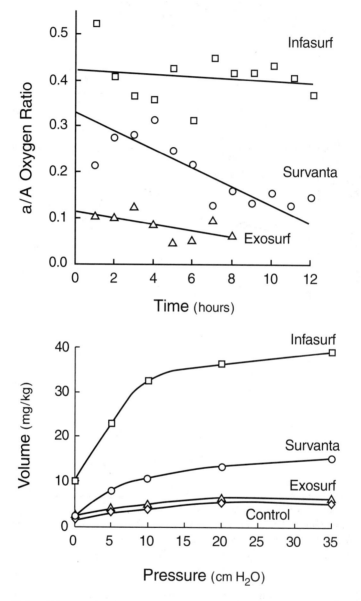

Figure 11-8   Effects of instillation of Infasurf, Survanta, and Exosurf in premature lambs *in vivo*. Premature lambs of 126 ± 1 days gestation received Exosurf, Infasurf, or Survanta in clinically recommended doses by tracheal instillation at birth (time zero). Activity was ordered as Infasurf > Survanta > Exosurf. (Top) Arterial/alveolar oxygen ratio vs time; (bottom) static P-V deflation curve at end-experiment. (Data are means from Ref. 176.)

Indirect Mechanisms of Exogenous Surfactant Activity (Incorporation into Recycling, Combination with Alveolar Surfactant). Studies on the physiological activity of exogenous surfactants in animal models of RDS in the preceding sections primarily assess effects related to the direct surface active function of these materials. Activity in some cases may also include a contribution from interactions with components of endogenous surfactant in the alveoli, particularly when older premature animals are studied. However, the acute models and examples discussed do not include long term functional benefits associated with the incorporation of exogenous surfactants into endogenous recycling pathways. It is well documented that the efficient and active recycling pathways in alveolar type II cells can incorporate components from exogenous surfactants into endogenous surfactant pools [30, 393, 496, 497, 502, 508, 515, 520, 524, 1060] (Chapter 6). This phenomenon gives exogenous surfactants the potential to have substantial indirect physiological benefits independent of their intrinsic surface activity. The impact of this may not be prominent in acute animal studies, but it almost certainly contributes to clinical outcomes in patients. The low direct activity of Exosurf in improving mechanics and function in premature lambs <130 days gestation suggests that at least some of its benefits in premature infants may occur by indirect mechanisms. The substantial amount of DPPC delivered to the lungs a single instilled dose of Exosurf could significantly enhance surfactant pools if only a fraction were taken up into type II cells and packaged with apoproteins. This mechanism would be consistent with the fact that some of the clinical benefits of Exosurf appear to be delayed relative to preparations such as Infasurf and Survanta (e.g., [480, 921, 1119]). ALEC, another protein-free synthetic surfactant composed of 7:3 DPPC:egg PG, also has delayed effects in premature infants [1077]. The effective activity of Exosurf (and ALEC) could also be increased in premature infants by combination with small amounts of surfactant apoproteins in the alveolar lumen [382]. Physiological improvements from these indirect mechanisms are less likely to occur if exogenous surfactants are aerosolized rather than instilled, since the amounts reaching the alveoli are not as great [25, 129, 920]. Although indirect benefits from interactions with endogenous surfactant in alveoli or type II cells may contribute to the clinical efficacy of exogenous surfactants, their direct effects on pulmonary mechanics and function as assessed in animal models are most relevant for surfactant replacement therapy.

VI. Overview of Animal Models of ARDS-Related Acute Lung Injury with Surfactant Dysfunction

Essentially all animal models of ARDS involve acute lung injury (ALI), which can be caused by a variety of methods in adult or full-term newborn animals ([114, 455, 620, 643, 784, 786, 912, 969, 1192] for review). Lung surfactant dysfunction or inactivation, with or without associated surfactant deficiency, has been demonstrated in a significant number of animal models of ALI. Mechanisms and characteristics of lung surfactant dysfunction are detailed in Chapter 9, and the pathophysiology of ARDS-related lung injury is described in Chapter 10. Representative animal models of ALI used in studying surfactant dysfunction and replacement in ARDS are given below. Subsequent sections provide specific examples illustrating the effects of exogenous surfactant therapy in improving pulmonary function and mechanics in several of these animal models.

A. Antibody-Induced Lung Injury

Antibodies against lung tissue are known to be present in the lungs under pathological conditions and to contribute to respiratory disease [485]. In addition, immune complexes or antibodies directed against lung tissue or components in lung surfactant can be infused or instilled in animals to cause ALI and surfactant dysfunction [283, 618, 836, 946, 1054]. Several animal models of this kind have been used to study surfactant inactivation and the mitigating effects of exogenous surfactant supplementation. For example, serum containing antilung antibodies injures the alveolocapillary membrane and causes severe ALI with edema and surfactant dysfunction if infused into guinea pigs [618]. ALI and surfactant dysfunction can also be generated in animals by intra-peritoneal injection of hybridoma cells that produce antibodies to surfactant protein SP-B [283, 1054].

B. Aspiration Lung Injury

Aspiration of caustic substances that damage airway and alveolar tissue as well as degrade or inactivate lung surfactant is another method of causing ALI in animals. Lung injury with surfactant dysfunction can be induced, for example, by aspiration of hydrochloric acid [586, 621, 1230], hydrochloride/pepsin [948] and meconium [16, 145, 823, 1047–1049]. Animal models of aspiration lung injury have direct clinical relevance, since aspiration of gastric contents is an important cause of ARDS in humans, and severe ARDS-related neonatal respiratory failure can accompany the aspiration of meconium by term infants at birth [606, 1061, 1075]. Exogenous surfactant therapy has been utilized to improve respiratory function in animals with meconium aspiration [16, 145, 823, 1047, 1049] and acid aspiration [586, 948, 1230].

C. Bacterial Infection/Toxin

Sepsis is a prevalent cause of clinical ARDS, and bacteria or bacterial toxins have been used to induce lung injury in a number of animal models.[2] *Escherichia coli* bacteria and lipopolysaccharide (LPS or endotoxin) infused intravascularly or instilled into the trachea are common inducers of ALI in animals [271, 330, 445, 677, 678, 773, 802, 1074]. *Pseudomonas aeruginosa* [100, 757, 860], *Pneu-*

---

2. For example, Refs. 45, 70, 100, 230, 271, 330, 445, 627, 677, 678, 734, 757, 773, 802, 860, 992, 995, 1074, 1191.

mocystis carinii [230, 992] and group B *Streptococcus* [70, 995], among others, have also been utilized. As in many other ARDS-related animal models, acute injury produced by bacteria or bacterial toxins includes a prominent component of pulmonary inflammation in addition to edema and surfactant dysfunction. Intrapulmonary shunting with decreased arterial oxygenation is also present, along with altered pulmonary mechanics with decreased compliance. Surfactant replacement interventions have been found to have positive effects in several animal models of bacterial lung injury as noted later [230, 677, 678, 773, 995, 1074].

D. Fatty Acid Lung Injury

Intravenous infusion of fluid free fatty acids such as oleic acid in animals severely damages the capillary endothelium and alveolar epithelium, leading to rapid and extensive edema formation [29, 379, 381, 988, 999, 1222, 1227]. Related injury to the alveolar epithelium in animals can also be caused by tracheal instillation of lysophosphatidylcholine, which disrupts membrane integrity similarly to fatty acids [774]. The widely studied oleic acid model of ALI has a significant component of surfactant inactivation both from biophysical interactions with inhibitors and from depletion of active large surfactant aggregates [379–381]. Oleic acid lung injury also exhibits prominent inflammatory pathology in addition to edema and surfactant dysfunction. Acute fatty acid lung injury in animals is directly relevant for the ARDS-related respiratory failure that can follow pancreatitis, which is associated with free fatty acids generated by pancreatic lipases released into the circulation and by increased activity in pulmonary endothelial lipoprotein lipase [687].

E. Hyperoxic Lung Injury

The significant pulmonary toxicity of high levels of oxygen has been known for decades [133, 204, 911]. The toxic effects of oxygen create a dilemma for treating patients in respiratory failure, but they also can be used to induce lung injury in animals so that mechanisms and therapies for ALI and ARDS can be examined. A substantial number of studies have investigated hyperoxic lung injury and oxygen-induced changes in pulmonary surfactant activity and metabolism in newborn and adult animals.[3] Depending on the severity and duration of exposure, hyperoxia can induce responses ranging from adaptive to lethal. Severe acute lung injury from prolonged exposure to high levels of oxygen (95–100%) includes

---

3. For example, Refs. 11, 58, 140, 182, 235, 275, 276, 450, 454, 457, 458, 478, 573, 623, 661, 700–704, 799, 817, 911, 1073, 1139, 1141, 1150–1152, 1172.

inflammation, edema, increased alveolocapillary membrane permeability, type II cell dysfunction, and surfactant dysfunction. Hyperoxia can also be used at less severe levels to generate chronic injury with fibrosis. Newborn animals have a significantly higher resistance to pulmonary oxygen toxicity than adult animals, in contrast to retinal oxygen toxicity, which is more severe in newborns. Hyperoxic lung injury exhibits some species variability in the details of its pathology, as is the case for many animal models of ALI. Hyperoxic lung injury can be mitigated by exogenous surfactant supplementation as described later [235, 451, 661, 703, 704, 799].

F. In Vivo Lavage

The use of *in vivo* lavage to deplete endogenous surfactant and induce lung injury in rats, guinea pigs, rabbits, and other adult mammals is a widely studied model for evaluating surfactant replacement interventions.[4] Animals are supported by mechanical ventilation and supplemental oxygen prior to and following *in vivo* lavage with normal saline. Lavage is typically done until arterial oxygenation is reduced below a predefined level, but the quantitative degree of surfactant deficiency is variable and is not measured. The *in vivo* lavage procedure together with mechanical ventilation of the lavaged lungs superimposes an acute injury onto the underlying surfactant deficiency. Although the *in vivo* lavage model involves a more significant component of surfactant deficiency than surfactant dysfunction, it is often studied as relevant for surfactant replacement interventions in ARDS.

G. Neurogenic Edema and Bilateral Vagotomy

Pulmonary edema can be caused neurogenically following cranial injury or by ligation of appropriate nerves [79, 165, 221, 322, 607]. In animals subjected to bilateral cervical vagotomy, the resulting edema and inflammatory lung injury have been shown to include a prominent component of surfactant dysfunction [79, 607]. Vagotomized animals have an increased alveolocapillary membrane permeability, and surfactant dysfunction is thought to result primarily from inactivation by the high content of plasma proteins in edema. Consistent with surfactant dysfunction, the lungs of animals injured with bilateral cervical vagotomy exhibit decreased compliance, atelectasis, and hyaline membrane formation. Exogenous surfactant therapy has been shown to improve pulmonary function and compliance in this animal model of ALI as described later [79].

---

4. For example, Refs. 46, 72, 328, 329, 374, 375, 507, 589, 601, 617, 619, 639, 741, 1104, 1137, 1140.

H. *N*-Nitroso-*N*-methylurethane (NNNMU) Injury

Subcutaneous injection of the nitrogenated urethane compound NNNMU gives rise to a progressive lung injury over several days that has a number of features relevant for ARDS [409, 637, 640, 641, 935, 1098]. NNNMU is known to have toxic effects on pulmonary cells including alveolar type II pneumocytes [935]. Animals with NNNMU injury have pulmonary edema with increased levels of plasma proteins and decreased levels of phospholipid in bronchoalveolar lavage. Arterial hypoxemia is also present in NNNMU-injured animals, and lung P-V compliance is decreased. Surface activity in bronchoalvelar lavage is reduced not only because of inactivation by plasma proteins and other constituents in edema but also by depletion of active large surfactant aggregates [640, 1098]. NNNMU-induced lung injury has been found to respond favorably to exogenous surfactant supplementation as noted in the following section [409, 637, 641, 1098].

I. Viral Infection

A significant number of viruses have the potential to cause severe acute inflammatory lung injury with edema and alveolocapillary membrane damage. Surfactant dysfunction can be one of many pathological components of this kind of viral-induced lung injury. Examples of viruses that can cause ARDS-related inflammatory lung injury with surfactant dysfunction in animals include influenza A [1101], respiratory syncytial virus [1110], and Sendai virus [1102, 1103]. The efficacy of exogenous surfactant supplementation in mitigating viral-induced pneumonia has also been studied [1101, 1103].

VII. Surfactant Replacement in Animal Models of Acute Lung Injury and ARDS

A number of the ARDS-related animal models in the preceding section have been shown to respond with improved pulmonary mechanics and acute respiratory function when exogenous surfactants are delivered to the lungs (Table 11-4). Because the complex pathophysiology of ALI and ARDS involves multiple aspects in addition to surfactant dysfunction (Chapter 10), exogenous surfactant therapy cannot be expected to completely normalize respiration. The prominent edema and inflammation characteristic of ARDS also make it more difficult to deliver exogenous surfactants to the alveoli. Despite these considerations, research in ARDS-related animal models indicates that surfactant supplementation strategies have the potential to generate meaningful improvements in lung function and mechanics in clinical therapy for this severe lung injury syndrome.

Table 11-4   Selected Animal Models of Acute Lung
Injury Shown to Respond to Exogenous Surfactant
Therapy

Acid aspiration pneumonia
Antibody-induced lung injury
Bacterial or endotoxin-induced lung injury
Bilateral vagotomy
Fatty acid lung injury
Hyperoxic lung injury
In vivo lung lavage
Meconium aspiration
N-Nitroso-*N*-methylurethane injury
Viral pneumonia

Representive studies of exogenous surfactant therapy in the tabulated animal models are: acid aspiration pneumonia [586, 621, 948,
1230], antibody- induced lung injury [618], bacterial or endo-toxin-
induced lung injury [230, 677, 678, 773, 995, 1074], bilateral
vagotomy [79], fatty acid lung injury [1227], hyperoxic lung injury
[235, 661, 703, 704, 799], *in vivo* lavage [46, 72, 328, 329, 374, 375,
507, 589, 617, 639, 741, 1137, 1140], meconium aspiration [16, 145,
823, 1047, 1049], NNNMU injury [409, 637, 641, 1098], and viral
pneumonia [1101, 1103].

Clinical surfactant therapy for ARDS, including its concurrent use with other
agents and interventions targeting different aspects of acute lung injury, is
discussed in Chapter 13. Examples illustrating responses of selected animal
models to exogenous surfactants are given below.

A. Example of Surfactant Therapy in Animals with Bilateral Cervical Vagotomy

Berry et al [79] studied exogenous surfactant replacement in adult rabbits injured
by bilateral cervical vagotomy. Severe respiratory distress developed in experimental animals within 4 hours of surgical ligation. Compared to sham-operated
controls, vagotomized animals had decreased compliance and significantly impaired arterial oxygenation despite mechanical ventilation with 100% oxygen.
These animals also exhibited histological evidence of alveolar edema, focal
hemorrhage, and hyaline membranes. The protein content and recovery of intra-
venously injected ^{125}I-albumin in bronchoalveolar lavage were substantially
higher in vagotomized animals, consistent with increased alveolar permeability

---

**Surfactant Replacement in Animal Models of
ARDS-Related Lung Injury**

Animal models of ARDS involve severe acute inflammatory lung injury and generally have a more prominent component of surfactant dysfunction than surfactant deficiency.

Research in animal models is essential for developing effective surfactant therapy for clinical ARDS, just as in the case of surfactant therapy for RDS.

A variety of studies in animal models of ALI and ARDS have found that active exogenous surfactants with the ability to resist inactivation can improve respiratory function and mechanics.

Although surfactant therapy can improve pulmonary function in ARDS, it is not able to abolish all of the pathological aspects of inflammatory lung injury.

(Table 11-5). Surfactant dysfunction was demonstrated by an increased minimum surface tension in lavage fluid (Table 11-5). However, surfactant separated away from inhibitory plasma proteins by centrifugation had normal activity, and lavage disaturated phosphatidylcholine levels were unchanged, suggesting that surfactant deficiency was not present. Vagotomized animals were significantly improved by tracheal instillation of lavaged sheep surfactant at the onset of respiratory distress (50 mg/kg in 6 ml/kg saline). Arterial oxygenation was increased, arterial carbon dioxide levels were decreased, and compliance was increased in surfactant-treated vagotomized animals compared to those given saline as a placebo (Figure 11-9). Minimum surface tension in lavage fluid was also significantly lower in surfactant-treated animals vs saline-treated animals (7.6 ± 2.2 vs 21.6 ± 1.1 mN/m). Although instillation of natural sheep surfactant improved respiratory function in vagotomized animals, its quantitative benefits were not as dramatic as found in animal models of RDS [79]. This was attributed in part to the difficulty of delivering exogenous surfactant by instillation to severely injured and atelectatic lungs [79]. Even so, the positive effects of surfactant supplementation in vagotomized rabbits indicates the potential utility of this therapeutic approach in ARDS.

Table 11-5 Bronchoalveolar Lavage Analysis Following Bilateral Cervical Vagotomy in Adult Rabbits[a]

Lavage parameter	Control rabbits	Vagotomized rabbits
Protein (mg/kg)	100 ± 20	285 ± 65
IV injected ^{125}I-albumin (% recovered)	0.20 ± 0.02	3.73 ± 1.02
Saturated PC (μmole/ml)	5.5 ± 1.4	5.7 ± 1.3
Minimum surface tension (mN/m), BAL	9.4 ± 2.2	19.6 ± 2.5
Minimum surface tension (mN/m), isolated surfactant	1.4 ± 1.4	1.9 ± 1.3

[a]Values are means ± SEM from Berry et al [79]. Minimum surface tension was measured on a Wilhelmy balance at 37°C. Isolated surfactant was pelleted from lavage by centrifugation at $7,000 \times g$ to obtain large active aggregates separate from inhibitory plasma proteins. Additional studies showed that instillation of natural surfactant significantly improved minimum surface tension and pulmonary function and mechanics in vagotomized animals (see text and Figure 11-9).

B. Example of Surfactant Therapy in Animals Injured by Antilung Serum

Another method of causing acute injury to the lungs is through immunologically active agents that activate the complement system or interact directly with components in lung tissue or surfactant. Lachmann et al [618] have shown that intravenous administration of antilung serum (ALS) in adult guinea pigs causes rapid and severe pulmonary injury. Respiratory failure occurs within minutes of ALS infusion and is characterized by decreased compliance, damage to the alveolocapillary membrane, and extensive interstitial and alveolar edema. Injured animals typically die within 30 minutes despite mechanical ventilation with 100% oxygen. Bronchoalveolar lavage from ALS-injured animals has increased levels of protein and impaired surface tension lowering properties compared to controls (Table 11-6). Decreased surface activity in lavage from injured animals is consistent with surfactant inactivation by plasma proteins and other constituents in edema. Reductions in lavage phospholipid content may also contribute to this effect, since surfactant is more sensitive to inhibition at lower concentration. Lachmann et al [618] found that the severe respiratory failure in ALS-injured animals could be significantly improved by intratracheal instillation of a high dose of lavaged bovine lung surfactant (280-350 mg phospholipid/kg body weight in 3–3.5 ml saline) (Table 11-7). Injured animals receiving surfactant had significantly increased arterial oxygenation and decreased ventilator mean airway pressure compared to those given an equal volume of saline-placebo. These improvements in lung function and mechanics were documented over a short time scale due to the

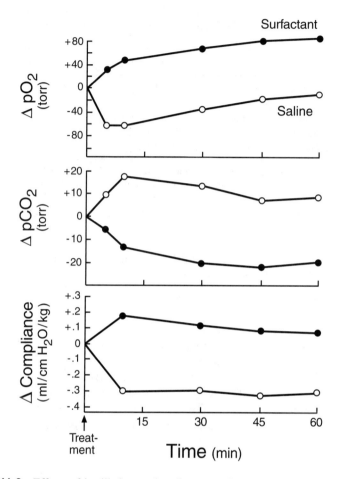

Figure 11-9 Effects of instilled natural surfactant on lung mechanics and function in vagotomized rabbits. Rabbits with acute pulmonary injury following bilateral cervical vagotomy had improved gas exchange and compliance from a single instilled dose of natural surfactant lavaged from adult sheep (50 mg lipid/kg in 6 ml/kg saline). Control rabbits received an equal volume of saline alone. Average time of instillation was 3.2 ± 0.5 hr after vagotomy. Data are expressed as changes relative to pretreatment baseline. See text for details. (Redrawn from Ref. 79.)

highly acute nature of the ALS injury model. Nevertheless, ALS lung injury does incorporate a number of the pathophysiologic features of ARDS and, because of its severity, represents a stringent test for the potential acute benefits of surfactant therapy.

Table 11-6 Bronchoalveolar Lavage Analyses after Antilung Serum (ALS) Infusion[a]

Lavage parameter	Control	ALS-injured
Total protein (mg/ml)	0.14 ± 0.003	0.61 ± 0.13
Protein/phospholipid (mg/μmol)	0.4 ± 0.1	2.8 ± 0.8
Total lavage phospholipid (μmol)	12.9 ± 3.09	7.84 ± 2.04
Total lavage saturated PC (μmol)	8.0 ± 2.2	4.6 ± 1.6
Minimum surface tension (mN/m)	7 ± 2	16 ± 1

[a]Data are means $\pm$ SD (n = 10–12) in lavage from control and ALS-injured adult guinea pigs from Lachmann et al [618]. Minimum surface tension was determined at minimum radius after 2 minutes pulsation on a bubble surfactometer (37°C, 20 cycles/min, 50% area compression). Results of surfactant treatment in ALS-injured animals are in Table 11-7.

C. Example of Exogenous Surfactant Therapy for Endotoxin Lung Injury in Animals

Because of the importance of sepsis as a cause of clinical ARDS, a number of studies have investigated surfactant therapy in animal models of lung injury from bacteria or bacterial toxins. As an example, Tashiro et al [1074] studied adult rats given *Escherichia coli* endotoxin by tracheal instillation (53 ± 19 mg/kg). A second dose of endotoxin was used in some animals in order to meet prospectively defined levels of oxygenation consistent with clinical ARDS ($PaO_2 < 200$ mmHg despite ventilation with 100% oxygen at a peak inspiratory pressure of 25 cm H_2O and a PEEP of 7.5 cm H_2O). After meeting criteria for ARDS, injured animals were randomly assigned to receive a modified porcine lung surfactant

Table 11-7 Pulmonary Function in Surfactant-Treated and Saline-Treated Guinea Pigs Infused with Antilung Serum (ALS)

	Pre-ALS baseline	Post-ALS infusion[a]	ALS-infused + saline[b]	ALS-infused + surfactant[c]
MAP (cm H_2O)	4 ± 0	21 ± 4	24 ± 4	13 ± 3
PaO_2 (torr)	322 ± 58	58 ± 17	52 ± 10	283 ± 47
$PaCO_2$ (torr)	37 ± 6	35 ± 11	38 ± 5	13 ± 6

Data are means $\pm$ SD (n = 7-8) from Ref. 618. [a]Measured 10 min after ALS infusion; [b]25 min after ALS infusion (15 min after saline instillation); [c]25 min after ALS infusion (15 min after surfactant instillation). MAP, mean airway pressure; PaO_2, arterial partial pressure of oxygen; $PaCO_2$, arterial partial pressure of carbon dioxide. Surfactant dose 280–350 mg/kg body weight. Animals were all breathing 100% O_2.

extract at a dose of 100 mg/kg in 2 ml/kg saline or air or saline as a placebo. Following assignment and treatment, mechanical ventilation with 100% oxygen was continued for 3 hours (or until death) while pulmonary function was monitored. Chest X-rays were obtained during the initial baseline period and just before treatment, as well as at the end of study prior to measurements of static P-V deflation mechanics [1074]. Rats receiving surfactant had significant improvements in arterial oxygenation compared to rats given air or saline (Figure 11-10). PaO_2 increased to 390 ± 116 mmHg within 15 minutes of surfactant treatment and stayed at this level throughout the period of study. In contrast, placebo-treated controls continued to be severely hypoxemic with PaO_2 values in the range of 100 mmHg. Surfactant-treated animals also had significantly improved static P-V mechanics compared to controls (Figure 11-10, right panel). Dynamic lung-thorax compliance was improved less substantially by surfactant treatment. Chest radiographs from surfactant-treated animals were consistent with decreased pulmonary edema, and the lungs of these animals also had better alveolar aeration on histological examination. These positive results again support the potential clinical utility of surfactant therapy in ARDS, although benefits reported for surfactant treatment in animals injured by intravenous rather than instilled endotoxin have been more modest [677, 678, 773].

D. Example of Exogenous Surfactant Therapy in Hyperoxic Lung Injury in Animals

A final example showing the potential benefits of exogenous surfactant supplementation in ARDS is hyperoxic lung injury in adult rabbits exposed to 100% oxygen [458, 661, 700, 701, 703, 704]. Alveolar permeability in these animals begins to increase after 48 hours in 100% oxygen and reaches a plateau by about 64 hours of exposure [458, 700, 701, 703]. Rabbits removed to room air at this time exhibit a progressive lung injury that peaks at about 24 hours after exposure. Arterial oxygenation in injured animals is decreased, and bronchoalveolar lavage has increased protein levels, decreased phospholipid levels, and elevated minimum surface tension consistent with surfactant dysfunction (Table 11-8). Pulmonary volumes and compliance are also decreased, but lung tissue forces are unchanged based on saline P-V measurements [458]. Acute hyperoxic injury in rabbits has been shown to be mitigated by instillation of exogenous surfactant [235, 661, 703, 704, 799]. In the protocol in Figure 11-11, CLSE was instilled intratracheally in a dose of 125 mg (~75 mg/kg) at the end of the 64 hour oxygen-exposure period and again 12 hours later [703, 704]. Surfactant administration did not improve oxygenation in injured animals breathing room air at 24 hours after exposure, but PaO_2 was significantly improved in the surfactant-treated group when breathing 100% oxygen during a 20 minute measurement period (Table 11-8). This behavior is consistent with a reduced right-to-left shunt

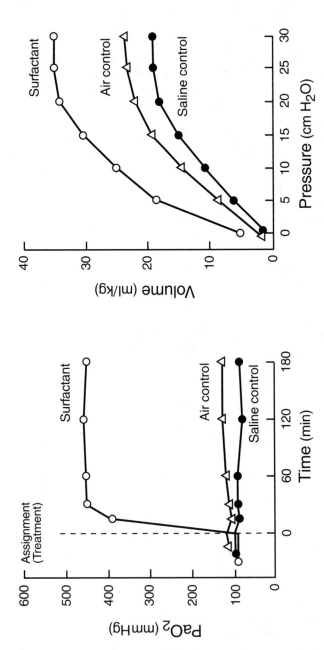

Figure 11-10 Effects of an extract of porcine lung lavage on pulmonary mechanics and function in endotoxin-injured rats. Rats injured by intratracheal *e. coli* endotoxin were randomized to receive an extract of porcine lung surfactant (100 mg/kg in 2 ml/kg saline) or placebo (saline or air). All animals were mechanically ventilated with 100% oxygen. Surfactant treatment significantly improved oxygenation and mechanics. Left panel: Partial pressure of arterial oxygen; right panel: quasi-static P-V deflation mechanics at end-experiment. (Data are means redrawn from Ref. 1074.)

Table 11-8 Pulmonary Changes in Acute Hyperoxic Injury in Adult Rabbits with and Without Surfactant Therapy

		64 hr in 100% O_2 + 24 hr in room air		
Parameter	Unexposed controls	Untreated exposed	Saline treated	Surfactant treated
PaO_2 (breathing air), mmHg	79 ± 3	54 ± 4	52 ± 6	54 ± 3
PaO_2 (breathing 100% O_2)	—	—	197 ± 55	446 ± 31
BAL phospholipid (μmoles/kg)	8.7 ± 1.6	4.5 ± 0.6	6 ± 1	34 ± 4
BAL protein (mg/kg)	7.4 ± 3.5	56 ± 4	60 ± 2	36 ± 3
Minimum BAL surface tension	< 1	30 ± 2	26 ± 2	< 1
Wet-to-dry lung weight	4.1 ± 0.3	6.3 ± 0.2	6.3 ± 0.3	5.6 ± 0.1

All values for surfactant-treated animals are significantly better than for saline-treated or untreated exposed animals except PaO_2 breathing air. Surfactant was CLSE given in two doses of 75 mg/kg according to the protocol in Figure 11-11.
Source: Data from Refs. 458, 704.

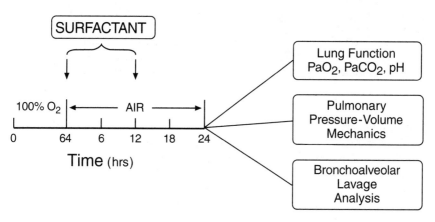

Figure 11-11 Protocol for exogenous surfactant replacement in rabbits with acute lung injury following 64 hours in 100% oxygen. Adult rabbits were exposed to 100% oxygen for 64 hours and then removed to room air. CLSE (125 mg total or 75 mg/kg) was instilled in the surfactant-treated group at the end of hyperoxic exposure and again 12 hours after exposure. The severity of ARDS-related lung injury at 24 hours after exposure was improved by exogenous surfactant therapy (Figure 11-12, Table 11-8). Protocol from Refs. 451, 703, 704.

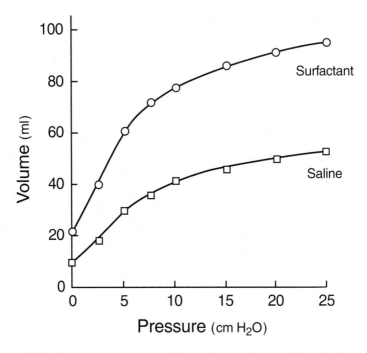

Figure 11-12 Improved pulmonary P-V mechanics from instillation of exogenous surfactant in hyperoxic lung injury. Deflation limbs of quasi-static P-V curves are shown for adult rabbits exposed to 100% oxygen and subsequently instilled with two doses of CLSE or an equal volume of saline alone (Figure 11-11). The static P-V deflation mechanics of CLSE-treated rabbits were greatly improved over saline-treated animals and were equivalent to unexposed, normoxic controls (not shown). (From Ref. 704.)

from pulmonary atelectasis in surfactant-treated animals [703, 704]. Surfactant administration also normalized minimum surface tension in lavage to <1 mN/m, and decreased pulmonary edema as shown by lower lavage protein levels and lung wet-to-dry weight ratios in treated animals (Table 11-8). Total lung capacity and deflation volumes were almost doubled in CLSE-treated vs saline-treated rabbits at 24 hours after exposure (Figure 11-12). Prophylactic administration of CLSE in adult rabbits prior to exposure to 100% oxygen has also been found to improve arterial oxygenation and survival time while decreasing histological evidence of lung injury and atelectasis [661]. While improvements from surfactant therapy in the animal models of ALI and ARDS discussed in this chapter are not as dramatic as in premature animals with RDS, they clearly indicate that active exogenous surfactants can be beneficial to lung function and mechanics in ARDS-related lung injury.

VIII. Chapter Summary

Research in animal models is invaluable for understanding lung surfactant function and for developing effective exogenous surfactant therapy for lung disease and injury. Physiological studies in animals provide a crucial link between laboratory research and clinical applications. Animal models of RDS primarily involve sufactant deficiency. These models include premature animals that are naturally deficient in endogenous surfactant, as well as adult animals or animal lungs depleted in endogenous surfactant by lavage. Examples of widely used animal models of surfactant deficiency described in this chapter are premature 27–28 day rabbit fetuses, premature lambs of gestation 120–135 days, and excised lavaged adult rat lungs. Studies in these and other surfactant-deficient animal models have been instrumental in the development of safe and effective clinical surfactant therapy for RDS. Lung surfactant therapy has also been widely studied in animal models of acute lung injury and surfactant dysfunction relevant for ARDS. Examples of such models include aspiration lung injury, antibody-induced lung injury, bilateral vagotomy, endotoxin- or sepsis-induced injury, viral pneumonia, fatty acid lung injury, *N*-nitroso-*N*-methylurethane (NNNMU)-induced lung injury, and hyperoxic lung injury. Exogenous surfactant supplementation has been shown to improve lung function and mechanics in multiple animal models of ARDS, although its effects are less pronounced than in RDS because of the complexity of acute inflammatory lung injury. Animal models vary in the details of their pathology, reproducibility, and relevance for human disease. The most meaningful and clinically relevant assessments of lung surfactant activity, dysfunction, and replacement depend on consistent results from several animal models together with correlations involving surface activity and composition. Exogenous surfactants and clinical surfactant therapies for RDS and ARDS are detailed in Chapters 12–14.

12

Clinical Surfactant Replacement Therapy for Neonatal RDS

I. Overview

This chapter summarizes the history and current status of clinical surfactant replacement therapy for RDS in premature infants. Coverage begins with early views and outcomes involving respiratory distress in premature infants in the 1950's and extends through successful clinical trials of surfactant therapy in the 1980's and 1990's. The role of lung surfactant deficiency in RDS or hyaline membrane disease was suggested over four decades ago, but the pathophysiology of this disease was debated for some time. Initial unsuccessful therapy with aerosolized DPPC in premature infants in the 1960's substantially delayed the development of clinically effective surfactant replacement interventions. Basic biophysical and animal model research emphasized in previous chapters was instrumental in establishing the foundation for successful exogenous surfactant therapy for premature infants. The results of clinical trials on the efficacy of surfactant therapy in premature infants are described in this chapter, along with ongoing efforts to optimize this treatment approach. Surfactant therapy in ARDS is covered in Chapter 13, and current and future clinical exogenous surfactants are detailed more fully in Chapter 14.

II. Mortality in Premature Infants in the Mid-1950's

The mortality in premature infants in 10 medical centers in New York City during the period 1955–1957 is shown in Table 12-1 [1001]. A great majority of premature infants below 1000 grams birth weight died within 3 days of birth. Premature infants weighing between 1001 and 1500 grams also had significant early mortality, although a majority survived. The primary therapy available at

Table 12-1 Inborn Premature Mortality at Ten New York City Medical Centers, 1955–1957

Birth weight cohort	Live births	Deaths		
		0–3 days	0–7 days	0–28 days
501–750 grams	278	262 (94.2%)	274 (98.6%)	274 (98.6%)
751–1000 grams	340	236 (69.4%)	256 (75.3%)	276 (81.2%)
1001–1500 grams	892	326 (36.5%)	350 (39.2%)	377 (42.3%)
Cumulative	1,510	824 (54.6%)	880 (58.3%)	927 (61.4%)

Source: Ref. 1001.

that time for premature infants was supplemental oxygen, and its use at high levels was restricted by concerns about retinal toxicity and the inability to measure arterial oxygen in patients. Nearly 90% of deaths in the first month in premature infants less than 1500 grams occurred within the first 3 days after birth (Table 12-1). Deaths beyond 3 days were relatively rare, partly because the most fragile infants succumbed by that time. From a modern perspective, it is clear that lung surfactant deficiency was a primary factor in most of these early deaths. In the textbook reporting the data in Table 12-1, intraventricular hemorrhage occurred in only 4/1000 autopsied premature infants, and necrotizing enterocolitis and bronchopulmonary dysplasia were not described [1001]. The majority of premature infants in Table 12-1 died of "idiopathic respiratory distress syndrome," the clinical manifestation of the pathologic diagnosis hyaline membrane disease (HMD).

III. Hyaline Membrane Disease Before Neonatal Intensive Care

The etiology, pathophysiology, and clinical characteristics of HMD (RDS) in premature infants are detailed in Chapter 10. HMD was identified as a specific disease entity in premature infants in the decade after World War II. In 1947, Gruenwald reported compelling physiologic studies indicating that the atelectasis in premature lungs at autopsy was due to high surface tension [357]. The association of HMD with surface tension abnormalities was again emphasized by Gruenwald in 1955 [356], and the relation of this disease to lung surfactant deficiency was suggested by Pattle [830] soon after the discovery of pulmonary surfactant. In 1959, Avery and Mead [34] published direct evidence showing that saline lung washings from preterm infants had much higher surface tension than similar material from full-term infants and older children who had died of other causes. The apparent clinical significance of lung surfactant provided a driving force for continuing research on its composition and properties. However, the role of surfactant deficiency as the primary causitive factor of HMD was not universally accepted, and the etiology and pathophysiology of this lung disease remained an open question in clinical medicine in the 1960's (e.g., [128, 129, 942, 1001]).

IV. Impact of Mechanical Ventilation and Intensive Care on HMD in Premature Infants

In the 1960's, technologies were introduced to allow sustained mechanical ventilation, monitoring of cardiopulmonary status, and measurement of blood gases in critically ill patients. These technologies had a dramatic impact on the prognosis of premature infants. Particularly important was the identification of techniques for using continuous positive airway pressure (CPAP) or positive end-expiratory pressure (PEEP) during mechanical ventilation of premature infants. The introduction of CPAP was the first effective clinical tool to manage the progressive atelectasis of RDS [343]. The impact of mechanical ventilation with CPAP or PEEP on the survival of premature infants was so dramatic that overall mortality rates for all newborns dropped by more than 50% in 20 years, from 18/1000 in 1965 to 7/1000 in 1985 [369, 765].

Just prior to the introduction of surfactant replacement therapy into general use in neonatology, an NIH-sponsored study reported on the outcome of extremely premature infants in-born at seven Level III perinatal centers in 1987–1988 [370]. The survival and mortality pattern for these premature infants was greatly altered from that found in the mid-1950's. Mortality in the first postnatal week was now almost exclusively in premature infants under 1000 grams birth weight, and a substantial percentage of even very small infants <750 gm birth weight survived the first week (Figure 12-1) [370]. For the total population of premature infants studied (1,765 infants of 500–1,500 gm birthweight), 81% survived to 1 week of age, 77% survived to 28 days, and 74% to discharge [370]. In contrast, less than 40% of infants of similar size survived to 28 days in 1955–1957, and only a tiny percentage of these were <750 grams (Table 12-1). Deaths in the first postnatal week, which were closely linked to RDS, accounted for the vast majority of neonatal deaths in very premature infants in the 1950's (Table 12-1). Despite the improved survival of premature infants in the 1980's, however, a nontrivial number of infants still died before leaving the hospital (Figure 12-1). These deaths occurred not only from lung disease but also from a variety of complications of prematurity and intensive care.

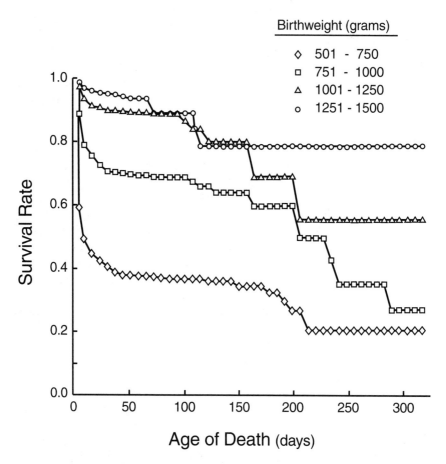

Figure 12-1 Survival rate is shown as a function of age for premature infants at seven centers in the National Institute of Child Health and Human Development Network in 1987–1988, just prior to the introduction of exogenous surfactant therapy into general clinical use in the United States. A total of 1765 infants was analyzed: 501–750 gm (n=349), 751–1000 gm (n=382), and 1,001–1,500 (n=1,034). (Redrawn from Ref. 370.)

V. Bronchopulmonary Dysplasia (BPD) in Premature Infants

One concomitant of neonatal intensive care was the appearance of the chronic lung disease BPD in many infants who survived severe respiratory failure by the use of mechanical ventilation. The decrease in early mortality after the introduction of CPAP and PEEP in intensive care was accompanied by an increased incidence of BPD among the survivors of RDS. BPD initially was described as a long term, life-threatening disease that primarily occurred only in patients who had severe respiratory failure in the first few days of life [781]. As survival improved in extremely premature infants <1,000 grams birth weight, a milder, self-resolving form of BPD became a frequent diagnosis in neonatal intensive care units [51]. In the population of premature infants surveyed by Hack et al [370], 26% of survivors of birth weight 500–1500 grams were receiving supplemental oxygen at 28 days, while only 8% were still on oxygen by 3 months. Persistent BPD (if defined as oxygen dependence at 3 months of age) tended to reflect birth weight as its strongest predictor rather than the incidence and severity of RDS. Oxygen dependence at 3 months of age was 15 times more common in survivors of birth weight 501–750 grams (36%) than in survivors of birth weight 1001–1500 grams (2.3%) [370]. The fact that BPD in a large number of premature infants in intensive care units in the 1980's was not directly associated with RDS has direct ramifications for the relevance of this chronic lung disease as an end-point in clinical surfactant trials as described later.

VI. Initial Unsuccessful Clinical Surfactant Therapy for RDS

Initial clinical studies of exogenous surfactant therapy were limited by the misconception that dipalmitoyl phosphatidylcholine (DPPC), also known as dipalmitoyl lecithin (DPL), was the active principle of pulmonary surfactant. Early compositional studies had identified this phospholipid compound, which is uncommon in cell membranes, as having a high content of surface active material lavaged or processed from the lungs [102, 103]. Although DPPC is the most abundant single component in endogenous surfactant and contributes significantly to its surface tension lowering ability, it also lacks several surface behaviors required for functional pulmonary surfactant (Chapters 3, 6, 8). In particular, the rigid DPPC molecule adsorbs very poorly to the air-water interface and also does not respread effectively in cycled surface films. As a result, DPPC alone is not highly active as an exogenous surfactant. Moreover, the aerosolization methodology used to deliver DPPC to premature infants in initial clinical trials was ineffective in achieving substantial alveolar concentrations of this material.

Robillard et al [920] in 1964 first attempted clinical surfactant replacement therapy by aerosolizing DPL (DPPC) into the incubators of premature infants with RDS. No control infants were studied, and empirical clinical observations measured the effectiveness of the therapy. Chest retractions were reported to be decreased and respiratory rate lowered after DPL aerosol treatment in the eight infants who survived but not in the three who died [920]. Although the authors were cautiously optimistic about the therapy, subsequent studies were not conducted. A second uncontrolled study with DPL aerosolized into the incubators of premature infants with RDS was conducted by investigators from the Cardiovascular Research Institute of the University of California at San Francisco in a large obstetrical hospital in Singapore [128, 129]. Measurements were made of lung volumes, dynamic compliance, and blood gases for 43 treatments to 15 infants. Improvements in compliance were reported in 79% of the treatments, but no changes in pre- and post-treatment blood gases were detected. Ten of the 15 infants died. The investigators went on to study DPL + acetylcholine therapy in three infants (one of whom died) and acetylcholine therapy alone in nine infants (four of whom died). Based on acute responses to acetylcholine, it was concluded that pulmonary vasospasm was the primary defect in RDS and that the surfactant deficiency was due to inadequate blood flow to alveolar cells. "It is our view that idiopathic respiratory distress of the newborn, or hyaline membrane disease, results from and primarily involves pulmonary ischemia and that secondary changes in structure, composition and physical properties of the lungs have tended to divert attention from it" [129]. The concept that DPL could "reinforce the natural alveolar lining" was also noted and ultimately proved insightful. However, the negative clinical findings and their overall interpretation by Chu et al [128, 129] contributed to a hiatus in further clinical studies of surfactant replacement.

VII. Basic Research Leading to Successful Surfactant Therapy for RDS

Several factors in the decade following initial unsuccessful attempts at surfactant therapy in premature infants laid the groundwork for the eventual success of this therapeutic modality in RDS. As described in Chapter 11, the ability of lung surfactant from adult animals, or active substitutes for it, to improve pulmonary mechanics and function when instilled into surfactant-deficient animals of the same or different species was firmly established during the 1970's and early 1980's.[1] The biophysical basis of lung surfactant function as necessitating interactions involving multiple constituents rather than simply DPPC also became firmly established and widely appreciated during this time [320, 569, 785, 788, 790, 792, 862, 1105] (Chapters 6, 8). This foundation of basic biophysical and physiological research supported the design and implementation of successful clinical studies of exogenous surfactant therapy for RDS in the 1980's. The technique of delivering exogenous surfactants to infants by intratracheal instillation was also defined in animal research by Enhorning, Robertson, and colleagues in

---

1. For example, Refs. 9, 74, 181, 227, 238–240, 242, 490, 492, 494, 498, 500, 511, 516, 522, 787, 913, 916.

Basic Research Supporting the Development of Clinical Surfactant Therapy for RDS

The essential ability of exogenous surfactant replacement to improve lung mechanics and function was extensively documented in surfactant-deficient animals in the 1970's and early 1980's.

Understanding about the surface activity of lung surfactant and the importance of its multiple functional lipid and protein components was established in basic biophysical research in the 1970's and 1980's.

The ability to deliver exogenous surfactants effectively to the lungs by direct airway instillation in premature animals with RDS was established in basic research.

Exogenous surfactants with substantial surface and physiological activity that could be safely formulated for human use were identified in basic biophysical and animal research.

---

the early 1970's [238, 239, 242]. Instillation remains the method of choice in today's surfactant therapy, although aerosol delivery continues to be investigated (e.g., [25, 205, 232, 639, 642, 677, 1222]).

VIII. Initial Positive Uncontrolled or Nonrandomized Surfactant Replacement Studies in RDS in the Early 1980's

Initial studies reporting the success of surfactant replacement therapy in premature infants with RDS were either uncontrolled or nonrandomized. In 1980, Fujiwara et al [289] reported an uncontrolled experience of surfactant replacement in 10 premature infants using an "artificial surfactant" made from an organic solvent extract of homogenized bovine lung tissue fortified with synthetic DPPC and phosphatidylglycerol (PG). Patients were diagnosed clinically with severe RDS before treatment, and the majority appeared to have a substantial beneficial response to instilled exogenous surfactant. This publication was followed in 1981 by another report including nine additional uncontrolled patients treated with the same surfactant preparation [285]. Also in 1981, the successful treatment of premature infants with RDS with a synthetic exogenous surfactant made of powdered 7:3 DPPC:PG and designated artificial lung expanding compound (ALEC) was reported in England [747]. Although this trial had a control group, the study was not randomized. Surfactant-treated patients were born during the work week, and the controls were born at night or on weekends so that different caregivers were involved. A subsequent uncontrolled study in 1983 by a different group of investigators did not find beneficial effects on respiratory status when 7:3 DPPC:PG was administered to infants with RDS [738].

Several additional uncontrolled studies in the early 1980's reported clinical benefits for exogenous surfactant therapy in premature infants. A chloroform:methanol extract of lavaged porcine lung surfactant called Surfactant CK was reported to improve respiration in seven premature infants in Japan [588, 780], and lung functional improvements were also reported for a related extract of lavaged bovine lung surfactant in six infants in Canada [1012]. Human surfactant derived from the amniotic fluid of term pregnancies at caesarean section was also reported to improve respiratory function in five premature infants in the United States [396]. In addition, beneficial effects on respiratory function in 31 premature infants were reported for Surfactant-TA [284], a modification of the original surfactant preparation of Fujiwara et al [289] that used DPPC, palmitic acid, and tripalmitin as additives rather than DPPC and PG. Thus between 1980 and 1984, uncontrolled or nonrandomized studies with at least five different exogenous surfactants had found apparently positive results in premature infants, setting the stage for prospective, randomized placebo-controlled trials.

IX. Successful Controlled Clinical Studies of Exogenous Surfactant Therapy for RDS

The efficacy of surfactant replacement therapy in premature infants was demonstrated in a series of randomized placebo-controlled clinical trials published from 1985 through the early 1990's.[2] These placebo-controlled trials demonstrated significant improvements in respiratory function and long term outcomes for premature infants treated with a number of different exogenous surfactants at birth (prophylactic therapy) or after a diagnosis of RDS (rescue therapy). Randomized, placebo-controlled trials of exogenous surfactant replacement in premature infants are tabulated in Table 12-2. Specific findings in clinical trials of surfactant therapy are reviewed in detail elsewhere [384, 518, 550, 725, 866, 867, 1016, 1018]. The overall message of the placebo-controlled trials in Table 12-2 is that exogenous surfactant replacement therapy is highly beneficial in improving

---

2. For example, Refs. 94, 153, 157, 222, 223, 243, 248, 288, 314, 335, 336, 394, 443, 469, 470, 553, 615, 622, 647, 662, 663, 728, 729, 851, 883, 990, 1017, 1077, 1181.

Table 12-2 Randomized Placebo-Controlled Clinical Trials of Exogenous Surfactant Therapy in RDS

Type of surfactant	Trial sites	Number of trials	Total patient number	Results
Synthetics				
ALEC	Great Britain	3	364	1 Positive, 2 negative
DPPC/HDL	Belfast	1	100	Negative
Exosurf	USA, Canada	9	2,678	All positive
Processed extracts of lung tissue				
Surfactant TA	Japan, USA	3	171	All positive
Survanta	USA, Europe	7	1,653	All positive
Curosurf	Europe	1	146	Positive
Extracts of lavaged surfactant				
bLES	Canada	3	329	All positive
CLSE, Infasurf	USA	2	92	Both positive
(SF-RI 1) Alveofact	Germany	1	69	Positive
Whole surfactant				
Human from amniotic fluid	USA, Finland	4	321	All positive
Totals		34	5,923	

References for randomized placebo-controlled trials are ALEC [1077, 1181], Alveofact (SF-RI 1) [335, 336], bLES (BLSE) [222, 223, 243], Curosurf [153], DPPC/HDL [385], Exosurf [94, 157, 248, 662, 663, 851], Human Surfactant [394, 622, 728, 729], Infasurf (CLSE) [553, 615, 990], Surfactant-TA [288, 314, 595, 596, 883], and Survanta [1, 443, 469, 470, 647, 1017].

respiratory function, decreasing pulmonary complications, and reducing mortality and morbidity in premature infants. This is also confirmed by statistical analysis of combined data from surfactant trials [384, 430, 550, 725, 867, 1016, 1018] and by analyzing neonatal mortality patterns after the introduction of surfactant therapy into clinical use [99, 406, 962, 963].

Although randomized, placebo-controlled clinical trials have been indispensible in estabishing the efficacy of surfactant replacement therapy, the number of such trials in Table 12-2 is large for a life-threatening disease. This reflects in part the significant number of exogenous surfactants that have been tested in controlled fashion. A number of different clinical exogenous surfactants have been found to be beneficial to respiratory function and outcome in premature infants, but these preparations are not equivalent in composition, surface properties, and physiological activity when examined in basic research (see Chapter 14 for more detailed discussion of exogenous surfactants). Additional factors including the involvement of pharmaceutical companies in exogenous surfactant development, as well as the endpoints chosen to define therapeutic efficacy, have also contributed to the number of placebo-controlled trials on exogenous surfactant therapy in premature infants.

X. Clinical Endpoints in Trials of Lung Surfactant Replacement Therapy for RDS

Both long-term and short-term endpoints have been used in clinical trials of surfactant replacement therapy. Decreases in mortality or improved survival without BPD, in particular, have been viewed as definitive endpoints to determine therapeutic efficacy. However, survival and long term outcomes such as chronic lung disease in very premature infants are influenced by multiple variables in addition to exogenous surfactant activity. Modern mechanical ventilation and neonatal intensive care are also quite effective in infants even in the absence of exogenous surfactant therapy. Only a small fraction of placebo-controlled trials of exogenous surfactant therapy have actually demonstrated differences in mortality [153, 157, 243, 443, 647, 663, 729, 1077] (Table 12-3). Most of these trials have involved relatively large numbers of patients (Table 12-3), but other trials of significant size using the same surfactants with equivalent protocols have found no differences in mortality. The two smallest trials showing mortality differences from surfactant therapy in premature infants had death rates that exceeded 50% in the placebo group [153, 729], more than twice as high as predicted [369, 370, 765]. Only a minority of placebo-controlled surfactant trials have also been able to demonstrate differences in survival without BPD, and only

Table 12-3 Placebo-Controlled Trials of Surfactant Therapy in Premature Infants Showing a Difference in Mortality

Surfactant study	Study type	Gestational age/weight	Study size	Percent mortality	
				Placebo	Surfactant
ALEC [1077]	Prophylaxis	25–29 wk	308	30%	19%
Human Surfactant [729]	Prophylaxis	24–29 wk	60	52%	23%
Survanta [443][a]	Prophylaxis	23–29 wk 600–1,250 g	430	19%	11%
Exosurf [157]	Prophylaxis	700–1,100 g	446	27%	16%
Curosurf [153]	Rescue	700–2,000 g	146	51%	31%
Survanta [647][a]	Rescue	600–1,750 g	798	27%	18%
Exosurf [663]	Rescue	700–1,350 g	419	29%	16%

[a]Refs. [443, 647] each combine two trials. An additional small trial by Enhorning et al [243] (not tabulated) showed a significant decrease in neonatal deaths but not total deaths in surfactant-treated vs control infants.

Current Status of Clinical Exogenous Surfactant Replacement Therapy for RDS

Uncontrolled or nonrandomized trials of surfactant therapy between 1980 and 1984 indicated that several exogenous surfactants could be beneficial to respiratory function in premature infants with RDS.

A large number of randomized, placebo-controlled clinical trials between 1985 and the early 1990's established the ability of multiple exogenous surfactants to improve lung function and long term outcomes including survival in premature infants.

Surfactant therapy in premature infants is now being optimized in terms of exogenous surfactant preparation, prophylactic vs treatment strategies, delivery methods, use with other agents, and other variables.

---

a few have shown a reduced incidence of BPD [662, 729]. As noted earlier, BPD and related chronic lung disease in many infants are associated more strongly with gestational age than with the severity of RDS. The clinical endpoints that are most directly related to lung surfactant activity are those involving respiratory function and the severity of acute RDS. Essentially all positive controlled trials of surfactant therapy have shown an improvement in one or more lung functional variables such as arterial oxygenation, required oxygen supplementation (FiO_2), or ventilator mean airway pressure (MAP). A strong independent association between respiratory variables like FiO_2 and MAP in the first 72 hours and the risk of death can be shown by statistical analysis of clinical trial data [605, 974]. Given enough patients, the impact of these variables emerges. Although mortality decreases are found in only a minority of clinical surfactant trials, the aggregate use of surfactant therapy has substantially improved survival in very premature infants (e.g., [99, 406, 963, 1016]).

XI. Clinical Trials Comparing Different Exogenous Surfactant Preparations

Although the basic efficacy of exogenous surfactant therapy in premature infants is established, several aspects of therapy are currently being optimized. One of the most important of these concerns the relative activity of different clinical exogenous surfactants. The ineffective use of DPPC in early clinical studies is just one example of the need to consider exogenous surfactant activity in therapeutic applications. A number of clinical studies have compared the efficacy of exogenous surfactant drugs in premature infants.[3] Results from eight prospective, randomized trials comparing exogenous surfactants in premature infants are given in Table 12-4. Comparison trials have demonstrated differences between exogenous surfactants in improving gas exchange and other pulmonary variables but not in reducing mortality or increasing survival without BPD (Table 12-4). Among four large comparison trials treating infants with RDS (rescue therapy), three found that Survanta and Infasurf were more active than Exosurf [471, 480, 1119] and one found that Infasurf was more active than Survanta [90]. In two large prophylaxis comparison studies, Infasurf was more active than Exosurf [481] and no significant differences were shown between Infasurf and Survanta [90] (Table 12-4). Infasurf, Survanta, Curosurf, and other exogenous surfactants containing lipids and proteins from endogenous surfactant are sometimes categorized as "natural surfactants." The results of clinical comparison studies indicate that "natural" surfactants are more efficacious in premature infants than current synthetic surfactants (Exosurf, ALEC) [127, 471, 480, 481, 921, 1119]. A similar conclusion is also found by statistical analysis of combined data from multiple surfactant trials (meta analysis) as noted in the next section. However, clinical and basic research also show that "natural surfactants" themselves vary in composition and activity and need to be considered individually as well as grouped together.

The activity differences found between exogenous surfactants in clinical trials are generally consistent with but smaller than those demonstrated in basic science studies (Chapters 11, 14). Clinical trials in general cannot discriminate mechanistic phenomena at the same level of quantitative specificity possible in basic research. The outcomes of infants in clinical surfactant trials are influenced by multiple aspects of prematurity and intensive care in addition to surface active function as noted above. An additional complicating factor in clinical surfactant comparisons is that exogenous surfactants affect patients not only through their intrinsic surface activity but also indirectly by incorporation into endogenous recycling pathways (Chapter 6). Exogenous surfactants can also be increased in activity by interacting with apoproteins or other components of endogenous surfactant in the lungs of some infants. Mechanisms such as these allow exogenous surfactants to enhance respiratory function and patient outcomes independent of their own surface and physiological activity. Activity differences in comparison studies where all patients receive an exogenous surfactant having some efficacy will by necessity be smaller than differences in placebo-controlled

---

3. For example, Refs. 89, 90, 127, 471, 480, 481, 921, 975, 1021, 1119.

Table 12-4 Randomized Clinical Comparison Trials of Exogenous Lung Surfactants in Premature Infants

Trial	Surfactant	N	Type	Population	Results[a]
Large studies					
Horbar [471]	Survanta vs Exosurf	617	Tx, nonbl	500–1500 g	Survanta: lower 0–72 hr average FiO_2 and MAP
Vt-Oxford [1119]	Survanta vs Exosurf	1,296	Tx, nonbl	501–1500 g	Survanta: fewer air leaks, lower FiO_2 at 72 hr, lower 0–72 hr average MAP
Hudak [480]	Infasurf vs Exosurf	1,126	Tx, bl	All with RDS	Infasurf: fewer air leaks, lower 0–72 hr FiO_2 and MAP[b]
Hudak [481]	Infasurf vs Exosurf	871	Plx, bl	<29 wk	Infasurf: less RDS, fewer air leaks, lower 0–72 hr average FiO_2 and MAP[b]
Bloom [90]	Infasurf vs Survanta	608	Tx, bl	<2000 g	Infasurf: lower: 0–72 hr average FiO_2 and MAP
	Infasurf vs Survanta	374	Plx, bl	<29 wks and <1250 g	No difference: all variables
Small studies					
Rollins [921]	Exosurf vs Curosurf	66	Tx, nonbl	All with RDS	Curosurf: Lower FiO_2 and improved a/A ratio at 24 hr
Sehgal [975]	Survanta vs Exosurf	41	Tx, nonbl	600–1750 g with RDS	No difference: all variables
Speer [1021]	Survanta vs Curosurf	73	Tx, nonbl	700–1500 g with RDS	Curosurf: lower FiO_2 until 24 hr

[a]All differences statistically significant, $P < 0.05$ or better.
[b]Studies of Hudak et al [480, 481] had a crossover design, with more Exosurf patients crossing over to Infasurf in both trials. All tabulated studies were prospective and randomized except Rollins et al [921]. Tx = treatment; Plx = prophylaxis; bl = blinded; nonbl = not blinded. None of these comparison trials showed statistically significant differences in mortality or BPD. See text for details.

surfactant trials. A substantial number of patients is thus required to demonstrate statistically significant differences between exogenous surfactants even in effects on pulmonary function (Table 12-4). Based on observed differences in functional activity, it can be hypothesized that sufficiently large comparison trials would demonstrate differences in long term outcomes including survival for some surfactants. However, the surface active properties and direct physiological activity of exogenous surfactants can be quantitated more specifically and in greater detail in basic research, where confounding clinical variables are not present. A similar argument for integrated basic and clinical research applies to ARDS as well as RDS as noted in Chapter 13. The use of basic science characterizations to help define the activity of therapeutic agents is not uncommon in clinical medicine. The activity of antibiotic drugs, for example, is routinely assayed under laboratory conditions rather than depending on controlled trials in severely ill septic patients with multifaceted pathology.

XII. Retrospective Assessments of Clinical Surfactant Trials and Exogenous Surfactants: Meta Analyses

A widely used method of analyzing therapeutic agents and patient outcomes from clinical trials is through retrospective statistical analysis of pooled data from multiple studies. Statistical analysis of this kind can be very useful in assessing the overall clinical or economic impact of a therapeutic modality. A number of retrospective meta analyses have been applied to data from clinical surfactant trials (e.g., [384, 430, 550, 725, 867, 1016, 1018]). Essentially all such analyses indicate that exogenous surfactant therapy has important benefits in premature infants. Lung function is improved, the severity of respiratory distress is decreased, complications such as pulmonary air leaks and pulmonary interstitial emphysema are decreased, and survival is increased. Meta analysis indicates that surfactant therapy does not decrease the incidence of several conditions including BPD and chronic lung disease, intraventricular hemorrhage, and retinopathy of prematurity. However, the incidence and severity of these conditions are not increased despite a larger percentage of surviving infants of extremely low birth weight. Meta analysis also consistently indicates that "natural" surfactants derived from animal lungs have greater efficacy as a group than current "synthetic" surfactants [384, 550, 1016, 1018] This finding agrees with the results of individual prospective surfactant comparison trials (Table 12-4). While meta analysis provides useful information on exogenous surfactants and the impact of surfactant therapy, it also has limitations. Retrospective pooling of data from trials using different patients, caregivers, protocols, and endpoints reduces overall resolving power. In addition, grouping results for therapeutic agents that vary in activity underestimates the efficacy of those that are the most active while overestimating the efficacy of others.

XIII. Prophylaxis versus Treatment (Rescue) Strategies for Exogenous Surfactant Therapy

Another important clinical issue is whether exogenous surfactant should be administered prophylactically to very premature infants or used primarily to treat established RDS ("rescue" therapy). Prophylactic administration of exogenous surfactant at birth has several potential advantages. Immediate surfactant replacement could minimize injury from ventilation of the surfactant-deficient lungs, and the distribution of exogenous surfactant might be improved by instillation while fetal lung liquid is still being absorbed. Prophylactic therapy at birth, however, requires exogenous surfactant delivery in conjunction with resuscitation and patient stabilization. Prophylactic therapy also means that exogenous surfactant will be given to some infants who would not develop RDS and thus receive an unnecessary drug. This latter concern has moderated as experience has shown that the toxicity and immunogenicity associated with short-term use of current clinical exogenous surfactants are very low [60, 1040-1042, 1161]. A number of randomized clinical trials have compared prophylactic and treatment strategies for exogenous surfactant therapy (Table 12-5). Given the numbers of patients required to show outcome differences in placebo-controlled surfactant trials, the three smaller studies in Table 12-5 appear underpowered to differentiate between two beneficial interventions [223, 228, 728]. The four larger studies in Table 12-5 all reported advantages for prophylaxis compared to rescue therapy [82, 551, 554, 810]. The major benefits of prophylactic therapy are found in the youngest and smallest infants <29 weeks gestation [82, 554, 810], although some benefits have also been reported in older infants [551]. Precise gestational age limits and prenatal indicators defining when prophylactic surfactant therapy is most useful have not yet been defined. Although there is not complete consensus among neonatologists, the low toxicity and potential benefits of prophylactic therapy appear to support its use in extremely premature infants. Prophylactic surfactant therapy has been found to be equally effective if given in the delivery room immediately after resuscitation and patient stabilization rather than prior to the first breath [556].

XIV. Delivery Methods for Exogenous Surfactant

Instillation has proved to be a surprisingly effective and practical method for delivering exogenous surfactants in the clinical setting. Direct endotracheal instillation is most common, but modifications such as bronchoscopic instillation have been utilized in patients with ARDS-related respiratory failure (Chapter 13). Instilled surface active material rapidly spreads and distributes to the periphery of the lung [189, 246, 355]. Spreading from the central airways toward the alveoli

Table 12-5 Clinical Trials Comparing Prophylaxis vs Rescue (Treatment) Strategies for Surfactant Replacement Therapy

Study	Surfactant	N	Patients	% Rescue-treated	Significant results[a]
Kendig [554]	Infasurf	479	<30 wk	58%	Prophylaxis: fewer deaths, less severe RDS, lower FiO_2 and MAP
Osiris [810]	Exosurf	2,690	"At risk"	73%	Prophylaxis: lower death and BPD; fewer air leaks
Kattwinkel [551]	Infasurf	1,248	29–32 wk	43%	Prophylaxis: fewer deaths, reduced severity of RDS and fewer with severe RDS, lower FiO_2 and MAP
Bevilacqua [82]	Curosurf	268	24–30 wk	40%	Prophylaxis: fewer with severe RDS, lower PIE
Egberts [228]	Curosurf	147	26–29 wk	43%	Prophylaxis: fewer severe RDS
Merritt [728]	Human	105	24–29 wk	76%	Prophylaxis: not different from rescue
Dunn [223]	bLES	122	<30 wk	52%	Rescue: fewer on O_2 at 28 d but not 36 d

[a]All results noted were statistically significant, $P < 0.05$ or better.

is promoted by surface tension gradients that drive transport from regions of high surfactant concentration to regions of lower surfactant concentration. Instillation allows comparatively large amounts of surfactant to be placed in the lungs, and a degree of inefficiency in spreading to the alveoli can be tolerated. Instilled surfactant doses of order 100 mg/kg body weight are typical, whereas <3.1 mg/kg of phospholipid (molecular weight 750) is required to cover the alveolar surface with a tightly packed surfactant monolayer [692] (Chapter 8). Excess instilled surfactant provides material for the hypophase and ultimately enters endogenous surfactant pools via recycling pathways to provide additional benefits to pulmonary function (Chapter 6).

Although exogenous surfactants can be delivered very effectively by instillation, several alternatives including aerosolization have been investigated. Phospholipid aerosols having stable particle sizes appropriate for alveolar deposition can be formed by ultrasonic or jet nebulization [692, 693, 1196], and exogenous surfactants have been aerosolized to animals and patients with RDS or ARDS-related lung injury [25, 205, 232, 639, 642, 677, 1222]. However, the theoretical potential of aerosols to improve the uniformity of alveolar deposition and reduce required surfactant dosages has not been replicated in practice. Aerosol technology to date has not been able to deliver exogenous surfactants to the alveoli as effectively as instillation. Enhancing the distribution of instilled exogenous surfactant with specialized modes of ventilation like jet ventilation [185, 186] or partial liquid ventilation [629, 630] is another area of clinical interest. The distribution of exogenous surfactant has also been improved by instilling larger fluid volumes or utilizing bronchoalveolar lavage in animals [46, 47, 327, 1104], although the use of these strategies is problematic in humans with edema and compromised lung function.

XV. Follow-up Analyses of Premature Infants and Surfactant Trials

Follow-up analyses assessing long term outcomes in treated patients and patient groups are an important aspect of clinical trial research. A number of follow-up studies have assessed neurodevelopmental and pulmonary outcomes in premature infants from clinical surfactant trials ([98, 99, 664, 725, 962] for review). A substantial percentage of very low birth weight infants are found to have minimal or only minor long term pulmonary or neurodevelopmental sequelae of prematurity and intensive care. However, as many as 20–30% of extremely premature infants have mild to moderate physical impairments or neurodevelopmental delays particularly early in life, and approximately 20% have even more severe disabilities ([98, 99] for review). These percentages tend to decrease significantly with age, although infants with the most severe impairments are often permanently affected. Despite the much greater numbers of extremely premature infants surviving due to surfactant therapy and advances in neonatal intensive care, the incidence of those with severe, permanent impairments has not undergone a proportional increase [99]. No assessments of long term outcome in premature infants have disputed the overall benefits and lifesaving nature of surfactant therapy.

As opposed to follow-up analyses documenting neurodevelopmental and other outcomes in premature infants, studies seeking to quantitate the persistence and impact of differences originally observed between patient groups in surfactant trials are much more difficult to carry out and interpret. Long term outcomes are affected by multiple uncontrolled social and enviromental variables not related to the original surfactant treatment, and the impact of these extraneous variables on outcome increases with time. Follow-up analyses also typically capture only a fraction of patients originally studied, further complicating meaningful group comparisons. Despite their complications, follow-up studies obviously address important issues about the impact of surfactant replacement therapy on the long term outcomes of premature infants and are a focus of continuing clinical investigation.

XVI. Chapter Summary

The majority of very premature infants in the 1950's died in the immediate perinatal period. Much of this early mortality resulted from hyaline membrane disease or RDS. Despite the early association of this disease with lung surfactant deficiency, clinically effective surfactant replacement therapy was not developed for decades. Surfactant therapy was unsuccessful when first attempted in premature infants in the 1960's, and extensive biophysical and animal research was necessary to establish a base of understanding for later effective clinical trials. The benefits of surfactant therapy in premature infants were first shown in uncontrolled or nonrandomized clinical trials from 1980 through 1984, and its efficacy was established in placebo-controlled trials from 1985 through the early 1990's. Essentially all clinical studies with active exogenous surfactants have found improvements in pulmonary function and the severity of acute RDS, which are closely related to lung surfactant activity. Surfactant therapy overall has significantly increased the survival of very premature infants, but only a minority of individual placebo-controlled trials have demonstrated decreased mortality and chronic lung disease. This reflects the efficacy of neonatal intensive care even without surfactant therapy, as well as the dependence of these outcomes on multiple factors in addition to surfactant activity.

Lung surfactant replacement therapy for RDS is currently being optimized in terms of exogenous surfactant preparation, treatment strategy, and other clinical variables. Several large trials support the prophylactic use of exogenous surfactant particularly in the youngest, smallest infants, although "rescue" therapy for established RDS is also beneficial in premature infants of all ages. Clinical research also demonstrates activity differences between exogenous surfactants. "Natural surfactants"containing apoproteins from animal lungs are more active as a group than current protein-free synthetic surfactants, and natural surfactants themselves also vary in composition and activity. Activity differences between exogenous surfactants in clinical trials are less apparent than found in basic research, where mechanistic differences can be examined specifically and in more detail (e.g., Chapter 14). While controlled trials are crucial for defining clinical efficacy, it is equally important to utilize information on exogenous surfactants from biophysical and animal research as described throughout this book. The history of surfactant replacement therapy for RDS reflects a partnership of integrated basic science and clinical investigation. This is also true of surfactant therapy for clinical ARDS as detailed in the next chapter.

13

Surfactant and Combined-Modality Therapies for Clinical ARDS and Acute Lung Injury

I. Overview

The widespread incidence and clinical severity of ARDS have led to significant interest in new therapies aimed at this complex lung injury syndrome. Exogenous surfactant replacement therapy is one such intervention. Lung surfactant dysfunction contributes to the pathology of ARDS, and lung surfactant can also be deficient if type II cell function is impaired. Biophysical and animal model research shows that lung surfactant dysfunction as well as deficiency can be reversed by exogenous supplementation with surfactants able to resist inactivation. This chapter summarizes the conceptual basis and current status of clinical exogenous surfactant therapy for ARDS-related respiratory failure. Also discussed are selected additional agents and interventions that address other aspects of acute lung injury and have the potential to be additive or synergistic if used in conjunction with exogenous surfactant in "multimodal" therapies for ARDS.

II. Overview of ARDS and Its Pathophysiology

Details about the definition, pathophysiology, and clinical features of ARDS are discussed in Chapter 10, and only a brief summary is given here. ARDS is a major cause of mortality and morbidity worldwide. Between 50,000 and 150,000 patients are diagnosed with ARDS annually in the United States alone [76, 483, 488, 599]. Infants and children, as well as adults, can be affected by ARDS-related respiratory failure [69, 191, 1084]. Despite striking advances in the sophistication of intensive care and mechanical ventilatory support over the past several decades, mortality in severely affected individuals remains at substantial levels of 30–70% [76, 191, 220, 428, 488, 599, 735, 1084]. Respiratory failure in

ARDS is the clinical manifestation of severe acute lung injury. This underlying injury induces a complex pathology including inflammation, edema, impaired gas exchange, tissue injury, and vascular dysfunction (Chapter 10). Clinical ARDS in many cases also includes sepsis and the involvement of organs outside the lungs. Lung surfactant dysfunction can occur in ARDS as a result of biophysical interactions with inhibitors present in edema or released through inflammation. Components in lung surfactant can also be degraded or altered chemically by phospholipases, proteases, or reactive oxidants. Surfactant deficiency secondary to type II cell injury or dedifferentiation can also occur in ARDS, although surfactant dysfunction is typically more prominent. The existence of lung surfactant dysfunction is well documented in patients with ARDS [344, 361, 398, 635, 858, 971, 1114] (Chapter 10).

III. Rationale for Surfactant Therapy in Clinical ARDS

Biophysical and animal model research provides a strong conceptual rationale for the use of exogenous surfactant therapy in clinical ARDS ([247, 402, 455, 643, 644, 771, 786, 798, 969, 1022] for review). The characteristics and mechanisms of lung surfactant dysfunction have been extensively defined in biophysical research, and it has been demonstrated that decreased surface activity can be reversed by raising surfactant concentration despite the continued presence of inhibitor substances (Chapter 9). Multiple animal models of acute lung injury and surfactant dysfunction have been shown to have improved respiratory function or

---

**Rationale for Exogenous Surfactant
Supplementation in ARDS**

Dysfunction of endogenous surfactant can occur in ARDS by several mechanisms and is known to contribute to its pathophysiology.

Biophysical and animal model research shows that surfactant dysfunction as well as deficiency can be overcome by raising the concentration of surface active material.

Exogenous surfactant supplementation in ARDS addresses only one aspect of a complex pathology, and additional complementary interventions may also be necessary.

mechanics following the delivery of active exogenous surfactants (Chapter 11). Exogenous surfactants with high inhibition resistance clearly have the potential to improve alveolar stability, increase lung volumes and compliance, and enhance gas exchange in clinical ARDS. As summarized in subsequent sections, exogenous surfactants have been found to improve pulmonary function and outcome in ARDS-related respiratory failure in term infants [31, 256, 563, 668] and children [616, 876, 1185, 1186]. Surfactant therapy in adults with ARDS has so far met with mixed success [25, 345, 412, 682, 782, 1023, 1135]. Surfactant therapy addresses only one aspect of the complex pathophysiology of ARDS, and its clinical benefits and impact on long term patient outcomes may be limited unless complementary interventions targeting other aspects of lung injury are also employed.

IV. Factors Complicating Clinical Assessments of Surfactant Therapy for ARDS

Several factors complicate clinical evaluations of exogenous surfactant therapy in ARDS. The fact that patient outcomes are influenced by a broad pathology makes it more difficult to assess the true physiological effectiveness of any individual intervention. Even if exogenous surfactant were able to normalize surface activity completely in ARDS, clinical improvements could be obscured by the remaining elements of disease. The severe edema in patients with ARDS also makes it difficult to deliver exogenous surfactants effectively to the alveoli. Clinical studies in ARDS are further complicated by a heterogeneous patient population of diverse age and physical condition. The severity and details of lung injury in individual patients depend to some degree on the initiating cause. Patients with ARDS also vary significantly in their nonpulmonary pathology, which in its most severe aspects can include systemic inflammatory response syndrome (SIRS), sepsis syndrome, multiple organ dysfunction syndrome (MODS), or multiorgan failure (MOF) [62, 76, 220, 483, 488, 548, 582, 599, 721, 1080] (Table 13-1). All these factors complicate study design and increase the numbers of patients needed to define efficacy and discern differences in long term outcomes. The difficulty of clinical research in ARDS argues strongly for continuing, detailed basic research on surfactant and combined-modality interventions to complement and facilitate clinical evaluations. The current status of surfactant therapy for ARDS is summarized below, followed by discussion of selected agents that could potentially be used along with exogenous surfactant in multimodal interventions.

Table 13-1 Spectrum of Clinical Respiratory Distress Syndromes

Syndrome	Primary process	Patient age	Associated conditions	Predisposing conditions
RDS	Surfactant deficiency	Premature infants	Complications of prematurity	Premature birth
ARDS/ALI (direct injury)	Direct toxic injury to lungs	Any age	Associated with the cause of lung injury	Aspiration Inhaled toxic gas Near drowning Pulmonary contusion Radiation pneumonitis
ARDS/ALI (indirect injury)	Injury through systemic inflammation	Any age	SIRS, MODS, MOF, and less severe multiorgan involvement	Blood transfusion Cardiopulmonary bypass Drug overdose Infection/sepsis Severe trauma Shock Uremia

Definitions, causes, pathophysiology, and features of clinical acute lung injury (ALI) and ARDS are discussed in Chapter 10. Patients with ARDS can have extensive disease outside the lungs as reflected, for example, in systemic inflammatory response syndrome (SIRS), sepsis syndrome, multiple organ dysfunction syndrome (MODS), and multiple organ failure (MOF). See text for details.

V. Surfactant Therapy in Full-Term Infants with ARDS-Related Respiratory Failure

Although ARDS was initially defined in adults, related respiratory disease can also affect infants and children. Term infants with respiratory failure from meconium aspiration or pulmonary infection are important examples of such patients. Meconium is known to inhibit lung surfactant activity *in vitro* [131, 750], and exogenous surfactant can improve pulmonary function in animals with meconium aspiration [16, 1047, 1049]. In an initial pilot trial in 1991, Auten et al [31] showed that tracheally instilled CLSE (Infasurf) improved oxygenation in 14 term infants with severe respiratory failure from meconium aspiration or pneumonia. The rapid increase in arterial/alveolar oxygen ratio after the initial dose of CLSE was similar to that previously reported for this exogenous surfactant in premature infants with RDS [190, 554] (Figure 13-1). Oxygenation Index ($OI = 100 \times$ mean airway pressure $\times FiO_2/PaO_2$) was also significantly improved in CLSE-treated term infants [31]. Eight infants received a second dose of CLSE, again leading to a substantial improvement in oxgenation [31]. A third dose was given only to three patients and resulted in minimal additional improvement. All 14 severely ill infants survived without developing air leak complications or chronic lung disease [31]. In another uncontrolled study, Khammash et al [563] reported acute improvements in lung function after instillation of bLES in 15 of 20 term infants with meconium aspiration. A later randomized controlled trial by Findlay et al [256] showed that instillation of three doses of Survanta (150 mg/kg) improved lung function and outcome in term infants with meconium-associated respiratory failure. Infants treated with Survanta had a significant reduction in the incidence of pneumothorax, duration of mechanical ventilation and oxygen therapy, and time of hospitalization [256] (Table 13-2). Fewer surfactant-treated infants also required extracorporeal membrane oxygenation (ECMO). Lotze et al [668] also reported favorable results using Survanta in a controlled trial in term infants referred for ECMO due to severe respiratory failure. Infants treated with four doses of Survanta (150 mg/kg) had improved pulmonary mechanics, decreased duration of ECMO treatment, and a lower overall incidence of complications after ECMO compared to a similar number of control infants [668]. Taken together, these studies strongly support the use of exogenous surfactant therapy in ARDS-related respiratory failure in infants.

VI. Surfactant Therapy in Children with Respiratory Failure

Several studies and case reports have shown that therapy with exogenous surfactant can have acute benefits in children with ARDS-related respiratory failure

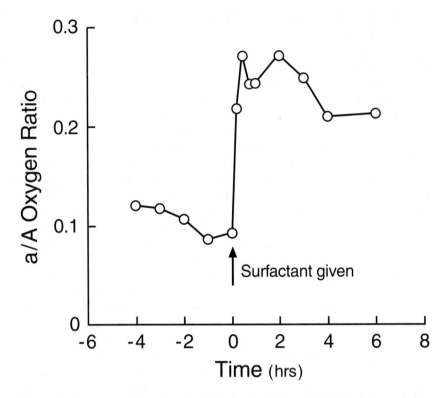

Figure 13-1 Improved lung function following instillation of CLSE in full-term infants with respiratory failure. The arterial/alveolar (a/A) oxygen ratio is shown before and after surfactant treatment in 14 term infants with severe respiratory failure from meconium aspiration or pneumonia. Surfactant treatment significantly improved oxygenation. CLSE dose was 90 mg/kg. (Redrawn from Ref. 31.)

[616, 876, 1185, 1186]. In an uncontrolled treatment study, Willson et al [1186] showed functional benefits from a single dose of Infasurf in 29 children admitted in respiratory failure to the pediatric intensive care unit (PICU) at six medical centers. Patients varied in age from 0.1 year through 16 years. The majority of patients (24/29) showed improved lung function, prospectively defined as a 25% decrease in Oxygenation Index, after instillation of Infasurf (70 mg/kg or 80 ml/m^2 body surface area) [824, 1186]. A subsequent randomized controlled trial assessed both functional and long term outcome differences in 42 children with respiratory failure at eight medical centers [1185]. Children receiving one or two doses of Infasurf (70 mg/kg or 80 ml/m^2) had significantly better OI values over the 50 hours following initial treatment (Figure 13-2). Mortality was not

Table 13-2 Improved Outcomes in Full-Term Infants with Meconium Aspiration Treated with Survanta

Outcome variable	Control group (n = 20)	Survanta-treated (n = 20)	P-value
Air leaks	5	0	0.024
Duration of mechanical ventilation (d)	10.8 ± 1.3	7.7 ± 0.7	0.047
Duration of O_2 therapy (d)	19.6 ± 2.6	13.0 ± 1.4	0.031
Duration of hospitalization (d)	24.3 ± 2.4	15.9 ± 1.2	0.003
ECMO requirement	6	1	0.037
Mortality < 28 d	0	0	NS
Discharge with O_2 therapy	8	6	NS

Surfactant-treated patients received three doses of Survanta (150 mg/kg in 6 ml instilled endotracheally over 20 minutes for each dose). Initial dose was at <6 hr, with others 6 hr apart. Control patients were instilled with air. Data are mean ± SEM.
Source: Ref. 256.

different between surfactant-treated and control groups in this small study, but several prospectively chosen outcome variables related to duration of disease were improved in patients receiving Infasurf (Table 13-3). The analysis in Table 13-3 excludes seven children who died or required ECMO, but including these patients in an intent-to-treat analysis did not alter study conclusions [1185]. Intent-to-treat analysis also indicated that children receiving Infasurf had a significant increase in "ventilator free days" during the first 14 days of hospitalization and a higher incidence of extubation by 72 hours [1185]. Although instillation of active exogenous surfactant has promising short term benefits in children with ARDS-related respiratory failure, larger controlled clinical trials are necessary to define in detail the impact of this intervention on long term outcomes including survival.

VII. Surfactant Therapy in Adults with ARDS

Studies of surfactant therapy in adults with ARDS have so far reported mixed success [25, 345, 412, 682, 782, 1023, 1135]. Several uncontrolled studies have found that instilled exogenous surfactant has acute functional benefits in adults. Surfactant-TA improved lung function in two adults with ARDS [782], and acute improvements in oxygenation were reported in six adults following bronchoscopic administration of Curosurf [1023]. Walmrath et al [1135] also found acute improvements in gas exchange following bronchoscopic instillation of high doses of Alveofact in 10 adults with sepsis and ARDS. A larger controlled trial tested the use of Survanta in 59 adults with ARDS from trauma, multiple blood

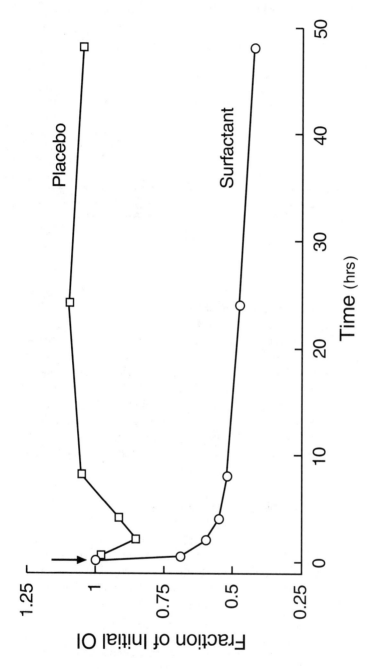

Figure 13-2 Improvements in oxygenation index after instillation of Infasurf in children with ARDS-related respiratory failure. Patients in eight pediatric intensive care units were randomized to Infasurf or control groups. Infasurf patients received 80 ml/m² body surface (70 mg/kg body weight) by tracheal instillation during hand ventilation with 100% oxygen (arrow). Control patients received hand ventilation and 100% oxygen alone. Ten of 21 surfactant-treated patients received a second dose 12 or more hours after the first. OI defined as $100 \times MAP \times FiO_2/PaO_2$, where MAP = mean airway pressure, FiO_2 = fraction of inspired oxygen, PaO_2 = arterial partial pressure of oxygen. (Data from Ref. 1185.)

Table 13-3 Long Term Outcomes in Pediatric Intensive Care Unit Patients Treated with Infasurf for Acute Respiratory Failure

Outcome variable[a]	Infasurf-treated (n = 17)	Control (n = 18)	P-value
Days of mechanical ventilation	9.0 ± 10.4	13.2 ± 8.4	0.03
Geometric mean	5.3	10.7	
95% CI	(3.1, 9.1)	(7.6, 15.1)	
Days in PICU	11.7 ± 11.6	16.7 ± 9.2	0.03
Geometric mean	8.1	14.4	
95% CI	(5.2, 12.7)	(10.9, 19.1)	
Days on supplemental O_2	13.5 ± 15	16.9 ± 11	0.06
Geometric mean	7.6	13.9	
95% CI	(4.3, 13.4)	(9.8, 19.6)	
Days in hospital	19.7 ± 15.9	23.1 ± 12.1	0.12
Geometric mean	14.0	20.4	
95% CI	(9.1, 21.5)	(15.8, 26.3)	

[a]Only patients surviving without ECMO are included in the tabulated variables. CI, confidence interval of geometric mean. Figure 13-2 shows lung functional improvements in Infasurf-treated vs control patients. See text for details. Data are mean ± SD.
Source: Ref. 1185.

transfusions, aspiration of gastric contents, or sepsis syndrome [345]. Survanta patients (n=43) were randomized to three groups to receive up to eight doses of 50 mg/kg, up to eight doses of 100 mg/kg, or up to four doses of 100 mg/kg. The latter group had a significantly decreased ventilator FiO_2 at 120 hours, but there were no significant differences from control in the other two Survanta-treated groups. Mortality in the group receiving up to four doses of 100 mg/kg was 19% compared to 44% in controls (P=0.075) [345]. Several additional clinical reports have found surfactant therapy to be ineffective in adult patients. The protein-free synthetic surfactant ALEC did not improve pulmonary function in two small studies in adults with ARDS or cardiopulmonary bypass [412, 682]. Moreover, a large randomized, controlled study in 725 patients with ARDS and sepsis reported no benefits following treatment with aerosolized Exosurf [25]. The negative results of this latter study, by far the largest now available on surfactant therapy in ARDS, were accompanied by an editorial in the *New England Journal of Medicine* stating that surfactant therapy was another on the long list of therapeutic interventions that have not altered the outcome of this condition [705]. This assessment fails to take into account that Exosurf (and ALEC as well) has much lower surface and physiological activity than other clinical exogenous surfactants in basic and clinical evaluations (Chapters 11, 12, 14). In addition, aerosol delivery of exogenous surfactants is not as effective as intratracheal or bronchoscopic instillation. Inferring that surfactant therapy is not beneficial in ARDS based on results for aerosolized Exosurf is reminiscent of what occurred after early unsuccessful attempts to treat premature infants with aerosolized DPPC in the 1960's [129]. The mistaken conclusion that RDS was globally nonresponsive to exogenous surfactants substantially delayed the development of effective clinical surfactant therapy.

VIII. Summary of Current Status of Clinical Surfactant Therapy for ARDS

The severe, multifaceted pathology of ARDS presents a challenge for exogenous surfactant therapy. Instillation of exogenous surfactant has been shown to be effective in treating term infants with ARDS-related respiratory failure from meconium aspiration. Exogenous surfactant therapy has also been found to have at least acute benefits in children with ARDS. The utility of this therapeutic approach in adults with ARDS is more uncertain. Meaningful testing of surfactant therapy in ARDS requires that the exogenous surfactants used have the highest surface activity and inhibition resistance and that they be delivered effectively to the alveoli. Several of the most active clinical surfactants have not yet had extensive evaluation in ARDS, particularly in adult patients. Moreover, as noted earlier, exogenous surfactant therapy may not have its most substantial impact on

---

Summary of Current Status of Lung Surfactant Therapy for Clinical ARDS

Instillation of several exogenous surfactants has been effective in improving respiratory function and outcome in term infants with respiratory failure from meconium aspiration.

Exogenous surfactant therapy has been found to have at least short term benefits in several studies in children with ARDS-related respiratory failure.

Clinical studies of surfactant therapy in adults with ARDS show limited success that may relate in part to ineffective delivery and to the activity of the exogenous surfactants used.

Exogenous surfactant therapy may be most effective when used in combination with complementary interventions targeting other pathophysiological aspects of ARDS.

patient outcomes in ARDS unless it is used in conjunction with complementary interventions aimed at inflammatory lung injury and vascular dysfunction. Subsequent sections describe representative therapeutic agents and interventions that have the potential to be additive or synergistic with exogenous surfactant therapy.

IX. Overview of Multimodal Therapy for Clinical ARDS

All patients with ARDS receive treatment for the underlying cause of lung injury plus respiratory support and other appropriate intensive care. In addition to exogenous surfactant, a variety of agents have been used to improve pulmonary function and antagonize inflammation and vascular dysfunction in ALI and ARDS. Examples of pharmacologic agents studied clinically in ARDS or sepsis syndrome include anticytokine antibodies and receptor antagonists; antibacterial antibodies; other anti-inflammatory drugs such as corticosteroids and pentoxifylline; antioxidants such as N-acetylcysteine (NAC); and vasoactive agents such as inhaled nitric oxide (INO), almitrine, and prostacyclin[1] (Table 13-4). Specialized modes of ventilation such as liquid ventilation and low tidal volume ventilation have also been studied clinically in ARDS-related respiratory failure, as have prone positioning and a number of other intervention strategies. Although not without benefits, few of these agents or interventions have had a substantial impact on survival in patients with ARDS. However, the conceptual rationale underlying the use of many of these therapies remains sound, and their relative lack of efficacy may reflect the multifaceted pathology of ARDS. To have a significant impact on patient outcomes, it may be necessary to target multiple aspects of inflammatory lung injury in a "multimodal" approach. Subsequent sections give examples of pharmacologic agents from current research that could potentially be used in conjunction with exogenous surfactant in treating ARDS (Table 13-4). Ventilatory interventions are also briefly noted. The agents and interventions discussed are representative only and with the exception of corticosteroids are primarily relevant for ALI and early phase ARDS. Many of the biochemical mediators and processes relevant for acute injury also affect chronic injury and late phase ARDS, but additional therapeutic targets associated with pulmonary remodeling, repair, and fibrosis also become important [80, 1153, 1224]. The necessity to include epithelial rescue in the therapy of ARDS is just

---

1. Examples of clinical studies with these agents are in Refs. 2–4, 38, 78, 85, 150, 193, 200, 202, 254, 262, 263, 302, 309, 324, 341, 465, 514, 538, 552, 626, 672, 717, 733, 763, 769, 801, 821, 822, 835, 928, 929, 1029, 1030, 1051, 1089, 1136, 1204, 1211, 1223, 1228, 1231. For additional review and discussion of agents and interventions for ARDS and sepsis see Refs. 27, 52, 69, 80, 252, 292, 377, 482, 548, 594, 833, 907, 980, 1076, 1170.

Table 13-4 Examples of Pharmacologic Agents for Potential Use with
Exogenous Surfactant in Multimodal Therapies for ARDS

Inhaled nitric oxide (INO)
Other vasoactive drugs (e.g., almitrine, prostacyclin)
Antibodies to inflammatory mediators (e.g., anti-TNFα, anti-IL8)
Antibodies to bacterial products (e.g., anti-endotoxin)
Blockers of inflammatory receptors or ligands (e.g., anti-CD40L, IL-1Ra)
Other anti-inflammatory drugs (e.g., pentoxifylline, corticosteroids)
Antioxidants (e.g., *N*-acetylcysteine, SODs, catalase)

Tabulated agents and categories are examples only. Many multimodal therapies are
also available for potential use in ARDS. See text for reference citations and details.
Abbreviations: TNF, tumor necrosis factor; IL-1Ra, interleukin 1 receptor antagonist;
IL-8, interleukin 8; SODs, superoxide dismutases; CD40L, CD 40 ligand.

one example [80]. A variety of new agents and interventions that could be utilized
in multimodal therapies for ARDS are also becoming available through continuing basic research and new medical technology.

Basic research has extensively investigated cellular and molecular mechanisms
in lung injury, but much less information is available on the efficacy of specific
combined-modality therapies in cell systems and animal models. Detailed information
of this kind would be very helpful in guiding clinical applications. Addressing multimodal therapies in basic research is nontrivial. Scientists prefer to design experiments
that minimize variables and maximize the interpretability of data. Animal models of
ALI and ARDS have a broad phenomenology even without the added complexity of
studying multimodal interventions. Examples of animal models of ALI and ARDS
include hyperoxia, aspiration lung injury, bilateral vagotomy, antibody-induced lung
injury, endotoxin- or sepsis-induced injury, influenza-induced pneumonia, radiation
pneumonitis, fatty acid infusion, *N*-nitroso-*N*-methylurethane (NNNMU)-induced lung
injury, and several others (see Chapter 11). Extrapolation of results found in animal
model and cellular research to human medicine is far from exact. Studies in cultured
cells or tissues *in vitro* never fully replicate the complex, interactive environment in
living organisms. Animal models *in vivo* also vary in the details of their pathology and
in how closely they approximate clinical ARDS (which itself is not homogeneous).
Despite these limitations, laboratory and animal studies are invaluable in addressing
the complexities of lung injury and the effects of specific agents in mitigating it.
Appropriately structured basic research can define and compare in detail the activity of
various combinations of agents and interventions, while continuing to investigate
cellular and molecular mechanisms at a level impossible in humans. Basic research can
also assess the consistency of data and interpretations across complementary experimental systems (physicochemical, cellular, and animal models) under standardized
conditions. Information can also be gained in basic research about detrimental as well

as positive effects from combinations of agents, something that would be highly undesirable in clinical research.

X. Inhaled Nitric Oxide (INO) and Other Vasoactive Agents for Use in ARDS

Nitric oxide is a naturally occurring vasoactive product that is made from L-arginine and is identical to endothelial-derived relaxing factor [489, 818, 819]. Nitric oxide is important in multiple processes *in vivo*, one of the most important being the stimulation of cyclic-GMP to promote calcium sequestration to generate vasodilation. The vasodiliatory activity of nitric oxide can be pharmacologic as well as physiologic. INO is active principally in ventilated lung regions because clinically insignificant concentrations diffuse into neighboring non-ventilated tissues. INO facilitates gas exchange by increasing blood flow to ventilated areas to improve ventilation/perfusion (V/Q) matching. INO has been used in the therapy of several respiratory diseases ([578, 768, 907, 1090] for review). One of the major clinical uses of INO is in pulmonary hypertension of the newborn [323, 444, 580, 910]. INO also reduces pulmonary artery pressure and vascular resistance in animal models of ALI [70, 277, 601, 870, 987, 988], and it has been widely used to treat infants, children, and adults with ARDS-related respiratory failure. INO improves oxygenation and lowers pulmonary artery pressure in adults with ARDS [200, 254, 302, 733, 821, 928, 929, 1089, 1211] and in infants and children with related acute respiratory failure [2, 193, 202, 324, 763, 769, 801]. Approximately two thirds of patients treated with therapeutic doses of INO show a positive response based on a 20% improvement in PaO_2/FiO_2 ratio and a reduction in pulmonary artery pressure. Combining INO therapy with the prone position to facilitate alveolar recruitment can further enhance oxygenation [309, 538, 821]. However, many patients become dependent on INO, necessitating prolonged ventilation, and approximately a third of patients have little or no response. Despite its benefits in reducing respiratory support, lowering right ventricular overload, and improving left ventricular output, INO has not increased survival in adults with ARDS. This failure to affect mortality has made the use of INO in ARDS the subject of debate [578, 706, 768, 907, 1090, 1221]. The clinical benefits of INO may be augmented substantially if it is used in conjunction with exogenous surfactant as noted below.

A. Rationale for Combining INO and Exogenous Surfactant Therapy

The rationale for combining INO with exogenous surfacant therapy stems from their complementary mechanistic actions. INO reaches ventilated lung units, where it dilates the accessible vasculature and improves V/Q matching. Surfactant improves ventilation by decreasing surface tension and enhancing alveolar stability and recruitment. Exogenous surfactant would theoretically increase the ventilated lung area accessible to INO, enhancing its impact on gas exchange. Conversely, the benefits of exogenous surfactant would be improved if INO were present to increase the perfusion of ventilated areas. The impact of exogenous surfactant is blunted if newly stabilized and recruited alveoli are not adequately perfused. High levels of INO could lead to oxidant-induced changes in lung surfactant, but detrimental effects on surface activity have not been found at the low levels of 5–20 ppm generally used clinically [386, 399, 699]. Additive improvements in lung function from the simultaneous use of INO and exogenous surfactant have been demonstrated in several animal models of surfactant deficiency or ALI [328, 410, 546, 882, 1227]. It has also been reported that exogenous surfactant can protect against the detrimental effects of an instilled NO synthase inhibitor that reduces the activity of endogenous NO [1219]. Further laboratory and animal research detailing the specific interactions and additivity of INO and exogenous surfactant will help assess the potential clinical utility of this combination of agents.

Vasodilatory Drugs Other Than INO. Although INO has received the most study as an inhaled pulmonary vasodilator in respiratory failure, another such drug is prostacyclin (PGI_2) [822, 945, 1136, 1231]. Prostacyclin is a microcirculatory vasodilator and inhibitor of platelet aggregation used for several indications in neonatal and adult intensive care medicine [945]. When aerosolized to patients with ARDS, its vasodilatory action in ventilated areas should in principle be similar to INO in improving V/Q matching without leading to systemic hypotension. Consistent with this interpretation, aerosolized prostacyclin has been found to give improvements in acute respiratory function equivalent to those of INO in several clinical studies in patients with ARDS [822, 1136, 1231]. Another vasodilatory prostaglandin, PGE_1, has also been found to give improvements similar to those of INO when delivered by aerosol to patients with ARDS [869]. These results suggest that aerosolized prostacyclin or similar drugs might be potential alternatives to INO in combined-modality regimens.

Vasoconstrictive Drugs Such as Almitrine in Conjunction with INO. Several studies have investigated the possibility of using INO together with vasoconstrictive agents in ARDS. The mechanistic rationale for this approach is that vasoconstrictive drugs could reinforce the natural hypoxic vasoconstriction of the pulmonary vasculature that occurs during respiratory failure. This vasoconstriction in theory could allow a larger fraction of pulmonary blood flow to be redirected by INO to ventilated areas, further improving V/Q matching. The use of vasoconstrictive agents also has the potential for negative responses, since inappropriate vasoconstriction could further impair an already compromised gas exchange process. Several clinical studies have shown that coadministration of INO and almitrine bismesylate, a selective pulmonary vasoconstrictor, has an additive effect in improving arterial oxygenation and reducing the level of required mechanical ventilatory support in patients with ARDS [309, 538, 672, 835, 1204]. These promising results with INO and almitrine provide additional evidence that rational combinations of vasoactive agents could be important elements in multimodal therapy for ARDS. Phenylephrine has also been reported to improve acute respiratory function in ARDS patients responding to INO [215], although the mechanisms involved have been questioned [1088]. Other vasoconstrictive agents such as norepinephrine [820] and prostaglandin PGF_{2_α} [585] have not been found to exhibit additivity with INO.

XI. Anti-Inflammatory Agents and Antioxidants for Use in ARDS

ALI and ARDS have substantial components of inflammatory and oxidant injury that can be antagonized pharmacologically. Agents directed against these aspects of lung injury have the potential for additivity with the pulmonary mechanical and functional activity of exogenous surfactant and vasoactive drugs like INO. Inflammatory and oxidant-induced changes in ARDS typically occur in a progressive cascade. Proteinaceous edema not only compromises gas exchange and contributes to surfactant inactivation but also promotes further inflammation. Injury to alveolar and airway cells from mechanical ventilation and supplemental oxygen can also worsen inflammation, as can the continuing host immune response to sepsis. Oxidant injury accompanies and interacts with other components of the inflammatory response. Reactive oxygen species including superoxide anion, hydrogen peroxide, hydroxyl radical, and a variety of others can cause cell and tissue damage and increase vascular permeability [116, 429]. Oxidant stress in ARDS is further exacerbated by the high inspired oxygen levels required clinically. Increased oxidant activity has been found in lavage fluid and in granulocytes and red blood cells from patients with ARDS [625, 816, 1063]. Antiinflammatory agents and antioxidants for potential use in ALI and ARDS are reviewed elsewhere (e.g., [27, 292, 482, 548, 594, 907, 1076]. Representative examples of agents that might give additive benefits with exogenous surfactant in ARDS are noted below.

A. Examples of Anti-Inflammatory Agents and Antioxidants

Clinically useful anti-inflammatory agents must reduce the pathologic part of the inflammatory response while allowing a physiologic component to remain to protect the organism. The normal balance of inflammatory and anti-inflammatory factors in the lung is disrupted in ALI and ARDS, with a shift in the direction of increased inflammation. Examples of antibodies to specific proinflammatory cytokines or mediators that could antagonize this shift include anti-TNFα [3, 4, 150, 1087], anti-IL-8 [101, 274, 1209], anti-CD40L [11, 12], IL-1Ra [262–264, 582, 800], and antibodies to endotoxin [341, 1228]. Two other agents with anti-inflammatory properties that have been used in ALI and ARDS are pentoxifylline [38, 626, 1029, 1030, 1223] and corticosteroids [85, 465, 552, 716–718, 803]. A variety of agents potentially capable of antagonizing oxidant injury in ARDS are also available including *N*-acetylcysteine (NAC) [78, 514, 1051] and antioxidant enzymes or related compounds [58, 188, 330, 331, 817, 908, 1073, 1139, 1141, 1172]. Details on several of these agents and their activities are provided in the following small type paragraphs.

Anti-Tumor Necrosis Factor Alpha (Anti-TNFα Antibodies). TNFα is an early and potent proinflammatory cytokine contributing to ALI and clinical ARDS (e.g., [26, 325, 427, 645, 674, 720, 751, 826, 1086]). The elevation of proinflammatory cytokines including TNFα in lavage, edema, or serum has been found to be a consistent predictor for an increased risk of developing ALI/ARDS or associated with poor outcomes in affected patients [26, 427, 718, 720, 826, 886]. The early release of TNFα can complicate its detection clinically. Peak plasma levels of TNFα were found to occur only 1 hour after IV administration of low dose endotoxin in human volunteers and were undetectable by 6–8 hr [734]. TNFα contributes to lung injury in animals with pulmonary infection, and anti-TNFα or TNFα receptor blockade has been shown to reduce the severity of lung injury in several animal models [325, 645, 1087, 1191]. When treated with monoclonal anti-TNF antibodies, patients with ARDS or sepsis have shown some acute benefits without significantly improved long term outcomes or reduced mortality [3, 4, 150]. Antibodies to TNFα have not yet been studied with exogenous surfactant or other agents in multimodal therapies for ARDS.

Anti-IL-8. IL-8 is a potent chemoattractant cytokine (chemokine) for neutrophils that has been widely studied as a marker for acute lung injury in high-risk patients (e.g., [63, 217, 333, 427, 608, 719, 720, 736, 737]). IL-8 levels are markedly elevated in pulmonary edema fluid from patients with ARDS compared to healthy volunteers or patients with hydrostatic edema [217, 608, 736, 737]. High IL-8 levels in lavage also correlate with increased mortality in ARDS patients [736] and with a high risk for development of ARDS [217]. IL-8 levels in lavage do not correlate with the persistence of ARDS [333], suggesting that it is more important in the pathogenesis of acute disease. IL-8 has been implicated in the early phase of ARDS through effects on neutrophil apoptosis [332]. Early treatment with anti-IL-8 antibodies has been found to reduce lung injury and mortality in animal models of acid aspiration [274] and endotoxemia [101, 1209], but antibodies to IL-8 have not been examined in research on multimodal interventions for ALI and ARDS.

Anti-CD40 Ligand (Anti-CD40L). CD40 is a 50 kDa receptor once thought to be expressed only on bone marrow–derived cells but now known also to be present on pulmonary fibroblasts [11, 12, 279, 976, 977]. Fibroblast CD40 serves as an activation structure for the synthesis of proinflammatory cytokines through interactions with CD40L, which is found on T lymphocytes and mast cells. A monoclonal anti-CD40L antibody termed MR1, which disrupts the CD40-CD40L interaction, has been shown to reduce the severity of hyperoxic lung injury and radiation-induced lung injury in mice [11, 12]. Intraperitoneal administration of MR1 into mice before or after the start of exposure to >95% oxygen reduced epithelial cell necrosis, edema, and influx of inflammatory cells. MR1 also substantially decreased the induction of cyclooxygenase-

2, a proinflammatory enzyme responsible for prostaglandin production [11]. Treatment with MR1 in radiation lung injury not only improved the severity of the acute phase of injury but also reduced fibrosis [12]. Anti-CD40L reagents for humans are in the testing phase by several pharmaceutical firms (e.g., for idiopathic thrombocytopenic purpura) but have not been utilized in clinical ARDS.

Pentoxifylline. This xanthine derivative is a phosphodiesterase inhibitor with multiple physiological effects (e.g., [229, 688, 1149, 1170] for review). As a therapeutic agent in chronic arterial occlusive disease and hypoperfusion syndromes, pentoxyfylline achieves many of its physiological benefits through vasodilation and other hemodynamic effects resulting in increased tissue oxygenation. Some of its major actions in ALI and sepsis appear to be anti-inflammatory and relate to its ability to raise cAMP levels [81, 251, 1031], inhibit free radical formation, and antagonize the production and actions of TNFα [48, 626, 1030, 1223, 1226]. Pentoxifylline has been shown to have beneficial effects in multiple animal models of ALI, shock, and sepsis [141, 271, 445, 505, 627, 652, 688, 994], as well as to reduce levels of TNFα and to improve cardiopulmonary function in patients with sepsis [38, 626, 1029, 1030, 1223]. The pharmacology of pentoxifylline is well characterized [1149], and it is safe for use in patients with ARDS [743]. However, pentoxifylline has not yet been evaluated for clinical efficacy in ARDS, particularly as part of multimodal interventions.

Corticosteroids. The anti-inflammatory effects of corticosteroids are well known, and systemic and topical use of these agents is found in a host of clinical settings. In ARDS, corticosteroids were initially studied during the early stages of disease and were found to be ineffective [75, 676]. There is growing evidence, however, that later prolonged treatment with corticosteroids can improve outcomes in patients with established (fibroproliferative phase) ARDS [85, 465, 552, 716-718, 803]. Corticosteroids would not be expected to be additive or useful in multimodal protocols with exogenous surfactant in the acute phases of ARDS. However, despite having a nontrivial toxicity with long-term use, they are potential candidates for targeting fibroproliferative pathology in combined-modality protocols that include this aspect of ARDS.

N-Acetylcysteine (NAC). NAC is an agonist for glutathione (GSH), an important pulmonary antioxidant. GSH has multiple biological actions ranging from protecting against reactive oxidants to participation in metabolic pathways including the synthesis of inflammatory mediators ([110, 113, 311, 429, 748] for review). Levels of GSH have been found to be reduced in lavage from patients with ALI/ARDS [816] and pulmonary fibrotic disorders [111]. NAC enhances GSH production by entering cells and providing cysteine, the rate-limiting amino acid in the synthesis of GSH [311, 748]. Administration of NAC can increase GSH and reduce the levels of proinflammatory cytokines in blood cells and plasma from patients with ARDS [625]. NAC also has direct antioxidant properties due to its SH group. Animal studies indicate that NAC treatment or pretreatment has a protective effect against hyperoxia, endotoxin administration, or glutathione blockers [77, 192, 623, 634, 940, 1132]. Treatment with NAC leads to improved respiratory function but not survival in adults with ARDS [78, 514, 1051]. Suter et al [1051] found that NAC-treated patients had an improved oxygenation ratio (PaO_2/FiO_2) and a reduction in the need for mechanical ventilation over the first 72 hr compared to a placebo group. A more recent double-blind, placebo-controlled study in 48 adults with ARDS found that treatment with NAC or with procysteine increased cardiac index and decreased the duration of lung injury [78]. NAC is ineffective in improving chronic lung disease in infants [84]. No adverse side effects have been reported from the use of NAC in ARDS, consistent with the broad experience using this drug as an antidote for acetaminophen overdose [311, 748].

    Superoxide Dismutase (SOD) and Catalase. SOD and catalase are important antioxidant enzymes with a broad biologic distribution ([116, 429, 1000] for review). There are three forms of SOD: cytoplasmic SOD (Cu,Zn-containing), mitochondrial SOD (Mn-containing), and extracellular SOD (Cu-containing) [116, 1000]. Levels of extracellular SOD are relatively high in lung and brain tissue [691, 813], and it has been speculated that this enzymatic form might be particularly important in pulmonary antioxidant defense [292]. However, all forms of SOD are active in catalyzing the conversion of superoxide anion to hydrogen peroxide. Hydrogen peroxide is subsequently reduced to water by catalase, a tetrameric, heme-containing protein [429, 1000]. A number of studies have suggested that antioxidant enzymes encapsulated in lipid vesicles (liposomes) or conjugated to polyethylene glycol to prolong half-life and increase entry into cells can protect against oxidant damage and mitigate the severity of ALI [58, 188, 817, 908, 1073, 1139, 1141, 1172]. The majority of studies have involved enzyme delivery by intratracheal instillation, but intraperitoneal injection and aerosolization have also been used. Several forms of recombinant human SOD are available for use in patients [188, 292, 908, 926]. A related agent is EUK-8, a synthetic low molecular weight compound having SOD-like and catalase-like activity [330, 331]. The use of antioxidant enzymes in multimodal therapies for ARDS has not been extensively studied, although the instillation of SOD with exogenous surfactant has been suggested as one approach to treating lung injury [187, 926]. The possibility that instilled exogenous surfactant might be inactivated by SOD has been reported but only at very high enzyme concentrations [373]. Other workers have found no adverse interaction between exogenous surfactant and human recombinant Cu,Zn-SOD [187].

XII. Specialized Modes of Ventilation for Treating ARDS

In addition to pharmacologic agents, other types of interventions can be utilized in multimodal regimens for treating ARDS. An important example of non-pharmacologic interventions is the use of specialized methods and strategies of mechanical ventilation. Mechanical ventilatory support is lifesaving in patients with respiratory failure, but it also contributes to lung injury and pulmonary complications during intensive care. A host of different techniques and strategies are utilized in an attempt to optimize gas exchange while minimizing lung injury and ventilator-associated complications in pediatric and adult patients with acute respiratory failure (e.g., [97, 195, 252, 377, 482, 833, 907, 982, 1075] for review). Examples of modes and types of ventilation include intermittent mandatory ventilation, assist/control volume- or pressure-limited ventilation with PEEP, low tidal volume ventilation, pressure-limited synchronous ventilation with PEEP, inverse ratio ventilation, high frequency jet ventilation, high frequency oscillatory ventilation, and liquid ventilation. A detailed discussion of methods and strategies of mechanical ventilation is outside the scope of coverage here. Liquid ventilation is described briefly below as an example of a ventilatory mode that could potentially enhance the physiological effects of agents such as INO and exogenous surfactant and facilitate the delivery and distribution of instilled surface active material.

Liquid Ventilation in ARDS-Related Respiratory Failure. Inert perfluorocarbon liquids with low surface tension and viscosity, and high oxygen and carbon dioxide solubility relative to water, have been used to improve oxygenation and reduce ventilatory pressures in several clinical settings ([195, 907, 982] for review) There are two primary forms of liquid ventilation: total or tidal liquid ventilation (TLV) and partial liquid ventilation (PLV), which is also called perfluorocarbon-assisted gas exchange (PAGE). TLV and PLV (PAGE) have been shown to improve compliance and gas exchange in surfactant-deficient animals [628, 981, 983, 1179, 1197] and also in animals with ALI and respiratory failure [438, 441, 442, 477, 770, 984, 1095–1097]. Several studies on the effects of liquid ventilation in patients with ARDS-related respiratory failure have also been done [439, 440, 629]. Clinical findings indicate that liquid ventilation can improve gas exchange and compliance in at least some patients with acute respiratory failure, although the efficacy and long term benefits of this therapy are not yet fully defined. Conceptually, the actions of liquid ventilation in facilitating alveolar recruitment and expansion could be additive with those of agents like INO in recruiting blood flow to ventilated areas (as described for INO and exogenous surfactant earlier). In addition, liquid ventilation could also potentially improve the delivery of exogenous surfactant to collapsed or edematous alveoli. Several studies in animal models of acute respiratory failure have reported that liquid ventilation can give additive benefits when utilized concurrently with INO [477, 1097, 1179] or with exogenous surfactant [630]. The possibility that other modes of ventilation such as high frequency jet ventilation could enhance the distribution of instilled exogenous surfactant has also been reported [185, 186].

XIII. Chapter Summary

Surfactant therapy for clinical ARDS and related acute respiratory failure is still under investigation. Basic biophysical and animal research indicates that this therapeutic approach has the potential to mitigate surfactant dysfunction in patients with ARDS. Surfactant therapy has been shown to be beneficial in term infants with meconium aspiration and to improve lung function in children with ARDS-related respiratory failure. Surfactant therapy for adults with ARDS has so far met with less success. Improvements in lung function have been reported following instillation of several exogenous surfactants in adult patients, but the response is much less pronounced than in infants with RDS and survival is unaffected. Moreover, studies with several synthetic exogenous surfactants have reported no benefits in adults with ARDS. In order to be effective in ARDS, exogenous surfactants must have optimal surface activity and inhibition resistance and must be delivered effectively to the alveoli. Surfactant therapy may also need to be used with other agents and interventions in a multimodal approach.

A variety of agents and interventions targeting inflammatory lung injury are available that could potentially be used in combined-modality therapies for ARDS. Examples noted in this chapter as complementary with exogenous surfactant include vasoactive agents such as INO, prostacyclin, and almitrine; anticytokine antibodies and receptor antagonists; pentoxifylline; antioxidants such as *N*-acetylcysteine, superoxide dismutase and catalase; and specialized modes of mechanical ventilation. Combined-modality therapies for ARDS require further study, including detailed basic science evaluations that examine mechanisms while defining the activities of specific combinations of agents in clinically relevant models. ARDS remains a severe clinical problem despite great advances in intensive care, and integrated basic and clinical research on multimodal interventions appears crucial to optimize its treatment.

14

Exogenous Lung Surfactants: Current and Future

I. Overview

The composition and activity of clinical exogenous surfactants currently in use to treat respiratory disease and injury are discussed in this chapter. Several exogenous surfactants used earlier in research on surfactant therapy in infants are also noted. In addition to detailing current surfactant preparations, new exogenous surfactants under active clinical investigation are described and potential additional approaches for future drug development are identified. Although clinical exogenous surfactants of high activity are currently available, most contain material from animal lungs. Active avenues of development for new clinical exogenous surfactants include the use of recombinant human sequence apoproteins or hydrophobic peptides combined with synthetic lipids. In addition, future synthetic surfactants could be formulated to contain novel components such as phospholipid analog molecules resistant to inflammation-induced enzymatic degradation. Some current clinical exogenous surfactants could also potentially be improved in activity by adding specific apoprotein or peptide components as discussed in this chapter.

II. Classification and Summary of Current Clinical Exogenous Surfactants

This section classifies clinical exogenous surfactants currently used worldwide to treat surfactant-related lung disease and gives an overview of their composition and characteristics. Also included are several additional exogenous surfactant preparations currently under active clinical evaluation for therapeutic use. Subsequent sections discuss the activity of clinical exogenous surfactants in basic

319

biophysical and animal research and present selected approaches for developing additional surfactant drugs or improving the activity of existing preparations. Additional details on the preparation, isolation, purification or synthesis of biological and non-biological surface active materials serving as components in clinical exogenous surfactants are given in Chapter 5.

Several classification systems are used for clinical exogenous lung surfactants. The categorization in Table 14-1 divides clinical exogenous surfactants into two broad groups: (I) preparations containing components of endogenous surfactant from animal lungs and (II) synthetic or recombinant surfactants that do not contain animal-derived material. Subdivisions within each of these categories then differentiate individual preparations based on their concept and composition. Category IA contains exogenous surfactants extracted in organic solvent from lavaged alveolar surfactant from animals (Alveofact, bLES, Infasurf). Categories IB and 1C contain exogenous surfactants processed from animal lung tissue with or without additional synthetic additives (Curosurf, Surfactant-TA, Survanta). Category IIA includes protein-free synthetic exgenous surfactants (Exosurf, ALEC), while categories IIB and IIC include preparations containing either synthetic peptides or recombinant human apoproteins. In addition to subdividing

Table 14-1 Clinical Exogenous Surfactants in Current Use Worldwide

I.   Preparations containing endogenous surfactant from animals
  A.   Organic solvent extracts of lavaged animal lung surfactant
       Alveofact (SF-RI 1)
       bLES
       Infasurf (CLSE)
  B.   Organic solvent extracts of processed animal lung tissue
       Curosurf
  C.   Supplemented organic solvent extracts of processed animal lung tissue
       Surfactant TA
       Survanta
II. Synthetic and recombinant exogenous lung surfactants
  A.   Protein-free synthetic surfactants
       ALEC
       Exosurf
  B.   Peptide-containing synthetic surfactants
       KL4[a]
  C.   Surfactants containing recombinant apoproteins
       Recombinant SP-C surfactant[a]

[a]Still under investigation for clinical efficacy. Manufacturers, composition, and characteristics of the tabulated clinical surfactant preparations are given in the text. Several additional exogenous surfactants such as human amniotic fluid surfactant have also been studied clinically but are not commercially available.

exogenous surfactants to discriminate their differences as in Table 14-1, the broader groupings themselves are used for convenience in the clinical literature. For example, the terms *natural surfactants* and *surfactant extracts* are commonly used to designate all the surfactants in category I since they are derived in whole or in part from endogenous surfactant extracted with organic solvent. A related broad term is *bovine surfactants*, which designates all surfactant preparations containing material from bovine lung lavage or lung tissue. Broad groupings of exogenous surfactants have obvious advantages for simplifying and summarizing clinical findings across multiple studies. At the same time, it is important to understand that the different surfactant preparations within these broader categories can vary substantially in their composition and activity. A brief summary of each of the clinical exogenous surfactants in Table 14-1 is given below.

ALEC [738, 746, 747, 1077, 1181]. ALEC (Britannia Pharmaceuticals, Redhill, Surrey, UK) is a protein-free synthetic exogenous surfactant composed of DPPC and egg PG in a 7:3 molar ratio. ALEC is an acronym for "artificial lung expanding compound," a name that also honors Dr. Alec Bangham, who was instrumental in developing this surfactant preparation [55, 747]. DPPC is the primary component responsible for surface tension lowering in ALEC, while egg PG is present to enhance adsorption and spreading [55]. ALEC is produced for clinical use as a fine, white powder that is mixed with sterile 0.15 M NaCl prior to delivery to patients. Mixing is at a temperature below 30°C in an attempt to produce a suspension with better spreading and adsorption after instillation.

Alveofact [60, 335–338]. Alveofact (SF-RI 1, Thomae GmbH, Biberach/ Riss, FGR) is a chloroform-methanol extract of surfactant obtained by bronchoalveolar lavage from intact cow lungs. Lung surfactant in lavage is pelleted by centrifugation, followed by extraction with chloroform:methanol to remove hydrophilic surfactant protein. The methodology for preparing Alveofact is similar to that used for Infasurf and bLES (see below), except for an added acetone extraction in the case of the latter preparation. Alveofact contains a weight distribution of approximately 99% phospholipids and neutral lipids, including 4% cholesterol plus 1% hydrophobic SP-B/SP-C [338]. Alveofact is formulated for clinical use as a sterile suspension in 0.15 M NaCl at a concentration of 45 mg/ml.

bLES [222, 223, 243, 1012]. bLES (bLES Biochemicals Inc, Ontario, Canada) is a chloroform:methanol extract of surfactant isolated by centrifugation from bronchoalveolar lavage of intact bovine lungs. bLES is an acronym for "bovine lipid extract surfactant." The general lavage, centrifugation, and chloroform:methanol extraction methods used to obtain bLES are similar to those for preparing Alveofact and Infasurf. An additional extraction with acetone is used to deplete cholesterol and other neutral lipids in the final preparation. This acetone extraction step was not included in the original form of this exogenous surfactant, which was called bovine lung surfactant extract (BLSE) and was used in clinical studies in infants in the early and mid-1980's by Enhorning, Possmayer, and colleagues [243, 1012, 1214]. After acetone extraction, bLES contains a weight distribution of approximately 98–99% phospholipid (79% of which is phosphatidylcholine) and 1% hydrophobic surfactant proteins. bLES is formulated for clinical use as a sterile suspension in 0.15 M NaCl and 1.5 mM $CaCl_2$ at a concentration of 25 mg phospholipid/ml.

Curosurf [82, 152, 153, 228, 777, 889, 915, 921, 1021]. Curosurf (Chiesi Farmaceutici, Parma, Italy) is a modified surfactant extract prepared from minced porcine lungs by washing, centrifugation at 1,000 and 3,000 × g, extraction with 2:1 chloroform:methanol, and liquid-gel affinity chromatography. The final phospholipid-rich fraction after chromatography contains approximately 99% polar lipids and 1% by weight of hydrophobic surfactant proteins. This fraction is sterilized by a high pressure filter system (0.45 and 0.2 μm filters), dried or lyophilized, and stored in vials prior to suspension in saline by sonication at a concentration of 80 mg/ml for clinical use. The distribution of phospholipids in Curosurf contains a higher percentage of tissue-derived phosphatidylethanolamine and sphingomyelin, and a lower percentage of PC, than found in lavaged lung surfactant. The phospholipid class distribution in Curosurf in weight percent has been reported to contain 65–75% PC [889, 915], plus a combined PE and SPH content of 12–22% [777, 889].

Exosurf [25, 94, 127, 157, 224, 248, 471, 480, 481, 662, 663, 810, 851, 975, 1119]. Exosurf (Glaxo-Wellcome, formerly Burroughs-Wellcome, Research Triangle Park, NC, USA) is a protein-free mixture of synthetic DPPC:hexadecanol:tyloxapol in a weight ratio of 1:0.111:0.075. DPPC is the primary surface tension lowering component of Exosurf, while the nonionic detergent tyloxapol and the C16:0 alcohol hexadecanol are present to enhance adsorption and spreading. Exosurf is formulated as a sterile, lyophilized powder in vacuum-sealed vials and is suspended in 8 ml of distilled water prior to clinical use. The 8 ml final sterile suspension contains 13.5 mg/ml DPPC, 1.5 mg/ml hexadecanol, and 1 mg/ml tyloxapol in 0.1M NaCl.

Infasurf [31, 89, 90, 190, 480, 481, 551, 553–556, 615, 990, 1185, 1186]. Infasurf (ONY, Inc, Amherst, NY, USA) is a chloroform:methanol extract of alveolar surfactant obtained by saline broncholalveolar lavage from intact calf lungs. Whole surfactant is pelleted by centrifugation at 10,000–12,000 × g prior to extraction with 2:1 chloroform methanol to give calf lung surfactant extract (CLSE). The preparation methods for CLSE are similar to those used for Alveofact and bLES (except for added acetone extraction in bLES as already noted). CLSE contains by weight about 93% phospholipid, 5% cholesterol and neutral lipids, and 1.5% protein that is a mixture of hydrophobic SP-B and SP-C. The weight distribution of phospholipid classes is approximately 83% PC, 6% PG, 4% (PI+PS), 3% PE, and 2% SPH (e.g., [383, 541, 555]). Desaturated PC accounts for slightly less than 50% of total phospholipid. The final Infasurf preparation for clinical use is formulated as a heat-sterilized suspension of CLSE in 0.15 M NaCl at a concentration of 35 mg phospholipid/ml.

KL4 [46, 142–145, 679, 730, 892, 1122, 1137]. KL4 is a synthetic exogenous surfactant composed of DPPC, palmitoyl-oleoyl phosphatidylglycerol (POPG), and palmitic acid combined with a 21 amino acid synthetic peptide containing repeating subunits with one lysine (K) and four leucine (L) residues. DPPC and POPG are present in a 3:1 ratio by weight, and palmitic acid has a content of 15% by weight relative to phospholipid. The hydrophobic peptide, which itself is sometimes called KL4, is present at 3% by weight relative to phospholipid. This hydrophobic peptide was designed to approximate the balance of hydrophobic and hydrophilic residues found in native SP-B [142, 143, 679]. KL4 surfactant is currently under clinical evaluation for use in infants. The general approach of forming synthetic exogenous surfactants by combining lipids with synthetic hydrophobic peptides is detailed in a subsequent section.

Recombinant SP-C surfactants [65, 184, 374, 375, 425, 497, 638]. Exogenous surfactants containing two forms of bacterially produced recombinant human SP-C are currently being evaluated for clinical use. The most widely studied of these preparations (Byk Gulden Pharmaceutical, Konstanz, Germany) contains a recombinant 34 amino acid analog of human SP-C that begins with Gly as N-terminal position 1 [184, 375, 497, 638]. Relative to the human sequence, positions 4 and 5 in the recombinant protein have Phe substituted for Cys and position 32 has Ile substituted for Met. This rSP-C protein is combined at 2% by weight with 7:3 (wt/wt) DPPC:POPG plus 5% by weight palmitic acid in the final surfactant preparation [184, 374, 375, 497, 638]. A second 34 amino acid form of recombinant human SP-C without substitutions relative to human sequence is also being investigated for use in clinical exogenous surfactants (Scios-Nova Inc, USA) [65, 425]. This latter recombinant SP-C is sometimes designated rSP-C(Cys)$_2$. The general approach of using recombinant human apoproteins as constituents in exogenous surfactants is discussed in more detail later.

Surfactant-TA [284, 288, 314, 595, 596, 883, 1062]. Surfactant-TA (Surfacten, Tokyo Tanabe, Tokyo, Japan) is a processed organic solvent extract of finely ground lung tissue that is supplemented with synthetic DPPC, palmitic acid, and tripalmitin. Ground lung tissue is initially treated with several differential centrifugation and flotation steps, extraction with ethyl acetate to reduce neutral lipids, and subsequent extraction of ethyl acetate–insoluble material with chloroform:methanol. This lung tissue extract is then supplemented with the above synthetic additives to give a final product containing approximately 84% phospholipids, 7% tripalmitin, 8% palmitic acid, and 1% protein [290]. The supplemented extract in chloroform is sterilized by high pressure filtration and lyophilized. Prior to instillation in patients, lyophilized Surfactant-TA is suspended in sterile 0.15 M NaCl at a concentration of 30 mg lipid/ml by hand swirling and extrusion through a small gauge needle.

Survanta [1, 90, 443, 469–471, 647, 975, 1017, 1021, 1119]. Survanta (Beractant, Abbott Laboratories, North Chicago, IL, USA, licensed from Tokyo Tanabe, Tokyo, Japan) is a processed, supplemented extract of bovine lung tissue prepared by similar methods and with identical synthetic additives to those used in Surfactant-TA. After supplementation with DPPC, palmitic acid, and tripalmitin, the total protein content of Survanta is slightly under 1% by weight. This protein contains only a small content of SP-B by ELISA measurement [403, 967]. The final form of Survanta for clinical use is produced as an autoclave-sterilized suspension in 0.15 M NaCl at a concentration of 25 mg phospholipid/ml.

A. Additional Exogenous Surfactants Studied Clinically in Premature Infants

In addition to the clinical exogenous surfactants above, several preparations have been investigated for therapeutic use in premature infants. *Human surfactant from amniotic fluid* has been utilized successfully in a number of clinical studies to treat or prevent RDS in premature infants (e.g., [225, 394, 396, 547, 622, 728, 729]). Amniotic fluid at term contains surfactant secreted from type II pneumocytes, along with additional nonsurfactant material derived from fetal tissue. It was recognized in the early 1980's that human surfactant could be processed from the amniotic fluid of normal term pregnancies by a combination of conventional and density gradient centrifugation and filtering under sterile technique [396, 950]. Human surfactant from amniotic fluid contains SP-A as well as hydrophobic lipids and proteins and cannot be heat sterilized after isolation without loss of activity. Human amniotic fluid surfactant was very important in the development of surfactant replacement therapy and was one of the initial exogenous surfactants used clinically in premature infants [396]. Amniotic fluid surfactant has subsequently been studied in both prophylactic and treatment protocols in premature infants in the United States and Europe [225, 394, 547, 622, 728, 729]. Due to considerations of source and processing, as well as the availability of other active exogenous surfactants, human surfactant from amniotic fluid is not produced commercially and is not currently in use outside a few locations including Finland. *Surfactant CK*, a chlorform-methanol extract of lavaged porcine lung surfactant, was one of the first exogenous surfactants used clinically. It was originally formulated as a sterile sonicated suspension in 0.15 M NaCl with added $CaCl_2$ and was shown to have beneficial effects on lung function in uncontrolled studies in premature infants in Japan in the early 1980's [588, 780]. The method of preparing Surfactant CK was later modified to include acetone precipitation and formulation in saline [587, 592, 914], but it has not been developed as a commercial pharmaceutical. *DPPC/HDL surfactant* was another exogenous surfactant preparation studied in early clinical research, but it was found to be ineffective in premature infants [385]. This surfactant was a mixture of synthetic DPPC with high density lipoprotein (HDL) isolated by ultracentrifugation of human serum. DPPC/HDL (10:1, w:w) was suspended by sonication in 5 ml of 0.15 M NaCl at a concentration of 6 mg phospholipid/ml and sterilized by irradiation prior to administration to infants [385].

III. Basic Science Comparisons of Activity in Clinical Exogenous Surfactants

Basic science research does not substitute for clinical evaluations, but it provides crucial complementary information in assessing mechanisms and functional details at a level impossible in patients. A number of randomized, multicenter trials in premature infants have compared the activity of clinical exogenous surfactants [89, 90, 127, 471, 480, 481, 921, 975, 1021, 1119] (Chapter 12). Resolving power in these clinical trials, however, is limited by the many pathophysiological variables other than lung surfactant activity that also contribute to patient outcome. In addition, modern mechanical ventilation technology and neonatal intensive care are highly beneficial in premature infants even in the absence of surfactant therapy. Significant numbers of patients are required to demonstrate differences in long term outcomes even in placebo-controlled trials of exogenous surfactant therapy, and outcome differences are necessarily smaller in comparison trials where all infants receive a surfactant drug with some beneficial effects (Chapter 12). Basic science research on the activity of exogenous surfactants includes compositional, biophysical, and physiological assessments. Basic research interpretations are most powerful when data on all these aspects of lung surfactants and their behavior are integrated together. Activity differences that are consistently displayed in correlated biochemical, biophysical, and animal research are indispensible in assessing and comparing exogenous surfactants. Basic research has identified substantial differences in the surface and physiological activity of clinical exogenous surfactants as illustrated in the following examples.

A. Examples of Biophysical Activity Differences Between Clinical Exogenous Surfactants

The differing source and composition of clinical exogenous surfactants summarized in the preceding section give rise to associated differences in surface activity. The surface properties of any surfactant material depend on the composition of the interfacial film it forms. The majority of biophysical studies suggest that current exogenous surfactants with the closest compositional analogy to lavaged alveolar surfactant tend to have the highest surface activity. The adsorption of several clinical exogenous surfactants from the work of Seeger et al [967] is shown in Figure 14-1. The surface tension reached after 0.2 minute of adsorption in a static bubble surfactometer is plotted against subphase surfactant concentration at 37°C. CLSE (Infasurf) and Alveofact were able to reach the lowest adsorption surface tensions at the smallest surfactant concentrations among the preparations studied. Figure 14-2 gives complete surface tension–time adsorption isotherms measured for Exosurf, Survanta, and CLSE in a study by Hall et al [382]. CLSE adsorbed almost identically to lavaged whole surfactant, while Survanta adsorbed to a higher equilibrium surface tension and Exosurf reached only a surface tension of 38 mN/m after 20 minutes of adsorption.

Clinical exogenous surfactants also vary in their ability to lower surface tension during dynamic compression and to resist biophysical inactivation by

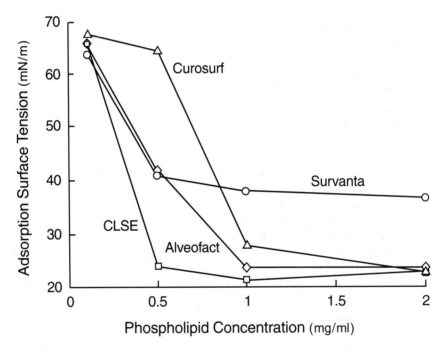

Figure 14-1   Adsorption surface tension as a function of phospholipid concentration in several clinical exogenous surfactants. Clinical exogenous surfactants were allowed to adsorb for 0.2 minute at different concentrations. Measurements were done in a bubble surfactometer in the static mode and contain a diffusion resistance. (Data are means redrawn from Ref. 967.)

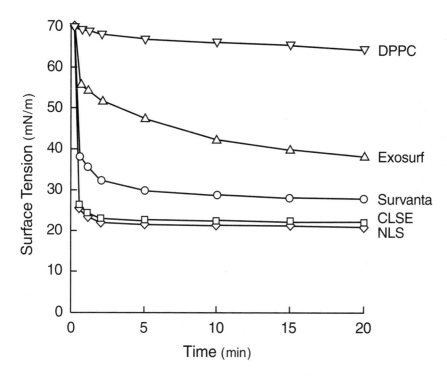

Figure 14-2 Adsorption isotherms of clinical exogenous surfactants. Surface tension–time adsorption isotherms at 37°C are shown for Exosurf, Survanta, and CLSE compared to DPPC and whole surfactant lavaged from calves (NLS). Surfactants were added at time zero to a stirred subphase of 0.15 M NaCl, 5 mM $CaCl_2$. Final surfactant concentration was 0.063 mg lipid/ml. (Data are means redrawn from Ref. 382.)

compounds such as blood proteins. Figure 14-3 shows data from Seeger et al [967] on the surface tension lowering ability of different exogenous surfactant preparations during dynamic cycling at physiologically relevant rate, area compression, temperature, and humidity. Suspensions of CLSE, Alveofact, and Curosurf reached minimum surface tensions <1 mN/m after 5 minutes of cycling, while Survanta reached a minimum of 4 mN/m (Figure 14-3). Similar dynamic activity studies with the pulsating bubble surfactometer have shown that Exosurf reaches minimum surface tensions near 30 mN/m even during prolonged cycling at a high concentration of 10 mg lipid/ml [382]. CLSE and Alveofact have also been found to resist inhibition by blood proteins such as fibrinogen or albumin more effectively than Survanta and Curosurf in dynamic surface activity studies [382, 967] (Figure 14-4). The relative sensitivity of exogenous surfactants to inactivation is particularly important for surfactant therapy in patients with ARDS-related lung injury (Chapter 13).

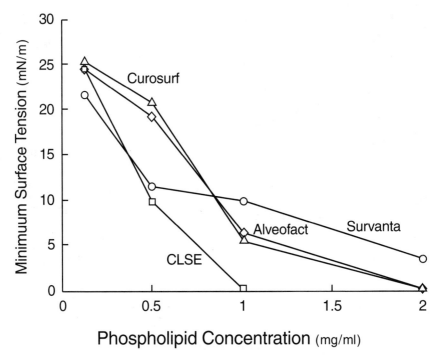

Phospholipid Concentration (mg/ml)

Figure 14-3 Dynamic surface tension lowering behavior of different clinical exogenous surfactants. The minimum surface tension after 5 minutes pulsation in a bubble surfactometer is plotted as a function of surfactant concentration (37°C, 20 cycles/min, 50% area compression). (Data are means redrawn from Ref. 967.)

B. Examples of Physiological Activity Differences Between Exogenous Surfactants in Animals

Differences in physiological as well as biophysical activity are found among clinical exogenous surfactants in basic research. The majority of available animal research indicates that exogenous surfactants with the greatest surface activity and inhibition resistance give better improvements in mechanics and function in surfactant-deficient or injured lungs. The activity of exogenous surfactants in animal models is discussed in detail in Chapter 11. One example of an animal model used in assessing the direct physiological activity of clinical exogenous surfactants involves studies of P-V deflation mechanics under standardized conditions in lavaged excised rat lungs (e.g., [74, 382, 452, 784]). Details of this model, which is FDA certified for the quality control manufacture of several clinical exogenous surfactants in the United States, are given in Chapter 11. The effects of Exosurf, Survanta, and CLSE (Infasurf) in improving P-V mechan-

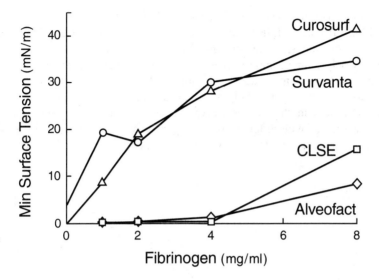

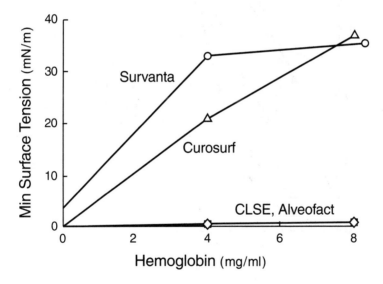

Figure 14-4 Resistance of exogenous surfactants to inhibition by fibrinogen and hemoglobin. Minimum surface tension after 5 minutes of pulsation in a bubble surfactometer is plotted against the concentration of the inhibitory proteins fibrinogen and hemoglobin. CLSE and Alveofact maintained an ability to reach low surface tension in the presence of higher inhibitor concentrations. Experiments were at 37°C, 20 cycles/min, and 50% area compression. Surfactant phospholipid concentration was uniform at 2 mg/ml. (Redrawn from Ref. 967.)

ics after instillation into lavaged surfactant-deficient rat lungs are shown in Figure 14-5 [382]. Instilled CLSE almost normalizes the P-V mechanics of the surfactant- deficient lungs, while Survanta has an intermediate effect and Exosurf produces only a small improvement. Research in several other animal models also indicates the same order of physiological activity for these three exogenous surfactants [176, 501, 740]. These physiological activity findings correlate with differences in the surface active properties of the surfactants described above, as well as with differences in their composition, particularly in terms of apoprotein content. Due to its origin from lavaged surfactant subjected only to

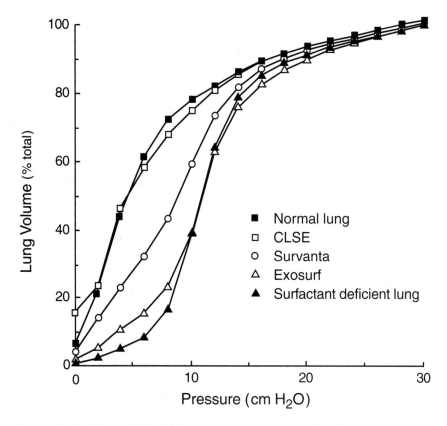

Figure 14-5 Effects of CLSE, Survanta, and Exosurf on P-V deflation mechanics in surfactant-deficient excised rat lungs. Exogenous surfactants were instilled into excised rat lungs depleted in endogenous surfactant by multiple lavage. Normal curve is immediately post-excision, and subsequent surfactant-deficient curve is after lavage. Surfactant dose was 20 mg phospholipid/2.5 ml saline for Infasurf and Survanta and 37.5 mg lipid/2.5 ml saline for Exosurf. Measurements were at 37°C. (Data are means redrawn from Ref. 382.)

chloroform:methanol extraction, CLSE (Infasurf) contains amounts and ratios of surfactant lipids and hydrophobic apoproteins near endogenous levels. In contrast, Exosurf contains neither of the two highly active hydrophobic apoproteins and Survanta has a very low content of SP-B [403, 967]. Survanta also has a reduced ratio of protein to phospholipid as a result of its synthetic additives [740]. The addition of hydrophobic surfactant proteins or related synthetic peptides to Exosurf and Survanta can improve their physiological activity in animals as described in a later section [382, 740, 1140].

IV. Overview of Approaches for New Surfactant Development or Increasing the Activity of Existing Preparations

There are various approaches for developing new clinical exogenous surfactants and only selected strategies are covered here. These include the modification of existing exogenous surfactants with specific additives to enhance activity, as well as the development of completely new preparations. Approaches under active investigation for new exogenous surfactants include combining synthetic phospholipids and fatty acids with recombinant human surfactant proteins or with synthetic apoprotein-related peptides (e.g., as in KL4 and recombinant SP-C surfactants noted earlier). Exogenous surfactants can also include novel synthetic molecules such as specialized phospholipid analogs, or they can utilize specific physical processing to enhance surface activity and inhibition resistance. Exoge-

---

Examples of Approaches for New Exogenous Lung Surfactants

New surfactants containing human recombinant apoproteins combined with synthetic phospholipids or other components.

New surfactants containing synthetic apoprotein-based hydrophobic peptides, or less specific peptides, combined with synthetic phospholipids.

Adding recombinant apoproteins or synthetic peptides to current clinical exogenous surfactants to increase activity.

New surfactants containing improved or novel phospholipid components such as active phospholipase-resistant synthetic phospholipid analogs.

nous surfactants manufactured *in vitro* have potential advantages in production, formulation, and quality control over animal-derived preparations. At the same time, they need to be assessed in light of the benefits and risks of current drugs. The high activity, low toxicity, and close compositional analogy to native surfactant of several existing clinical exogenous surfactants provide significant challenges for new drug development. Details on several approaches for developing new exogenous surfactants are given in subsequent sections.

V. Exogenous Surfactants Containing Recombinant Surfactant Apoproteins

The formulation of exogenous surfactants containing phospholipids combined with human sequence surfactant proteins produced by recombinant DNA technology is an active area of drug development. SP-A [367, 1130, 1131], SP-B [1208], SP-C [425, 947, 1113], and SP-D [166] can all be produced by recombinant techniques for use in exogenous surfactants (Chapter 5). Recombinant SP-B and SP-C, or related synthetic peptides described in the next section, have been emphasized in research because of the known high surface activity of the hydrophobic proteins in endogenous surfactant. The general validity of the approach of combining recombinant hydrophobic surfactant proteins with lipids to form active exogenous surfactants is apparent from basic research. The ability of SP-B and SP-C purified from endogenous surfactant to substantially increase the surface activity of phospholipids is well documented in biophysical studies (Chapter 8). Recombinant forms of SP-C have thus far received the most study as components in exogenous surfactants for therapeutic use [65, 184, 374, 375, 425, 638, 967, 968, 973]. The high surface activity of several synthetic surfactants containing recombinant SP-C, or recombinant SP-C plus purified animal SP-B, has been demonstrated *in vitro* [967, 968, 973]. Synthetic exogenous surfactants containing phospholipids plus recombinant SP-C have also been shown to improve lung function and mechanics in premature and adult animals [65, 184, 374, 375, 425, 638]. The uptake of recombinant SP-C has also been studied in several animal species [497, 856]. Clinical evaluations of several recombinant SP-C preparations are currently in progress as noted earlier.

One relevant molecular modification for recombinant SP-C (and synthetic SP-C peptides below) involves cysteine-linked acylation, which is present at residues 5 and 6 in native human SP-C. As discussed in Chapter 8, palmitoylation in native SP-C affects its structural behavior and may influence surface active interactions with phospholipids [37, 535, 536, 841, 973, 1112, 1142]. Acylation increases the percentage of α-helical secondary structure in SP-C [535, 536, 1112, 1142], including effects on the region of sequence containing its two positively charged amino acids. Palmitic chains in SP-C also could affect surface activity by direct interactions with fatty chains in DPPC and other phospholipids. The quantitative importance of acylation in the surface active function of SP-C is still somewhat uncertain. Even if acylation contributes significantly to surface activity in native SP-C, it may be less important in recombinant or synthetic forms of this protein, particularly those containing amino acid substitutions. Acylation does influence the structure and physical properties of some forms of recombinant SP-C [160] but does not significantly increase the physiological activity of synthetic surfactants containing rSP-C(Cys)$_2$ in premature rabbits [425]. Synthetic surfactants containing phospholipids plus 2–10% nonacylated recombinant SP-C have been shown to have excellent overall surface activity, although they are less effective in resisting plasma protein– induced inhibition than surfactant preparations containing native SP-B [968, 973]. Palmitoylated recombinant SP-C has been produced in the baculovirus expression system [1113] but has not yet been studied for activity in comparison to nonacylated SP-C in exogenous surfactants.

VI. Synthetic Exogenous Surfactants Containing Synthetic Peptides

Synthetic peptides with varying degrees of homology to lung surfactant apo-proteins have been studied for their biophysical behavior and surface active interactions with phospholipids in exogenous surfactants.[1] Synthetic peptides of almost any primary amino acid sequence of reasonable length can be prepared by solid phase synthesis and related techniques (Chapter 5). For applications involving exogenous lung surfactants, attention has focused largely on hydrophobic or amphipathic peptides. A host of synthetic peptides related to SP-B and SP-C, including full-length and regional human sequence peptides, have been studied in research applications (Tables 14-2, 14-3). Exogenous surfactants containing synthetic SP-B peptides as in Table 14-2 have been of particular interest because of the greater effectiveness of native SP-B relative to SP-C in increasing phospholipid surface activity (Chapter 8). Exogenous surfactants containing synthetic SP-C peptides or mixtures of SP-B/SP-C peptides have also been extensively investigated, and some studies involving regional SP-A peptides have been done. Less specific synthetic hydrophobic peptides have also been investigated as constituents in exogenous surfactants, with KL4 being one prominent example described earlier. KL4 has been shown to improve lung function in premature animals [730, 892] and in initial clinical studies in premature infants [144]. Amphipathic α-helical peptides based on sequences such as Leu-Leu-Glu-Lys-Leu-Leu-Glu-Lys(Trp)-Leu-Lys, as well as nonspecific hydrophobic peptides,

---

1. For example, Refs. 18, 19, 36, 106, 139, 142–144, 160, 249, 334, 425, 532, 535, 653, 654, 656, 666, 667, 679, 711–713, 730, 775, 895, 939, 1064, 1118, 1122, 1137, 1138, 1140, 1154, 1155, 1225.

Table 14-2 Examples of Human Sequence SP-B
Synthetic Peptides Studied in Research and
Development

SP-B$_{1-7}$[a]	SP-B$_{11-25}$	SP-B$_{40-60}$
SP-B$_{1-9}$	SP-B$_{14-25}$	SP-B$_{41-55}$
SP-B$_{1-15}$	SP-B$_{15-25}$	SP-B$_{49-66}$
SP-B$_{1-20}$	SP-B$_{20-60}$	SP-B$_{51-65}$
SP-B$_{1-25}$	SP-B$_{21-35}$	SP-B$_{53-78}$
SP-B$_{1-60}$	SP-B$_{27-78}$	SP-B$_{59-80}$
SP-B$_{1-78}$	SP-B$_{31-45}$	SP-B$_{61-75}$
SP-B$_{8-25}$	SP-B$_{35-46}$	SP-B$_{71-81}$

[a]Amino acid (AA) numbering is from the N-terminus based on
the human SP-B sequence in Chapter 8 (Table 8-6). Hydrophobic peptides based on nonhuman SP-B sequences or incorporating specific AA substitutions are also available (e.g., [18, 106,
249]). Also studied in research are SP-C peptides (Table 14-3)
and less specific synthetic peptides.
Source: Refs. 18, 36, 334, 895, 939, 1154, 1155.

have also been studied in exogenous surfactants [711, 713, 1118, 1225]. Current
research demonstrates that exogenous surfactants containing lipids and full-length
or regional SP-B/C peptides or less specific peptides can have significant surface
and physiological activity. However, activity in general is not as high as found in
comparable mixtures containing purified native apoproteins even when full-
length synthetic SP-B and SP-C peptides are used ([462, 529, 530, 532] for
review). The development and optimization of peptide-based synthetic exogenous
surfactants are subjects of ongoing basic research.

An important and poorly understood factor in the activity of peptide-based
exogenous surfactants involves secondary and higher structure and its impact on
peptide surface activity (e.g., [462, 530, 532, 711]). The functional importance of
structural modificatons such as acylation in SP-C peptides is also not well defined as
noted above. The structure and conformation of proteins determine their interactions
with other molecules. The conformation of full-length and regional SP-B and SP-C
peptides synthesized by solid state chemistry can be quite different from that of the
native apoproteins. Surfactant apoproteins are all derived from larger precursors *in vivo*,
and post-translational processing and intracellular packaging mediate their structural
associations with lipids. In contrast, peptides in exogenous surfactants are typically
synthesized in a lipid-free environment and mixed with phospholipids *in vitro*. The
secondary and tertiary structure of these peptides is solvent dependent and also varies
with the amount and character of associated lipid. Specific core lipidation during

peptide synthesis may be required to maintain desired structural behavior and aid subsequent combination with synthetic phospholipids. The extreme hydrophobicity of many synthetic peptides further complicates their handling and characterization. Information is still incomplete about specific conformational features needed to maximize the activity of peptides in exogenous surfactants. Although α-helical regions appear to influence some surface active behavior in native SP-B and SP-C (Chapter 8), the precise balance of structural domains, amphipathicity, acylation, aggregation, and oligomerization that optimizes interactions between peptides and phospholipid bilayers and films is not yet known. Possible synergy between SP-A, SP-B, and SP-C peptides in exogenous surfactants has also not been studied in detail.

VII. Addition of Recombinant Apoproteins or Synthetic Peptides to Existing Exogenous Surfactants

Surfactant apoproteins are highly interactive with phospholipids in bilayers and films and significantly enhance adsorption and dynamic film behavior in endogenous surfactant (Chapter 8). Recombinant apoproteins or related synthetic peptides can be used not only as components in new exogenous surfactants but also as additives to improve the surface activity or inhibition resistance of existing preparations. Several examples illustrating this approach are given below.

Table 14-3 Examples of SP-C Synthetic Hydrophobic Peptides Used in Exogenous Surfactant Research

Full-length SP-C synthetic peptides	*Regional SP-C peptides*
e.g., SP-C$_{1-35}$, human	e.g., SP-C$_{1-10}$, human
SP-C$_{2-35}$, human	SP-C$_{5-35}$, human
SP-C$_{1-35}$, porcine	SP-C$_{7-35}$, human
SP-C$_{1-35}$, with covalent C16:0	SP-C$_{1-12}$, porcine
SP-C$_{1-35}$, with covalent C2–C20	SP-C$_{1-17}$, porcine
	SP-C$_{1-21}$, porcine
	SP-C$_{1-31}$, canine
Modified full-length SP-C peptides	*Modified regional SP-C peptides*
e.g., SP-C with specific amino acid substitutions or incorporation of helical sequences such as from bacteriorhodopsin	e.g., Regional SP-C peptides with specific AA substitutions

*Amino acid numbering is from the N-terminus based on the sequence of human SP-C in Chapter 8 (Table 8-6). Nonhuman sequences are closely homologous.
Source: Refs. 18, 19, 139, 425, 535, 932, 1137, 1140.

A. Adding SP-A to Improve Activity in Exogenous Surfactants

Both recombinant human sequence SP-A [367, 1130, 1131] and SP-A peptides [712, 1138] are available for possible use in exogenous surfactants. SP-A is known to improve several aspects of phospholipid surface activity including adsorption and dynamic film behavior [226, 421, 422, 576, 894, 927, 960, 1070, 1215], and it is synergistic with SP-B in forming tubular myelin [1053, 1184]. Addition of purified SP-A has also been shown to improve the ability of organic solvent surfactant extracts containing hydrophobic apoproteins to resist inactivation by plasma proteins [149, 1116, 1220]. Although the addition of recombinant SP-A or SP-A peptides to exogenous surfactants has potential benefits particularly for increasing inhibition resistance, it would also increase costs and has some associated complications. Aside from economic considerations, SP-A is much more sensitive than the hydrophobic apoproteins to environmental variables such as heating, which is widely used in sterilizing exogenous surfactant suspensions. This problem could be less important for regional SP-A peptides containing only lipid-associated amphipathic domains rather than the whole apoprotein [712, 1138]. The presence of SP-A in exogenous surfactants could also affect surfactant uptake and secretion by type II cells since this apoprotein plays a regulatory role in these processes (Chapter 6). A decreased alveolar half-life of instilled exogenous surfactant because of more rapid uptake into type II cells could reduce its duration of effect. As opposed to animal-derived SP-A, the antigenicity of recombinant human SP-A or synthetic human sequence SP-A peptides in exogenous surfactants is unlikely to be a practical concern.

B. Example of Adding SP-B/C to Exosurf to Improve Activity

As shown earlier, Exosurf has relatively low surface activity and physiological effects on P-V mechanics in surfactant-deficient lungs compared to several apoprotein-containing exogenous surfactants. If hydrophobic surfactant apoproteins purified from calf lung surfactant are combined with Exosurf *in vitro* at a content of 1% by weight, overall surface activity is greatly improved. Mixtures of Exosurf + 1% SP-B/C are able to reach minimum surface tensions <1 mN/m in pulsating bubble experiments [382]. Mixtures of Exosurf + 1% SP-B/C also give a much more substantial improvement in P-V deflation mechanics when instilled into surfactant-deficient excised lungs relative to Exosurf alone (Figure 14-6) [382]. These results indicate that tyloxapol and hexadecanol in Exosurf are not as effective as SP-B/C in enhancing surface and physiological activity in this exogenous surfactant. They also support the possibility that Exosurf could be increased in activity in the lungs of some premature infants by combination with small amounts of endogenous apoprotein. In a surfactant development sense, supplementation of Exosurf with SP-B/C recombinant proteins or synthetic

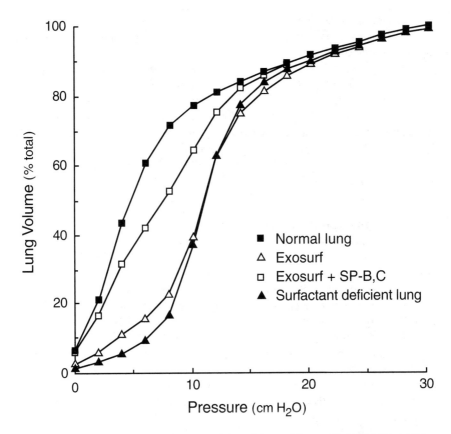

Figure 14-6 Improved physiological activity of Exosurf from added SP-B/C. Addition of purified bovine SP-B/C (1% by weight) to Exosurf gave a substantial improvement in its activity in restoring surfactant-deficient P-V deflation mechanics when instilled into lavaged, excised rat lungs at 37°C. Surfactant dose was 37.5 mg lipid/2.5 ml saline. (Data are means redrawn from Ref. 382.)

peptides in a modified exogenous surfactant could give a preparation with enhanced clinical efficacy.

C. Example of Adding SP-B to Survanta to Improve Activity

ELISA measurements indicate that the SP-B content of Survanta is very small [403, 967]. The effects of adding supplemental SP-B to Survanta were studied in 27-day-gestation rabbit fetuses by Mizuno et al [740]. Infasurf and whole sheep surfactant, which contain much larger amounts of SP-B than Survanta, were used in activity comparisons. Instillation of Infasurf or Survanta to 27-day gestation fetal rabbits improved P-V mechanics compared to untreated controls, but Infasurf and whole sheep surfactant had a significantly larger beneficial effect (Figure 14-7). Addition of 2.0% by weight of SP-B to Survanta significantly improved its activity in fetal rabbit lungs, giving activity equivalent to that of natural sheep surfactant (Figure 14-7, bottom). Addition of 0.5% by weight SP-B to Survanta did not improve activity (data not shown) [740]. These results indicate that SP-B in Survanta is functionally deficient and that activity could be improved by adding supplemental SP-B in a modified exogenous surfactant. Findings consistent with this have also been reported by Walther et al [1140], who showed that the activity of Survanta in a rat model of surfactant deficiency could be improved by the addition of synthetic SP-B peptides.

VIII. Phospholipid Components in Synthetic Exogenous Surfactants

Although protein/peptide constituents in exogenous lung surfactants are highly important in activity, phospholipid constituents should not be ignored. The majority of synthetic exogenous surfactants investigated to date have contents of DPPC that significantly exceed those found in endogenous lung surfactant. DPPC is the most prevalent phospholipid component in endogenous lung surfactant, but it accounts for only about one third of total surfactant phospholipid (Chapter 8). Non-DPPC phospholipids are important functionally in endogenous surfactant, and this is true by extension for exogenous surfactants as well. More comprehensive and detailed studies of the activity of phospholipids other than DPPC in synthetic exogenous surfactants may be very useful in optimizing activity. A variety of phosphatidylcholines other than DPPC are found in endogenous surfactant, including both disaturated and unsaturated molecular species (Chapter 8). Mixed chain disaturated compounds like palmitoyl-myristoyl-PC and myristoyl-palmitoyl-PC have gel-to- liquid-crystal transitions slightly below body temperature [123, 979] and are just two examples of phospholipids that might be included in synthetic exogenous surfactants. The headgroup region of phospholipids also affects their biophysical behavior and surface active interactions. In addition to zwitterionic phospholipids, anionic phospholipids may have specific molecular interactions with positively charged peptides or proteins in synthetic surfactants. Current synthetic exogenous surfactants utilize several anionic phospholipid constituents including POPG and egg-PG. However, the activity of PG compounds such as dipalmitoyl PG, which has fully miscible chains for interactions with DPPC, has not been examined in detail. A large number of additional synthetic glycerophospholipids are also available for potential use as components in exogenous surfactants.

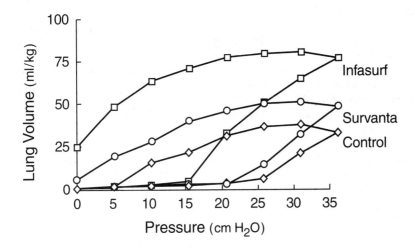

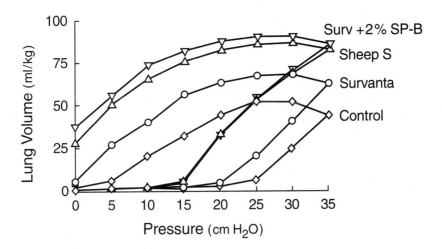

Figure 14-7 Effect of supplementation of Survanta with SP-B on physiological activity in 27-day gestation rabbit fetuses. (top) Rabbit fetuses treated with Survanta, Infasurf, and untreated controls; (bottom) rabbit fetuses treated with Survanta + SP-B (2% by wt), natural surfactant from adult sheep (Sheep S), or untreated controls. Surfactants were all instilled at a dose of 100 mg/kg and static P-V deflation mechanics were measured after 15 minutes of mechanical ventilation. (Data are means redrawn from Ref. 740.)

The miscibility of phospholipids is an important factor for their activity in synthetic exogenous surfactants. Phospholipids interact most readily if they form miscible or partially miscible surface films. The miscibility of phospholipid compounds is significantly influenced by the similarity of their fatty chains in length and saturation [22, 301, 1007] (Chapters 2, 3). Phospholipids with saturated chains of equal length, for example, tend to form miscible, tightly packed films. Headgroup characteristics also affect miscibility and film behavior, both through effects on packing and as a result of intermolecular interactions involving electrostatic forces or hydrogen bonding. Combining zwitterionic DPPC with one or more unsaturated anionic PG compounds is common in synthetic exogenous surfactants, but a variety of other possibilities exist to contribute additional molecular interactions. The balance of fluid and rigid phospholipids in synthetic exogenous surfactants is potentially just as important to function as in endogenous surfactant (Chapter 8). Saturated phospholipids in general generate better surface tension lowering than fluid phospholipids, while the latter improve spreading and adsorption. Saturated compounds such as DPPC and DPPG, for example, incorporate zwitterionic-anionic headgroup interactions with a high degree of chain miscibility and surface tension lowering ability. Interactions involving fluid phospholipids in synthetic surfactants can be generated by monoenoic unsaturated PC and PG compounds containing C16:0-C16:1, C16:0-C18:1, or related chains or by phospholipids with higher degrees of unsaturation. An alternative for fluid, liquid crystal phase phospholipids at body temperature is the use of compounds containing shorter saturated chains such as C14:0-C16:0 as noted above. Novel phospholipid-like synthetic compounds with specific phase and physical behavior could also be utilized in synthetic exogenous surfactants as described in the following section.

IX. Novel Phospholipase-Resistant Phospholipid and Phosphonolipid Analogs in Synthetic Exogenous Surfactants

In addition to glycerophospholipids as in endogenous surfactant, synthetic surfactants can contain novel phospholipid analogs designed with specific physicochemical properties. Examples of relevant analogs are diether phospholipids like dihexadecyl-PC (DHPC), as well as a series of novel C16:0 diether phosphonolipids with different N-headgroups synthesized by Turcotte, Notter, and co-workers (Chapter 5).[2] The structure of a diether phosphonate analog of DPPC, which has been designated DEPN-8 in racemic form, is shown in Figure 14-8. Several diether phospholipids and phosphonolipids including DHPC and DEPN-8 have been found to have significantly better adsorption and film respreading than DPPC, while retaining the ability to reduce surface tension to <1 mN/m under dynamic cycling [657, 658, 1092, 1093] (Table 14-4, Figure 14-9). The change from ester to ether linkages in these compounds increases their gel to liquid

---

2. For example, Refs. 207, 566, 651, 657–659, 673, 1006, 1091–1093.

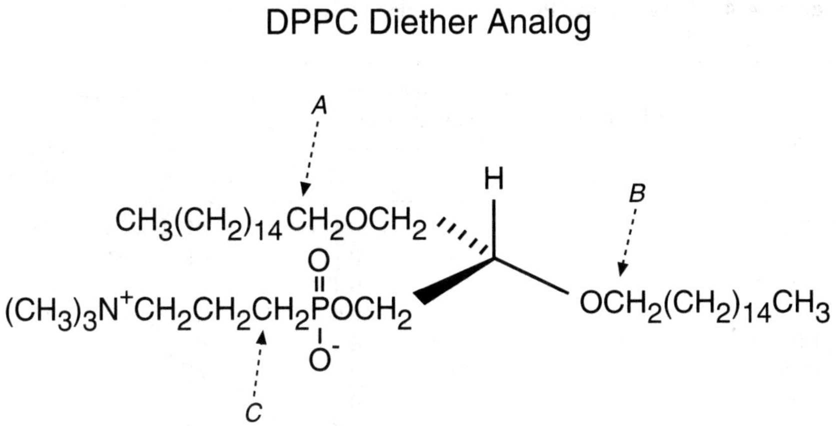

DPPC Diether Analog

Figure 14-8 Molecular structure of a diether phosphonolipid analog of DPPC. The analog molecule has saturated C16 chains like DPPC but ether rather than ester linkages with the glycerol backbone (sites A and B). A methylene group is also substituted for oxygen in the headgroup phosphate (site C). Analogs can be synthesized with varying absolute conformation (the molecule shown has an S conformation opposite to the R conformation of pulmonary DPPC). The racemic (R)(S) form of this compound has been designated as DEPN-8 in published studies (see text). Structures of additional phospholipid analog compounds are shown in Chapter 5. (Redrawn from Ref. 1093.)

crystal transition temperatures relative to glycerophospholipids [566, 659, 673, 1006, 1093]. However, the greater hydrophobicity and more flexible nature of ether vs ester bonds facilitate fatty chain interactions without a steric penalty from increased packing density, correlating with improved film respreading. The improved spreading and adsorption of diether phospholipids and phosphonolipids may also be related to the fact that at least some of these compounds are able to form interdigitated rather normal opposed bilayers [566, 1006].

An interesting feature of diether phospholipids and phosphonolipids as components in synthetic exogenous surfactants is that they lack sites for cleavage by phospholipases A_1 and A_2. Phosphonolipids also lack a cleavage site for phospholipase D and, due to chiral and/or steric considerations, may be more resistant to the action of phospholipase C [651]. The ability to resist phospholipase degradation could make analog-containing surfactants attractive for use in treating inflammatory lung injury. Resistance to degradation would not only maintain surface activity but also decrease the production of toxic, inhibitory byproducts such as lyso-PC and fluid free fatty acids. The molecular characteristics of diether analogs could also potentially be detrimental in terms of altered metabolic interactions with type II cells or other cells in the lungs. However, it

Table 14-4 Surface Active Properties of Selected Synthetic Diether Phospholipids
and Phosphonolipids

Surfactant compound	Adsorption surface tension	Respreading ratio for cycles		Minimum surface tension	
		2/1	7/1	Wilhelmy balance	Pulsating bubble
DPPC	68	0.37	0.11	<1	32
DHPC	65	0.84	0.14	<1	<1
DEPN-8	58	0.66	0.49	<1	<1
DEPN-9	46	0.81	0.66	<1	<1
DEPN-10	60	0.67	0.22	<1	4
DEPN-11	59	0.83	0.54	<1	2
DEPN-12	56	0.83	0.79	4	<1
DEPN-13	44	0.69	0.61	7	2

Abbreviations: DEPN, diether phosphonolipids (see Figure 5-7 for molecular structures); DHPC, dihexadecyl-PC; DPPC, dipalmitoyl-PC. Adsorption surface tensions (mN/m) were measured after 20 minutes in a dish with a stirred subphase (37°C, surfactant concentration 0.063 mg/ml). Minimum surface tensions from the bubble apparatus were after 20 minutes pulsation (37°C, 20 cycles/min, surfactant concentration 1 mg/ml). Minimum surface tensions and respreading ratios from the Wilhelmy balance were for films spread to 15 Å^2/molecule at 23°C (compression ratio 4.34:1, rate 5 min/cycle). A collapse plateau ratio of 1 is perfect respreading and 0 is no respreading (see Chapter 2).
Source: Data from Refs. 657, 658, 1092.

has been shown that diether phospholipids are able to enter surfactant recycling pathways in type II cells [508, 899–901], and they have not been found to have short-term pulmonary toxicity in animals [207]. Synthetic exogenous surfactants containing purified sheep SP-B/C combined with DHPC or with DEPnC (DEPN-8) have been shown to improve P-V mechanics in 27-day gestation rabbit fetuses as effectively as a chloroform:methanol extract of sheep surfactant [207]. Diether analogs are just one example of a spectrum of synthetic phospholipid-like components that could be utilized in future exogenous surfactant development.

In addition to serving as possible components of exogenous surfactants, synthetic phospholipid analogs are highly useful in probing the molecular structural basis of surface behaviors like spreading and adsorption. The ability of diether phospholipids like DHPC [566] and DEPN-8 [1006] to form interdigitated bilayers, and its possible correlation with improved adsorption, has been noted above. Analog compounds have also been studied in dynamically cycled surface films to elucidate headgroup-related factors contributing to improved film respreading [657, 658]. In-

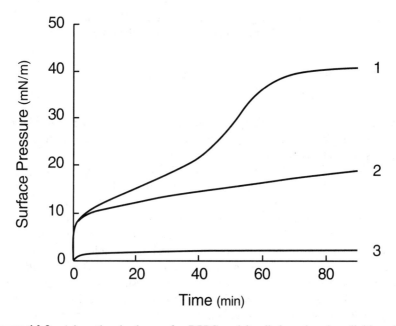

Figure 14-9 Adsorption isotherms for DPPC and its diether phosphonolipid analog DEPN-8. The molecular substitutions in the diether phosphonolipid analog in Figure 14-8 result in greatly enhanced adsorption relative to DPPC. Curve 1, DEPN-8, concentration 0.12 mg/ml; curve 2, DEPN-8, concentration 0.06 mg/ml; curve 3, DPPC, concentration 0.24 mg/ml. Data at 37°. See text for details. (Redrawn from Ref. 1092.)

creased fluidity in phospholipid fatty chains is a major contributor to improved respreading in lung surfactant films (Chapter 8). However, phospholipid headgroups also interact in films to affect respreading and other surface properties. Dynamic surface pressure–area studies of a number of phospholipid and phosphonolipid compounds with equivalent saturated C16 acyl chains demonstrate that those with more flexible ether linkages have consistently better respreading than those with ester or amide linkages [657, 658]. Increased hydrogen bonding between phospholipid headgroups also correlates strongly with improved dynamic respreading in cycled surface films, and a cationic headgroup charge due to a neutral (protonated) phosphate/phosphonate group tends to decrease respreading for some compounds [657, 658]. In terms of dynamic surface tension lowering, films of saturated diether analogs with choline headgroups generally exhibit very low minimum surface tensions over a broad range of pH, while films of compounds with strong hydrogen bonding headgroups like ethanolamine have higher minimum surface tensions that vary more substantially with pH [657, 658]. Gel to liquid crystal transition temperatures in phosphonolipid analogs increase as intermolecular hydrogen bonding increases and as headgroup electrostatic repulsion and hydration decrease [659]. Continuing surface and molecular biophysical studies with synthetic phospholipid analog compounds may further elucide mechanistic contributions to adsorption and film behaviors of interest for pulmonary surfactants.

X. Additional Strategies in Exogenous Surfactant Development

A variety of physicochemical approaches for new surfactant development other than the examples covered in this chapter are possible. For instance, an alternative to the use of proteins or peptides in synthetic exogenous surfactants is the addition of emulsifiers [353, 710], fatty acids [146, 459, 1066], or fatty alcohols [224] to enhance the adsorption or spreading of phospholipids. Similarly, the phosphonolipid analogs discussed above are only one example of a host of novel constituents that could be synthesized *de novo* for use in exogenous surfactants. Continuing advances in the active areas of synthetic bioorganic chemistry and surface science may in future allow the design of biologically compatible surfactant substances with even greater surface activity. Surface activity in exogenous surfactants could also be enhanced by using special physical dispersion methods to generate phospholipid microstructures with enhanced adsorption [843, 1213] or by dehydration to improve spreading behavior [55]. It is clear, however, that basic research described in this and other chapters has already identified a number of active clinical exogenous surfactants and is progressing along productive avenues toward the development of others.

XI. Chapter Summary

A number of clinical exogenous surfactants are currently used worldwide. These include organic solvent extracts of alveolar surfactant lavaged from intact animal lungs (Alveofact, bLES, Infasurf), supplemented or unsupplemented organic solvent extracts of surfactant processed from animal lung tissue (Curosurf, Surfactant-TA, Survanta), and protein-free synthetic lung surfactants (Exosurf, ALEC). Several additional exogenous surfactant preparations are also under active investigation for clinical efficacy (KL4 and recombinant SP-C surfactant). All exogenous surfactants approved for clinical use have been found to be beneficial to lung function or outcome in premature infants. At the same time, these exogenous surfactants vary in concept, composition, and activity. Basic research suggests that exogenous surfactants with the closest compositional analogy to endogenous surfactant exhibit correspondingly high surface and physiological activity.

The development of new exogenous surfactants depends to some extent on the activity, benefits, and risks of existing preparations, as well as on economic considerations. Developing new exogenous surfactants of high activity and low toxicity at reasonable cost provides a significant research challenge. Two active areas of new surfactant development involve combining synthetic lipids with human sequence recombinant apoproteins or with related synthetic peptides. It may also be possible to improve the activity of some current clinical exogenous surfactants by supplementation with apoproteins or synthetic peptides. In addition, new synthetic exogenous surfactants could be prepared using novel components such as phospholipid analogs that are resistant to enzymatic cleavage during inflammatory lung injury. Regardless of future drug development, existing basic science research on endogenous and exogenous lung surfactants and their function detailed throughout this book has already had important benefits. This basic research, combined and integrated with extensive clinical trial evaluations of surfactant replacement interventions, has had a major impact on the treatment of premature infants and other patients suffering from diseases of lung surfactant deficiency and dysfunction.

Glossary

Adsorption. The name given to the process whereby surfactant molecules enter the interface from the liquid subphase to form a surface film.

ALEC (artificial lung expanding compound). A clinical exogenous lung surfactant containing synthetic DPPC and egg phosphatidylglycerol in a 7:3 ratio. Produced by Britannia Pharmaceuticals, Redhill, Surrey, United Kingdom.

ALI (acute lung injury). A complex inflammatory condition, often including surfactant dysfunction, induced in the lungs by a variety of causes. Clinical ALI is defined from specific criteria associated with respiratory failure in affected patients.

Alveofact (SF-RI 1). A clinical exogenous lung surfactant obtained by chloroform-methanol extraction of bovine lung surfactant from bronchoalveolar lavage. Produced by Thomae GmbH, Biberach, Germany.

Apoprotein. A specific protein with a functional importance in a biological system. There are four lung surfactant apoproteins (SP-A, B, C, D), the first three of which are highly important to surface active function.

ARDS. The acute (formerly adult) respiratory distress syndrome, a clinical manifestation of lung injury with severe respiratory failure that can affect patients of all ages from term infants to adults. All patients with ARDS have clinical ALI, but with more severely compromised respiratory function.

Atelectasis. Collapse of the alveolar airsacs in the lung.

bLES (bovine lipid extract surfactant). A clinical exogenous lung surfactant made by chloroform-methanol and acetone extraction of lung surfactant lavaged from cows. Produced by bLES Biochemicals, Inc, Ontario, Canada.

Bovine lung surfactants. A nonspecific grouping of clinical exogenous surfactants sometimes used in the literature to include all preparations containing material from bovine lungs.

Bulk (bulk phase). Refers to a three-dimensional, macroscopic phase.

Captive bubble surfactometer. An instrument for measuring the adsorption and/or dynamic surface tension lowering of dispersed surfactants.

Chemokines. Chemotactic cytokines that attract inflammatory cells.

CLSE (calf lung surfactant extract). Surface active material made by chloroform-methanol extraction of lung surfactant lavaged from calves. CLSE is the substance of the clinical surfactant Infasurf.

Collapse (film collapse). For surfactant films, refers to compression to the point where there is not enough surface area to accommodate all the surfactant molecules in a single monomolecular layer.

Compliance. The change in volume generated by a change in pressure in the lungs ($\Delta V/\Delta P$).

Curosurf. A clinical exogenous lung surfactant obtained by a combination of centrifugation, chloroform-methanol extraction and affinity chromatography from minced porcine lung tissue. Produced by Chiesi Farmaceutici, Parma, Italy.

Cytokines. A class of cell-active proteins produced in inflammation.

Detergent. Can be defined generally as any cleansing agent or compound that emulsifies fats and oils. More restrictive definitions differentiate detergents chemically from soaps, although their functions overlap. All detergents are surface active.

DPPC (dipalmitoyl phosphatidylcholine). A phosphatidylcholine compound with two saturated palmitic acid (C16:0) chains. The most prevalent phospholipid in endogenous lung surfactant.

DSPC. Disaturated phosphatidylcholine (PC); the sum of all PC compounds in lung surfactant having two saturated acyl chains.

Dynamic surface activity. The surface tension lowering ability of a surfactant film under dynamic (nonequilibrium) compression. Lung surfactants reduce surface tension to lower values when compressed dynamically at rapid physiologic rates.

Endogenous lung surfactant. Surface active material made by the type II pneumocyte in the mammalian lung. Also called whole or natural surfactant. Removed most directly from the alveoli by bronchoalveolar lavage.

Exogenous lung surfactant. Any surfactant delivered to the lungs of patients or animals from outside the body. Exogenous surfactants have a wide range of concept, composition, and activity.

Exosurf. A clinical exogenous lung surfactant containing synthetic DPPC, tyloxapol, and hexadecanol. Produced by Glaxo Wellcome (formerly Burroughs-Wellcome), Research Triangle Park, North Carolina, USA.

FRC (functional residual capacity). The amount of air remaining in the lungs following a forced expiration.

Gel to liquid crystal transition temperature (T_c). The temperature at which hydrated phospholipid bilayers undergo a transition from the rigid lamellar gel phase to the more fluid lamellar liquid-crystal phase.

Growth factors. Proteins that promote cell division, differentiation, or growth. Along with other mediators (e.g., cytokines, chemokines), growth factors affect normal development, inflammation, and injury.

Human amniotic fluid surfactant. An exogenous surfactant containing lipids and hydrophobic and hydrophilic surfactant proteins obtained from the amniotic fluid of normal term pregnancies.

Hyaline membrane disease. Another name for the respiratory distress syndrome (RDS) of premature infants. The name derives from the appearance of shiny protein-rich layers in the alveoli of affected infants at autopsy.

Hysteresis. Path-dependent behavior. For surfactant films, hysteresis refers to a difference in surface tension–area behavior between compression and expansion. For lungs, hysteresis refers to a difference in P-V behavior between inflation and deflation.

Infasurf. A clinical exogenous lung surfactant obtained by chloroform-methanol extraction of lung surfactant lavaged from calves (CLSE). Produced by ONY, Inc, Amherst, NY & Forest Pharmaceuticals, St. Louis, MO, USA.

Interface. A boundary between any two bulk phases (liquid-gas, liquid-solid, etc.). For lung surfactants, the air-water interface is of primary interest.

Interfacial tension. A force existing at any interface, related thermodynamically to the change of free energy with surface area. The interfacial tension at a liquid-gas interface is called surface tension.

KL4. A clinical exogenous surfactant containing synthetic phospholipids, palmitic acid, and a synthetic hydrophobic peptide with repeated sequences of lysine (K) plus four leucine (L) residues.

Langmuir-Blodgett film (multilayer). A multilayered film deposited on a slide or other support drawn repeatedly through a film at the air-water interface. These deposited films are often studied in molecular biophysical research.

Langmuir trough. A surface balance based on the design of Irving Langmuir in the early part of the 20th century that uses surface barriers and a float-torsion balance arrangement to measure surface pressure-area behavior in compressed surfactant films.

Liposome. A vesicular structure containing phospholipid bilayers and enclosing a volume in the aqueous phase. Liposomal vesicles are unilamellar or multilamellar depending on their number of bilayers and vary in size from nanometers to microns in diameter.

Lung extract. A nonspecific term in the early lung surfactant literature describing any surface active material obtained (i.e., "extracted") from lungs, usually by saline but also sometimes by organic solvents.

Lung surfactant extract. Typically used to designate a hydrophobic organic solvent extract of lung surfactant or processed lung tissue that contains surfactant lipids and hydrophobic surfactant apoproteins. The composition of such extracts varies depending on source and preparation methods.

Lung tissue extract. Generally used to refer to a hydrophobic organic solvent extract of lung tissue that contains cellular and surfactant lipids and hydrophobic proteins.

Mesophases. Intermediate phases between the solid crystal and the isotropic liquid formed as a function of temperature by hydrated phospholipids.

Monolayer. A surface film that is one molecule thick, also called a monomolecular film.

Natural surfactant (endogenous surfactant, whole surfactant) Refers to alveolar or lamellar body surfactant containing all functional lipids and apoproteins.

Natural surfactants. A grouping of clinical exogenous surfactants sometimes used in the literature to include all preparations containing material from animal lungs.

Peptides (polypeptides). Small proteins usually less than 100 amino acids in length. Peptides of varying homology to lung surfactant proteins can be synthesized by solid-state chemistry or other techniques for use in basic research and as components in exogenous surfactants.

Phase. A region that is homogeneous in all of its properties.

Phospholipid or phosphonolipid analogs. Synthesized compounds with structural analogy to glycerophospholipids. Examples are phospholipids or phosphonolipids with ether- or amide-linked acyl moieties and/or specific N-headgroup differences relative to PC, PG, etc.

Phospholipids (glycerophospholipids). Molecules with two fatty acyl chains that are ester-linked to a glycerol backbone attached to a headgroup containing a phosphate (PO_4) moiety plus additional atoms that determine the phospholipid class.

Phospholipid classes. Phospholipid compounds with a specific headgroup. Examples of phospholipid classes are phosphatidylcholine (PC), phosphatidylethanolamine (PE), phosphatidylglycerol (PG), phosphatidylinositol (PI), and phosphatidylserine (PS).

Phosphosphingolipids. Compounds with equivalent headgroups to glycerophospholipids but having a hydrophobic group that is a ceramide. Sphingomyelin, found in low content in lung surfactant and higher concentration in cell membranes, is one example.

Pulsating (oscillating) bubble surfactometer. An instrument for measuring the overall surface activity of lung surfactants accounting for both adsorption and dynamic surface tension lowering ability.

RDS. The neonatal respiratory distress syndrome, a surfactant-deficient disease of premature infants. Synonymous with hyaline membrane disease.

Recombinant apoproteins. Surfactant proteins SP-A, B, C, and D produced by genetic engineering techniques using recombinant DNA (typically human sequence).

Recombinant SP-C surfactants. Exogenous surfactants containing synthetic lipids plus human sequence or modified human sequence recombinant SP-C. Surfactants containing two forms of recombinant SP-C are currently under study for clinical use (Byk Gulden Pharmaceutical, Germany; Scios-Nova, USA).

Respreading (dynamic respreading). Refers to the process whereby surfactant molecules squeezed out of a surfactant film during compression reenter and reintegrate back into the film during expansion. Respreading occurs from dynamically generated structures in the interfacial region, as opposed to adsorption from the bulk subphase.

Soap. A surface active salt of a fatty acid and a metal, typically sodium or potassium.

SP-A, SP-B, SP-C, SP-D. Specific proteins (apoproteins) in endogenous lung surfactant. SP-A, SP-B and SP-C are important contributors to biophysical function through interactions with surfactant lipids.

Squeeze out (selective squeeze out). The process where surfactant molecules with a lower affinity for the interface are ejected from a multicomponent surface film during compression.

Subphase. The bulk liquid phase existing below a surface film; also called the liquid subphase.

Subtypes (surfactant subtypes, subfractions). Subpopulations of aggregates of different size/density obtained by centrifugation from a suspension of lung surfactant. Large aggregate subtypes are more surface active than small aggregate subtypes. These subtypes are thought to reflect surfactant in different stages of alveolar processing.

Surface balance. A device for measuring the surface tension lowering behavior of a surfactant film during compression and expansion, usually in terms of a surface pressure-area (π-A) or surface tension–area (σ-A) isotherm.

Surface film. A film, typically containing surface active molecules, that exists at a surface or interface.

Surface phase. A two-dimensional interfacial phase, sometimes defined by analogy to three dimensional bulk phases (solids, liquids, gases).

Surface potential. A property measuring the effect of a surfactant film on the electrical potential at an interface (typically in units of millivolts).

Surface pressure. The amount by which a surfactant lowers surface tension in a film. Has the same units as surface tension.

Surface pressure–area (π-A) isotherm. Describes how a surfactant film lowers surface tension as a function of surface area at fixed temperature. Surface area A is given either as a percent of maximum area or as an inverse surface concentration in Å^2/molecule (1 Å = 0.1 nm).

Surface tension. The interfacial tension at a liquid-gas interface, with units of force/length or free energy/area (typically dynes/cm or mN/m). It is defined thermodynamically as the partial derivative of the Gibbs or Helmholtz free energy with area and is related to the work necessary to change the surface area of the system.

Surface tension–area (σ-A) isotherm. Equivalent to a surface pressure–area (π-A) isotherm for a compressed or cycled surfactant film, but expressed in terms of surface tension instead of surface pressure.

Surface viscosity. A property measuring the resistance of a surfactant film to flow or deformation in analogy to bulk phase viscosity; both surface shear viscosity and surface dilational viscosity can be defined (gm/sec, surface poise).

Surfactant. A contraction for "surface active agent." Any substance with an energetic preference to adsorb and form a film at an interface. All surfactants act to reduce interfacial tension.

Surfactant film. An interfacial film that lowers surface tension as a function of the concentration of surfactant molecules within it (or equivalently as a function of surface area).

Surfactant-TA. A clinical exogenous surfactant made by organic solvent extraction of processed bovine lung tissue supplemented with synthetic DPPC, tripalmitin, and palmitic acid. Produced by Tokyo Tanabe, Tokyo, Japan.

Survanta. A clinical exogenous surfactant made by organic solvent extraction of processed bovine lung tissue supplemented with synthetic DPPC, tripalmitin, and palmitic acid (as in Surfactant-TA). Produced by Abbott Laboratories, N. Chicago, Illinois, USA.

TLC (total lung capacity). The volume of the lungs after a forced inspiration to maximal pressure.

Tubular myelin. A distinctive large aggregate microstructure in endogenous surfactant that appears as a cross-hatched network of phospholipid bilayers under electron microscopy. Tubular myelin formation requires SP-A, SP-B, and calcium, and has been associated with more rapid adsorption in whole surfactant.

Whole surfactant (natural surfactant). Generally refers to endogenous lung surfactant containing all its functional lipids and SP-A, SP-B, and SP-C.

Wilhelmy surface balance. A surface balance that utilizes a hanging slide (Wilhelmy's method) to measure surface tension during the compression and expansion of a surfactant film.

References

Authors' names are presented here without diacritical marks.

1. Abbott/Ross Laboratories. Survanta Investigators Brochure. Columbus, OH (1989).
2. Abman SH, Griebel JL, Parker DK, Schmidt JM, Swanton D, Kinsella JP. Acute effects of inhaled nitric oxide in children with severe hypoxemic respiratory failure. J Pediatr 124:881–888 (1994).
3. Abraham E, Anzueto A, Gutierrez G, Tessler S, San Pedro G, Wunderink R, Dal Nogare A, Nasraway S, Berman S, Cooney R, Levy H, Baughman R, Rumbak M, Light RB, Poole L, et al. Double-blind randomized controlled trial of monoclonal antibody to human tumour necrosis factor in treatment of septic shock. Lancet 351:929–933 (1998).
4. Abraham E, Wunderink R, Silverman H, Perl TM, Nasraway S, Levy H, Bone R, Wenzel RP, Balk R, Alfred R, Pennington JE, Wherry JC. Efficacy and safety of monoclonal antibody to human tissue necrosis factor alpha in patients with sepsis syndrome: A randomized, controlled, double-blind, multicenter clinical trial. JAMA 273:934–941 (1995).
5. Adam NK. The Physics and Chemistry of Surfaces. New York: Dover Publications (1968).
6. Adams FH, Enhorning GA. Surface properties of lung extracts. I. A dynamic alveolar model. Acta Physiol Scand 68:23–27 (1966).
7. Adams FH, Fujiwara T. Surfactant in fetal lamb tracheal fluid. J Pediatr 63:537–542 (1963).
8. Adams FH, Fujiwara T, Emmanouilides GC, Raiha N. Lung phospholipid of the human fetus and infants with and without hyaline membrane disease. J Pediatr 77:833–841 (1970).
9. Adams FH, Towers B, Osher AB, Ikegami M, Fujiwara T, Nozak M. Effect of tracheal instillation of natural surfactant in premature lambs. I. Clinical and autopsy findings. Pediatr Res 12:841–848 (1978).
10. Adamson AW, Gast AP. Physical Chemistry of Surfaces. 6th ed. New York: Wiley-Interscience (1997).
11. Adawi A, Zhang Y, Baggs R, Finkelstein J, Phipps R. Disruption of the CD40-

CD40 ligand system prevents an oxygen-induced respiratory distress syndrome. Am J Pathol 152:651–657 (1998).

12. Adawi A, Zhang Y, Baggs R, Rubin P, Williams J, Finkelstein JN, Phipps R. Blockage of CD40-CD40 ligand interactions protects against radiation-induced pulmonary inflammation and fibrosis. Clin Immunol Immunopath 89:222–230 (1998).

13. Ahuja A, Oh N, Chao W, Spragg RG, Smith RM. Inhibition of the human neutrophil respiratory burst by native and synthetic surfactant. Am J Respir Cell Mol Biol 14:496–503 (1996).

14. Albert RK, Lakshminarayan S, Hildebrandt J, Kirk W, Butler J. Increased surface tension favors pulmonary edema formation in anesthetized dogs lungs. J Clin Invest 63:1015–1018 (1979).

15. Alberts B, Bray D, Lewis J, Raff M, Roberts K, Watson JD. Molecular Biology of the Cell. New York: Garland Publishing (1994).

16. Al-Mateen KB, Dailey K, Grimes MM, Gutcher GR. Improved oxygenation with exogenous surfactant administration in experimental meconium aspiration syndrome. Pediatr Pulmonol 17:75–80 (1994).

17. Ames BN. Assay in inorganic phosphate, total phosphate and phosphatases. Methods Enzymol 8:115–118 (1966).

18. Amirkhanian JD, Bruni R, Waring AJ, Navar C, Taeusch HW. Full length synthetic surfactant protein SP-B and SP-C reduce surfactant inactivation by serum. Biochim Biophys Acta 1168:315–320 (1993).

19. Amirkhanian JD, Bruni R, Waring AJ, Taeusch HW. Inhibition of mixtures of surfactant lipids and synthetic sequences of surfactant proteins SP-B and SP-C. Biochim Biophys Acta 1096:355–360 (1991).

20. Amirkhanian JD, Merritt TA. Inhibitory effects of oxyradicals on surfactant function: Utilizing in vitro Fenton reaction. Lung 176:63–72 (1998).

21. Amrein M, Schenk M, von Nahman A, Sieber M, Reichelt R. A novel force-sensing arrangement for combined scanning force/scanning tunneling microscopy applied to biological objects. J Microsc 179:261–265 (1995).

22. Ansell GB, Hawthorne JN, Dawson RMCE. Form and Function of Phospholipids. New York: Elsevier (1973).

23. Ansfield MJ, Benson BJ. Identification of the immunosuppresive components of canine pulmonary surface active material. J Immunol 125:1093–1098 (1980).

24. Antal JM, Divis LT, Erzurum SC, Wiedemann HP, Thomassen MJ. Surfactant suppresses NF-κB activation in human monocytic cells. Am J Respir Cell Mol Biol 14:374–379 (1996).

25. Anzueto A, Baughman RP, Guntupalli KK, Weg JG, Wiedemann HP, Raventos AA, Lemaire F, Long W, Zaccardelli DS, Pattishall EN (Exosurf ARDS Sepsis Study Group). Aerosolized surfactant in adults with sepsis-induced acute respiratory distress syndrome. N Engl J Med 334:1417–1421 (1996).

26. Armstrong L, Millar AB. Relative production of tumour necrosis factor α and interleukin 10 in adult respiratory distress syndrome. Thorax 52:442–446 (1997).

27. Artigas A, Bernard GR, Carlet J, Dreyfuss D, Gattinoni L, Hudson L, Lamy M, Marini JJ, Matthay MA, Pinsky MR, Spragg R, Suter PM (Consensus Committee). The American-European consensus conference on ARDS. Part 2: Ventilatory,

pharmacologic, supportive therapy, study design strategies and issues related to recovery and remodeling. Intensive Care Med 24:378–398 (1998).

28. Ashbaugh DG, Bigelow DB, Petty TL, Levine BE. Acute respiratory distress in adults. Lancet 2:319–323 (1967).

29. Ashbaugh DG, Uzawa T. Respiratory and hemodynamic changes after injection of free fatty acids. J Surg Res 8:417–423 (1968).

30. Ashton MR, Postle AD, Hall MA, Austin NC, Smith DE, Normand ICS. Turnover of exogenous artifical surfactant. Arch Dis Child 67:383–387 (1992).

31. Auten RL, Notter RH, Kendig JW, Davis JM, Shapiro DL. Surfactant treatment of full-term newborns with respiratory failure. Pediatrics 87:101–107 (1991).

32. Auten RL, Watkins RH, Shapiro DL, Horowitz S. Surfactant apoprotein A (SP-A) is synthesized in airways cells. Am J Respir Cell Mol Biol 3:491–496 (1990).

33. Avery ME, Fletcher BD, Williams RG. The Lung and Its Disorders in the Newborn Infant. Philadelphia: WB Saunders (1981).

34. Avery ME, Mead J. Surface properties in relation to atelectasis and hyaline membrane disease. Am J Dis Child 97:517–523 (1959).

35. Baatz JE, Elledge B, Whitsett JA. Surfactant protein SP B induces ordering at the surface of model membrane bilayers. Biochemistry 29:6714–6720 (1990).

36. Baatz JE, Sarin V, Absolom DR, Baxter C, Whitsett JA. Effects of surfactant associated protein SP B synthetic analogues on the structure and surface activity of model membrane bilayers. Chem Phys Lipids 60:163–178 (1991).

37. Baatz JE, Smyth KL, Whitsett JA, Baxter C, Absolom DR. Structure and functions of a dimeric form of surfactant protein SP-C: A Fourier transform infra-red and surfactometry study. Chem Phys Lipids 63:91–104 (1992).

38. Bacher A, Mayer N, Klimscha W, Oismuller C, Steltzer H, Hammerle A. Effects of pentoxifylline on hemodynamics and oxygenation in septic and nonseptic patients. Crit Care Med 25:795–800 (1997).

39. Bachofen H, Gehr P, Weibel ER. Alterations of mechanical properties and morphology in excised rabbit lungs rinsed with a detergent. J Appl Physiol 47:1002–1010 (1979).

40. Bachofen H, Hildebrandt J. Area analysis of pressure-volume hysteresis in mammalian lungs. J Appl Physiol 30:493–497 (1971).

41. Bachofen H, Hildebrandt J, Bachofen M. Pressure-volume curves of air- and liquid-filled excised lungs—surface tension in situ. J Appl Physiol 29:422–431 (1970).

42. Bachofen H, Schurch S, Possmayer P. Disturbance of alveolar lining layer: Effects on alveolar microstructure. J Appl Physiol 76:1983–1992 (1994).

43. Bachofen H, Schurch S, Urbinelli M, Weibel ER. Relations among alveolar surface tension, surface area, volume, and recoil pressure. J Appl Physiol 62:1878–1887 (1987).

44. Bachofen H, Wilson TA. Micromechanics of the acinus and alveolar walls. In: Crystal RG, West JB, Weiberl ER, Barnes PJ, Eds. The Lung: Scientific Foundations. 2nd ed. Philadelphia: Lippincott-Raven, pp. 1159–1167 (1997).

45. Bachofen M, Weibel ER. Alterations of the gas exchange apparatus in adult respiratory insufficiency associated with septicemia. Am Rev Respir Dis 116:589–615 (1977).

46. Balaraman V, Meister J, Ku TL, Sood SL, Tam E, Killeen J, Uyehara CFT, Egan E, Easa D. Lavage administration of dilute surfactants after acute lung injury in neonatal piglets. Am J Respir Crit Care Med 158:12–17 (1998).

47. Balaraman V, Sood SL, Finn KC, Hashiro G, Uyehara CFT, Easa D. Physiologic response and lung distribution of lavage vs bolus Exosurf in piglets with acute lung injury. Am J Respir Crit Care Med 153:1838–1843 (1996).

48. Balibrea-Cantero JL, Arias-Diaz J, Garcia C, Torres-Melero J, Simon C, Rodriguez JM, Vara E. Effect of pentoxifylline on the inhibition of surfactant synthesis induced by TNF-α in human type II pneumocytes. Am J Respir Crit Care Med 149:699–706 (1994).

49. Ballard PL. Hormonal regulation of pulmonary surfactant. Endocr Rev 10:165–181 (1989).

50. Ballard PL, Hawgood S, Liley H, Wellenstein G, Gonzales LW, Benson B, Cordell B, White RT. Regulation of pulmonary surfactant apoprotein SP28–36 gene in fetal human lung. Proc Natl Acad Sci USA 83:9527–9531 (1986).

51. Bancalari E. Neonatal chronic lung disease. In: Fanaroff AA, Martin RJE, Eds. Neonatal Perinatal Medicine. St. Louis: Mosby, pp. 1074–1089 (1997).

52. Bancalari E, Bidegain M. Respiratory disorders of the newborn. In: Taussig LM, Landau LI, Le Souef PN, Morgan WJ, Martinez FD, Sly PD, Eds. Pediatric Respiratory Medicine. St. Louis: Mosby, pp. 464–488 (1999).

53. Bangham AD. Lung surfactant: How it does and does not work. Lung 165:17–25 (1987).

54. Bangham AD, Hill MW, Miller NGA. Preparation and use of liposome as models of biological membranes. In: Korm EDE, Ed. Methods in Membrane Biology. New York: Plenum Press, pp. 1–68 (1974).

55. Bangham AD, Morley CJ, Phillips MD. The physical properties of an effective lung surfactant. Biochim Biophys Acta 573:552–556 (1979).

56. Baritussio A, Pettenazzo A, Benevento M, Alberti A, Gamba P. Surfactant protein C is recycled from the alveoli to the lamellar bodies. Am J Physiol 263:L607–L611 (1992).

57. Baritussio AM, Benevento A, Pettenazzo A, Bruni R, Santucci D, Dalzoppo P, Barcaglioni P, Crepaldi G. The life cycle of low molecular weight protein of surfactant (SP-C) in 3-day-old rabbits. Biochim Biophys Acta 1006:19–25 (1989).

58. Barnard ML, Baker RR, Matalon S. Mitigation of oxidant injury to lung micro-vasculature by intratracheal instillation of antioxidant enzymes. Am J Physiol 265:L340–L345 (1993).

59. Barnwell CN. Fundamentals of Molecular Spectroscopy. 3rd ed. London: McGraw-Hill (1983).

60. Bartmann P, Bamberger U, Pohlandt F, Gortner L. Immunogenicity and immunomodulatory activity of bovine surfactant (SF-RI 1). Acta Paediatr 81:383–388 (1992).

61. Batenburg JJ. Surfactant phospholipids: Synthesis and storage. Am J Physiol 262:L367–L385 (1992).

62. Baue AE, Durham R, Faist E. Systemic inflammatory response syndrome (SIRS), multiple organ dysfunction (MODS), multiple organ failure (MOF): Are we winning the battle? Shock 10:79–89 (1998).

63. Baughman RP, Gunther KL, Rashkin MC, Keeton DA, Pattishall EN. Changes in the inflammatory response of the lung during acute respiratory distress syndrome: Prognostic indicators. Am J Respir Crit Care Med 154:76–81 (1996).

64. Belai Y, Hernandez-Juviel JM, Bruni R, Waring AJ, Walther FJ. Addition of α1-antitrypsin to surfactant improves oxygenation in surfactant-deficient rats. Am J Respir Crit Care Med 159:917–923 (1999).

65. Benson BJ. Genetically engineered human pulmonary surfactant. Clin Perinatol 20:791–811 (1993).

66. Benson BJ, Hawgood S, Schilling J, Clements J, Damm D, Cordell B, White RT. Structure of canine pulmonary surfactant apoprotein: cDNA and complete amino acid sequence. Proc Natl Acad Sci USA 82:6379–6383 (1985).

67. Benson BJ, Williams MC, Sueishi K, Goerke J, Sargeant T. Role of calcium ions on the structure and function of pulmonary surfactant. Biochim Biophys Acta 793:18–27 (1984).

68. Beppu O, Clements JA, Goerke J. Phosphatidylglycerol-deficient lung surfactant has normal properties. J Appl Physiol 55:496–502 (1983).

69. Berger HM, Moison RMW, Van Zoeren-Grobben D. The ins and outs of respiratory distress syndrome in babies and adults. J R Coll Physicians (Lond) 28:24–33 (1994).

70. Berger JI, Gibson RL, Redding GJ, Standaert TA, Clarke WR, Truog WE. Effect of inhaled nitric oxide during group B streptococcal sepsis in piglets. Am Rev Respir Dis 147:1080–1086 (1993).

71. Berggren P, Curstedt T, Grossmann G, Nilsson R, Robertson B. Physiological activity of pulmonary surfactant with low protein content: Effect of enrichment with synthetic phospholipids. Exp Lung Res 8:29–51 (1985).

72. Berggren P, Lachmann B, Curstedt T, Grossmann G, Robertson B. Gas exchange and lung morphology after surfactant replacement in experimental adult respiratory distress induced by repeated lung lavage. Acta Anaesthesiol Scand 30:321–328 (1986).

73. Bermel MS. Evaluation of synthetic lung surfactants [Ph.D. Dissertation]. Rochester, NY: University of Rochester (1988).

74. Bermel MS, McBride JT, Notter RH. Lavaged excised rat lungs as a model of surfactant deficiency. Lung 162:99–113 (1984).

75. Bernard GR, Luce JM, Sprung CL, Rinaldo JE, Tate RM, Sibbald WJ, Kariman K, Higgins S, Bradley R, Metz CA, Harris TR, Brigham KL. High dose corticosteroids in patients with the adult respiratory distress syndrome. N Engl J Med 317:1565–1570 (1987).

76. Bernard GR, Artigas A, Brigham KL, Carlet J, Falke K, Hudson L, Lamy M, Legall JR, Morris A, Spragg R. The American-European Consensus Conference on ARDS: Definitions, mechanisms, relevant outcomes, and clinical trial coordination. Am J Respir Crit Care Med 149:818–824 (1994).

77. Bernard GR, Lucht WD, Niedermeyer ME, Snapper JR, Ogletree ML, Brigham KL. Effect of N-acetylcysteine on the pulmonary response to endotoxin in awake sheep and upon in vitro granulocyte function. J Clin Invest 73:1772–1784 (1984).

78. Bernard GR, Wheeler AP, Arons MM, Morris PE, Paz HL, Russell JA, Wright PA,

the Antioxidant in ARDS Study Group. A trial of antioxidants *N*-acetylcysteine and procysteine in ARDS. Chest 112:164–172 (1997).

79. Berry D, Ikegami M, Jobe A. Respiratory distress and surfactant inhibition following vagotomy in rabbits. J Appl Physiol 61:1741–1748 (1986).

80. Berthiaume Y, Lesur O, Dagenais A. Treatment of adult respiratory distress syndrome: Plea for rescue therapy of the alveolar epithelium. Thorax 54:150–160 (1999).

81. Bessler H, Gilgal R, Djaidetti M, Zahavi I. Effect of pentoxifylline on the phagocyte activity, cAMP levels and superoxide production by monocytes and polymorphonuclear cells. J Leukoc Biol 40:747–754 (1986).

82. Bevilacqua G, Parmigiani S, Robertson B (for trial participants). Prophylaxis of respiratory distress syndrome by treatment with modified porcine surfactant at birth: A multicenter prospective randomized trial. J Pediatr Med 24:609–620 (1996).

83. Bhamidipati SP, Hamilton JA. Interactions of lyso 1-palmitoylphosphatidylcholine with phospholipids: A ^{13}C and ^{31}P NMR study. Biochemistry 34:5666–5677 (1995).

84. Bibi H, Seifert B, Oulette M, Belik J. Intratracheal *N*-acetylcysteine use in infants with chronic lung disease. Acta Paediatr 81:335–339 (1992).

85. Biffl WL, Moore FA, Moore EE, Haenel JB, McIntyre RC, Burch JM. Are corticosteroids salvage therapy for refractory acute respiratory distress syndrome? Am J Surg 170:591–595 (1995).

86. Birdi KS. Lipid and Biopolymer Monolayers at Liquid Interfaces. New York: Plenum Press (1989).

87. Blanch L, Joseph D, Fernandez R, Mas A, Valles J, Diaz E, Baigorri F, Artigas A. Hemodynamic and gas exchange responses to inhalation of nitric oxide in patients with the acute respiratory distress syndrome and in hypoxemic patients with chronic obstructive pulmonary disease. Intensive Care Med 23:51–57 (1997).

88. Bligh EG, Dyer WJ. A rapid method of total lipid extraction and purification. Can J Biochem Physiol 37:911–917 (1959).

89. Bloom BT, Delmore P, Rose T, Rawlins T. Human and calf lung surfactant: A comparison. Neonat Intensive Care March/April:31–35 (1993).

90. Bloom BT, Kattwinkel J, Hall RT, Delmore PM, Egan EA, Trout JR, Malloy MH, Brown DR, Holzman IR, Coghill CH, Carlo WA, Pramanik AK, McCaffree MA, Toubas PL, Laudert S, et al. Comparison of Infasurf (calf lung surfactant extract) to Survanta (Beractant) in the treatment and prevention of RDS. Pediatrics 100:31–38 (1997).

91. Body DR. The phospholipid composition of pig lung surfactant. Lipids 6:625–629 (1971).

92. Boggaram V, Qing K, Mendelson CR. The major apoprotein of rabbit pulmonary surfactant. Elucidation of primary sequence and cyclic AMP and developmental regulation. J Biol Chem 263:2939–2947 (1988).

93. Bondurant S, Miller DA. A method for producing surface active extracts of mammalian lungs. J Appl Physiol 17:167–168 (1962).

94. Bose C, Corbet A, Bose G, Garcia-Prats J, Lombardy L, Wold D, Donlon D, Long

W. Improved outcome at 28 days of age for very low birth weight infants treated with a single dose of a synthetic surfactant. J Pediatr 117:947–953 (1990).

95. Boston RW, Geller F, Smith CA. Arterial blood gas tensions and acid base balance in the management of respiratory distress syndrome. J Pediatr 68:74–89 (1966).

96. Bowden DH. Alveolar response to injury. Thorax 2:357–375 (1981).

97. Boynton BR, Carlo WA, Jobe AH, Eds. New Therapies for Neonatal Respiratory Failure. New York: Cambridge University Press (1994).

98. Bregman J. Developmental outcome in very low birthweight infants. Pediatr Clin North Am 45:673–690 (1998).

99. Bregman J, Kimberlin LVS. Developmental outcome in extremely premature infants: Impact for surfactant. Pediatr Clin North Am 40:937–953 (1993).

100. Brigham K, Woolverton WC, Blake CH, Staub NC. Increased sheep lung vascular permeability caused by *Pseudomonas* bacteremia. J Clin Invest 54:792–804 (1974).

101. Broaddus VC, Boylan AM, Hoeffel JM, Kim KJ, Sadik M, Chuntharapai A, Hebert CA. Neutralization of IL-8 inhibits neutrophil influx in a rabbit model of endotoxin-induced injury. J Immunol 152:2960–2967 (1994).

102. Brown ES. Chemical identification of a pulmonary surface active agent. Fed Proc 21:438 (1962).

103. Brown ES. Isolation and assay of dipalmitoyl lecithin in lung extracts. Am J Physiol 207:402–406 (1964).

104. Brown ES. Lung area from surface tension effects. Proc Soc Exp Biol Med 95:168–170 (1957).

105. Brown LA, Wood LH. Stimulation of surfactant secretion by vasopressin in primary cultures of adult rat type II pneumocytes. J Appl Physiol 1001:76–81 (1989).

106. Bruni R, Taeusch HW, Waring AJ. Surfactant protein B-lipid interactions of synthetic peptides representing the amino-terminal amphipathic domain. Proc Nat Acad Sci USA 88:7451–7455 (1991).

107. Bruns G, Stroh H, Veldman GM, Latt SA, Floros J. The 35kD pulmonary surfactant-associated protein is encoded on chromosome 10. Hum Genet 76:58–62 (1987).

108. Cadenhead DA, Phillips MC. Molecular interactions in mixed monolayers. Adv Chem Ser 84:131–148 (1968).

109. Cajal Y, Dodia C, Fisher AB, Jain MK. Calcium-triggered selective intermembrane exchange of phospholipids by the lung surfactant protein SP-A. Biochemistry 37:12178–12188 (1998).

110. Cantin AM, Bein R. Glutathione and inflammatory disorders of the lung. Lung 169:123–138 (1991).

111. Cantin AM, Hubbard RC, Crystal RG. Glutathione deficiency in the epithelial tract in idiopathic pulmonary fibrosis. Am Rev Respir Dis 139:370–372 (1989).

112. Cantin AM, Larivee P, Begin RO. Extracellular glutathione suppresses human lung fibroblast proliferation. Am J Respir Cell Mol Biol 3:79–85 (1990).

113. Cantin AM, North SL, Hubbard RC, Crystal RG. Normal alveolar epithelial lining fluid contains high levels of glutathione. J Appl Physiol 63:152–157 (1987).

114. Cantor JO, Ed. Handbook of Animal Models of Pulmonary Disease. Vol I, II. Boca Raton, FL: CRC Press (1989).

115. Carlton DP, Cho SC, Davis P, Lont M, Bland RD. Surfactant treatment at birth reduces lung vascular injury and edema in preterm lambs. Pediatr Res 37:265–270 (1995).

116. Chabot F, Mitchell JA, Gutteridge JMC, Evans TW. Reactive oxygen species in acute lung injury. Eur Respir J 11:745–757 (1998).

117. Chang CH, Franses EI. Dynamic tension behavior of aqueous octanol solutions under constant-area and pulsating-area conditions. Chem Eng Sci 49:313–325 (1994).

118. Chang CH, Franses EI. Modified Langmuir-Hinshelwood kinetics for dynamic adsorption of surfactants at the air-water interface. Colloids Surf 69:189–201 (1992).

119. Chang R, Nir S, Poulain FR. Analysis of binding and membrane destabilization of phospholipid membranes by surfactant apoprotein B. Biochim Biophys Acta 1371:254–264 (1998).

120. Chapman D. Physical chemistry of phospholipids. In: Ansell GB, Hawthorne JN, Dawson RMCE, Eds. Form and Function of Phospholipids. New York: Elsevier, pp. 117–141 (1973).

121. Charafeddine L, D'Angio CT, Richards JL, Stripp BR, Finkelstein JN, Orlowski CC, LoMonaco MB, Paxhia A, Ryan RM. Hyperoxia increases keratinocyte growth factor mRNA expression in neonatal rabbit lung. Am J Physiol 20:L105–L113 (1999).

122. Chen PS, Toribara TY, Huber W. Microdetermination of phosphorus. Anal Chem 28:1756–1758 (1956).

123. Chen SC, Sturtevant JM. Thermotropic behavior of bilayers formed from mixed-chain phosphatidylcholines. Biochemistry 20:713–718 (1981).

124. Chevalier G, Collet AJ. In vivo incorporation of choline-^{3}H, leucine-^{3}H and galactose ^{3}H in alveolar type II pneumocytes in relation to surfactant synthesis. A quantitative radioautographic study in mouse by electron microscopy. Anat Rec 174:289–310 (1972).

125. Chollet-Martin S, Gatecel C, Kermarrec N, Gougerot-Pocidalo MA, Payen DM. Alveolar neutrophil functions and cytokine levels in patients with the adult respiratory distress syndrome during nitric oixde inhalation. Am J Respir Crit Care Med 153:985–990 (1996).

126. Chollet-Martin S, Jourdain B, Gibert C, Elbim C, Chastre J, Gougerot-Pocidalo MA. Interactions between neutrophils and cytokines in blood and alveolar spaces during ARDS. Am J Respir Crit Care Med 153:594–601 (1996).

127. Choukroun ML, Llanas B, Apere H, Fayon M, Galperine RI, Guenard H, Demarquez JL. Pulmonary mechanics in ventilated preterm infants with respiratory distress syndrome after exogenous surfactant administration: A comparison between two surfactant preparations. Pediatr Pulmonol 18:273–298 (1994).

128. Chu J, Clements JA, Cotton EK, Klaus MH, Sweet AY, Thomas MA, Tooley WH. The pulmonary hypoperfusion syndrome. Pediatrics 35:733–742 (1965).

129. Chu J, Clements JA, Cotton EK, Klaus MH, Sweet AY, Tooley WH. Neonatal

pulmonary ischemia. Clinical and physiologic studies. Pediatrics 40:709–782 (1967).

130. Chung JB, Hannemann RE, Franses EI. Surface analysis of lipid layers at air-water interfaces. Langmuir 6:1647–1655 (1990).

131. Clark DA, Nieman GF, Thompson JE, Paskanik AM, Rokhar JE, Bredenberg CE. Surfactant displacement by meconium free fatty acids: An alternative explanation for atelectasis in meconium aspiration syndrome. J Pediatr 110:765–770 (1987).

132. Clark JC, Wert SB, Bachurski CJ, Stahlman MT, Stripp BR, Weaver TE, Whitsett JA. Targeted disruption of the surfactant protein B gene disrupts surfactant homeostasis, causing respiratory failure in newborn mice. Proc Natl Acad Sci USA 92:7794–7798 (1995).

133. Clark JM, Lambertsen CJ. Pulmonary oxygen toxicity: A review. Pharmacol Rev 23:37–133 (1971).

134. Claypool WD, Wang DL, Chandler A, Fisher AB. An ethanol/ether soluble apoprotein from rat lung surfactant augments liposomes uptake by isolated granular pneumocytes. J Clin Invest 74:677–684 (1984).

135. Clements JA. Surface tension of lung extracts. Proc Soc Exp Biol Med 95:170–172 (1957).

136. Clements JA, Brown ES, Johnson RP. Pulmonary surface tension and the mucus lining of the lungs: Some theoretical considerations. J Appl Physiol 12:262–268 (1958).

137. Clements JA, Hustead RF, Johnson RP, Gribetz I. Pulmonary surface tension and alveolar stability. J Appl Physiol 16:444–450 (1961).

138. Clements JA, Tierney DF. Alveolar instability associated with altered surface tension. In: Handbook of Physiology, Respiration. Bethesda, MD: American Physiological Society, pp. 1565–1583 (1965).

139. Clercx A, Vandenbussche G, Curstedt T, Johansson J, Jornvall H, Ruysschaert J-M. Structural and functional importance of the C-terminal part of the pulmonary surfactant polypeptide SP-C. Eur J Biochem 229:465–472 (1995).

140. Coalson JJ, King RJ, Winter VT, Prihoda TJ, Anzueto AR, Peters JI, Johanson WG. O_2- and pneumonia-induced lung injury. I. Pathological and morphometric studies. J Appl Physiol 67:346–356 (1989).

141. Coccia MT, Waxman K, Soliman MH, Tominaga G, Pinderski L. Pentoxifylline improves survival following hemorrhagic shock. Crit Care Med 17:36–38 (1989).

142. Cochrane CG, Revak SD. Protein-phospholipid interactions in pulmonary surfactant. Chest 105:57S-62S (1994).

143. Cochrane CG, Revak SD. Pulmonary surfactant protein B (SP-B): Structure-function relationships. Science 254:566–568 (1991).

144. Cochrane CG, Revak SD, Merritt TA, Heldt GP, Hallman M, Cunningham MD, Easa D, Pramanik A, Edwards DK, Alberts MS. The efficacy and safety of KL4-surfactant in preterm infants with respiratory distress syndrome. Am J Respir Crit Care Med 153:404–410 (1996).

145. Cochrane CG, Revak SD, Merritt TA, Schraufstatter U, Hoch RC, Henderson C, Andersson S, Takamori H, Oades ZG. Bronchoalveolar lavage with KL4-surfactant in models of meconium aspiration syndrome. Pediatr Res 44:705–715 (1998).

146. Cockshutt A, Absolom DR, Possmayer F. The role of palmitic acid in pulmonary

surfactant: Enhancement of surface activity and the prevention of inhibition by plasma proteins. Biochim Biophys Acta 1085:248–256 (1991).

147. Cockshutt A, Possmayer F. Lysophosphatidylcholine sensitizes lipid extracts of pulmonary surfactant to inhibition by plasma proteins. Biochim Biophys Acta 1086:63–71 (1991).

148. Cockshutt AM, Possmayer F. Metabolism of surfactant lipids and proteins in the developing lung. In: Robertson B, van Golde LMG, Batenburg JJ, Eds. Pulmonary Surfactant: From Molecular Biology to Clinical Practice. Amsterdam: Elsevier, pp. 339–378 (1992).

149. Cockshutt AM, Weitz J, Possmayer F. Pulmonary surfactant-associated protein A enhances the surface activity of lipid extract surfactant and reverses inhibition by blood proteins in vitro. Biochemistry 19:8424–8429 (1990).

150. Cohen J, Carlet J. INTERSEPT: An international, multicenter, placebo controlled trial of monclonal antibody to human tumor necrosis factor-alpha in patients with sepsis. Crit Care Med 24:1431–1440 (1996).

151. Coleman M, Dehner LP, Sibley RK, Burke BA, L'Heureux PR, Thompson TR. Pulmonary alveolar proteinosis: An uncommon cause of chronic neonatal respiratory distress. Am Rev Respir Dis 121:583–386 (1980).

152. Collaborative European Multicentre Study Group. Factors influencing the clinical response to surfactant replacement therapy in babies with severe respiratory distress syndrome. Eur J Pediatr 150:433–439 (1991).

153. Collaborative European Multicentre Study Group. Surfactant replacement therapy for severe respiratory distress syndrome; an international randomized clinical trial. Pediatrics 82:683–691 (1988).

154. Collaborative Group on Antenatal Steroid Therapy. Effect of antenatal dexamethasone administration on the prevention of respiratory distress syndrome. Am J Obstet Gynecol 141:276–286 (1981).

155. Comroe J. Physiology of Respiration. 2nd ed. Chicago: Year Book Medical Publishers (1974).

156. Coolbear KP, Berde CB, Keough KMW. Gel to liquid-crystalline phase transitions of aqueous dispersions of polysaturated mixed-acid phosphatidylcholines. Biochemistry 22:1466–1473 (1983).

157. Corbet A, Bucciarelli R, Goldman S, Mammel M, Wold D, Long W. Decreased mortality rate among small premature infants treated at birth with a single dose of synthetic surfactant: A multicenter controlled trial. J Pediatr 118:595–605 (1991).

158. Crapo JD, Barry BE, Gehr P, Bachofen M, Weibel ER. Cell number and cell characteristics of the normal human lung. Am Rev Respir Dis 125:332–337 (1982).

159. Crapo JD, Peters-Golden M, Marsh-Salin J, Shelbourne JS. Pathologic changes in the lungs of oxygen-adapted rats. Lab Invest 39:640–653 (1978).

160. Creuwels LA, Demel RA, van Golde LMG, Benson BJ, Haagsman HP. Effect of acylation on structure and function of surfactant protein C at the air-liquid interface. J Biol Chem 268:26752–26758 (1993).

161. Creuwels LA, Demel RA, van Golde LMG, Haagsman HP. Characterization of a dimeric canine form of surfactant protein C (SP-C). Biochim Biophys Acta 1254:326–332 (1995).

162. Creuwels LAJM, Boer EH, Demel RA, van Golde LMG, Haagsman HP. Neutral-

ization of the positive charges of surfactant protein C: Effects on structure and function. J Biol Chem 270:16225–16229 (1995).

163. Creuwels LAJM, van Golde LMG, Haagsman HP. The pulmonary surfactant system—Biochemical and clinical aspects. Lung 175:1–39 (1997).

164. Creuwels LAJM, van Golde LMG, Haagsman HP. Surfactant protein B—Effects on lipid domain formation and intermembrane lipid flow. Biochim Biophys Acta 1285:1–8 (1996).

165. Crittenden DJ, Beckman DL. Traumatic head injury and pulmonary damage. J Trauma 22:766–769 (1982).

166. Crouch E, Chang D, Rust K, Persson A, Heuser J. Recombinant pulmonary surfactant protein D. Post-translational modification and molecular assembly. J Biol Chem 269:15808–15813 (1994).

167. Crouch E, Parghi D, Kuan S-F, Persson A. Surfactant protein D: Subcellular localization in non-ciliated bronchiolar epithelial cells. Am J Physiol 263:L60–L66 (1992).

168. Crouch E, Persson A, Chang D, Heuser J. Molecular structure of pulmonary surfactant protein D (SP-D). J Biol Chem 269:17311–17319 (1994).

169. Crouch E, Persson A, Chang D, Parghi D. Increased accumulation in silica-induced pulmonary lipoproteinosis. Am J Pathol 139:765–776 (1991).

170. Crouch E, Rust K, Marienchek W, Parghi D, Chang D, Persson A. Developmental expression of pulmonary surfactant protein D (SP-D). Am J Respir Cell Mol Biol 5:13–18 (1991).

171. Crouch E, Rust K, Veile R, Donis-Keller H, Grosso L. Genomic organization of human surfactant protein D (SP-D). SP-D is encoded on chromosome 10q22.2-23.1. J Biol Chem 268:2976–2983 (1993).

172. Crouch EC. Collectins and pulmonary host defense. Am J Respir Cell Mol Biol 19:177–201 (1998).

173. Crowley P, Chalmers I, Keirse MJNC. The effects of corticosteroid administration before preterm delivery: An overview of the evidence from controlled trials. Br J Obstet Gynaecol 97:11–25 (1990).

174. Cruz A, Casals C, Plasencia I, Marsh D, Perez-Gil J. Depth profiles of pulmonary surfactant protein B in phosphatidylcholine bilayers, studied by fluorescence and electron spin resonance spectroscopy. Biochemistry 37:9488–9496 (1998).

175. Crystal RG, West JB, Weibel ER, Barnes PJ, Eds. The Lung: Scientific Foundations. Vol. 1, 2. Philadelphia: Lippincott Raven (1997).

176. Cummings JJ, Holm BA, Hudak ML, Hudak BB, Ferguson WH, Egan EA. A controlled clinical comparison of four different surfactant preparations in surfactant-deficient preterm lambs. Am Rev Respir Dis 145:999–1004 (1992).

177. Curatolo W. The effects of ethylene glycol and dimethyl sulfoxide on cerebroside metastability. Biochim Biophys Acta 817:134–138 (1985).

178. Curstedt T, Johansson J, Barros-Soderling J, Robertson B, Nilsson G, Westberg M, Jornvall H. Low-molecular-mass surfactant protein type I. The primary structure of a hydrophobic 8 kDA polypeptide with eight half-cystine residues. Eur J Biochem 172:521–525 (1988).

179. Curstedt T, Johansson J, Persson P, Eklund A, Robertson B, Lowenadler B, Jornvall H. Hydrophobic surfactant-associated polypeptides: SP-C is a lipopeptide with 2

palmitoylated cysteine residues, whereas SP-B lacks covalently linked fatty acyl groups. Proc Natl Acad Sci USA 87:2985–2989 (1990).

180. Curstedt T, Jörnvall H, Robertson B, Bergman T, Berggren P. Two hydrophobic low-molecular-mass protein fractions of pulmonary surfactant: Characterization and biophysical activity. Eur J Biochem 168:255–262 (1987).

181. Cutz E, Enhorning G, Robertson B, Sherwood WG, Hill DE. Hyaline membrane disease: Effect of surfactant prophylaxis on lung morphology in premature primates. Am J Pathol 92:581–590 (1978).

182. D'Angio CT, Finkelstein JN, LoMonaco MB, Paxhia A, Wright SA, Baggs RB, Notter RH, Ryan RM. Changes in surfactant protein gene expression in a neonatal rabbit model of hyperoxia-induced fibrosis. Am J Phsyiol 272:L720–L730 (1997).

183. Davies JT, Rideal EK. Interfacial Phenomena. New York: Academic Press (1963).

184. Davis AJ, Jobe AH, Hafner D, Ikegami M. Lung function in premature lambs and rabbits treated with a recombinant SP-C surfactant. Am J Respir Crit Care Med 157:553–559 (1998).

185. Davis JM, Notter RH. Lung surfactant replacement for neonatal pathology other than primary repiratory distress syndrome. In: Boynton B, Carlo W, Jobe A, Eds. New Therapies for Neonatal Respiratory Failure: A Physiologic Approach. Cambridge: Cambridge University Press, pp. 81–92 (1994).

186. Davis JM, Richter SE, Kendig JW, Notter RH. High frequency jet ventilation and surfactant treatment of newborns in servere respiratory failure. Pediatr Pulmonol 13:108–112 (1992).

187. Davis JM, Rosenfeld WN, Koo HC, Gonenne A. Pharmacologic interactions of exogenous lung surfactant and recombinant human Cu/Zn superoxide dismutase. Pediatr Res 35:37–40 (1994).

188. Davis JM, Rosenfeld WN, Sanders RJ, Gonenne A. Prophylactic effects of recombinant human superoxide dismutase in neonatal lung injury. J Appl Physiol 74:2234–2241 (1993).

189. Davis JM, Russ GA, Metlay L, Dickerson B, Greenspan BS. Short-term distribution kinetics in intratracheally administered exogenous lung surfactant. Pediatr Res 31:445–450 (1992).

190. Davis JM, Vaness-Meehan K, Notter RH, Bhutani VK, Kendig JW, Shapiro DL. Changes in pulmonary mechanics after the administration of surfactant to infants with respiratory distress syndrome. N Engl J Med 319:476–479 (1988).

191. Davis SL, Furman DP, Costarino AT. Adult respiratory distress syndrome in children: Associated disease, clinical course, and predictors of death. J Pediatr 123:35–45 (1993).

192. Davreux CJ, Soric I, Nathens AB, Watson RWG, McGilvray ID, Suntres ZE, Shek PN, Rotstein OD. N-Acetylcysteine attenuates acute lung injury in the rat. Shock 8:432–438 (1997).

193. Day RW, Guarin M, Lynch JM, Vernon DD, Mean JM. Inhaled nitric oxide in children with severe lung disease: Results of acute and prolonged therapy with two concentrations. Crit Care Med 24:215–221 (1996).

194. Day RW, Lynch JM, White KS, Ward RM. Acute response to inhaled nitric oxide in newborns with respiratory failure and pulmonary hypertension. Pediatrics 98:698–705 (1996).

195. Day SE, Gedeit RG. Liquid ventilation. Clin Perinatol 25:711–722 (1998).
196. De Sanctis GT, Tomkiewicz RP, Rubin BK, Schurch S, King M. Exogenous surfactant enhances mucociliary clearance in the anaesthesized dog. Eur Respir J 7:1616–1621 (1994).
197. Deamer DW, Uster PS. Liposome preparation: Methods and mechanism. In: Ostro MJ, Ed. Liposomes. New York: Marcel Dekker, pp. 27–51 (1983).
198. Defay R, Prigogine I. Surface Tension and Adsorption. New York: John Wiley & Sons (1966).
199. DeLemos RA, Shermeta D, Knelson J, Kotas RV, Avery ME. Acceleration of appearance of pulmonary surfactant in the fetal lamb by administration of corticosteroids. Am Rev Respir Dis 102:459–461 (1970).
200. Dellinger RP, Zimmerman JL, Taylor RW, Straube RC, Hauser DL, Criner GJ, Davis K, Hyers TM, Papadakos P (and INO in ARDS Study Group). Effects of inhaled nitric oxide in patients with acute respiratory distress syndrome: Results of a randomized phase II trial. Crit Care Med 26:15–23 (1998).
201. deMello DE, Nogee LM, Heyman S, Krous HF, Hussain M, Merritt TA, Hsueh W, Haas JE, Heidelberger K, Schumacher R, Colten HR. Molecular and phenotypic variability in the congenital alveolar proteinosis syndrome associated with inherited surfactant protein B deficiency. J Pediatr 125:43–50 (1994).
202. Demirakca S, Dotsch J, Knotche C, Magsaam J, Reiter HL, Bauer J, Kuehl PG. Inhaled nitric oxide in neonatal and pediatric acute respiratory distress syndrome: Dose response, prolonged inhalation, and weaning. Crit Care Med 24:1913–1919 (1996).
203. Denbigh K. Principles of Chemical Equilibrium. Cambridge, England: Cambridge University Press (1981).
204. Deneke SM, Fanburg BL. Normobaric oxygen toxicity of the lung. N Engl J Med 303:76–86 (1980).
205. Dijk PH, Heikamp A, Bambang-Oetomo S. Surfactant nebulization versus instillation during high frequency ventilation in surfactant-deficient rabbits. Pediatr Res 44:699–704 (1998).
206. Discher BM, Maloney KM, Schief WR, Grainger DW, Vogel V, Hall SB. Lateral phase separation in interfacial films of pulmonary surfactant. Biophys J 71:2583–2590 (1996).
207. Dizon-Co L, Ikegami M, Ueda T, Jobe AH, Lin WH, Turcotte JG, Notter RH, Rider ED. In vivo function of surfactants containing PC analogs. Am J Respir Crit Care Med 150:918–923 (1994).
208. Dluhy RA, Cornell DG. In situ measurement of the infrared spectra of insoluble monolayers at the air-water interface. J Phys Chem 89:3195–3197 (1985).
209. Dluhy RA, Mitchell ML, Pettinski T, Beers J. Design and interfacing of an automated Langmuir-type film balance to an FT-IR spectrometer. Appl Spectrosc 42:1289–1293 (1988).
210. Dluhy RA, Reilly KE, Hunt RD, Mitchell ML, Mautone AJ, Mendelsohn R. Infrared spectroscopic investigations of pulmonary surfactant. Surface film transitions at the air-water interface and bulk phase thermotropism. Biophys J 56:1173–1181 (1989).
211. Dluhy RA, Stephens SM, Widayati S, Williams AD. Vibrational spectroscopy of

biophysical monolayers. Applications of IR and Raman spectroscopy to bio-membrane model systems at interfaces. Spectrochim Acta A 51:1413–1447 (1995).

212. Dobbs LG. Isolation and culture of alveolar type II cells. Am J Physiol 258:L134–L147 (1990).

213. Dobbs LG, Gonzales RF, Marinari LA, Mescher EJ, Hawgood S. The role of calcium in the secretion of surfactant by rat alveolar type II epithelial cells. Biochim Biophys Acta 877:305–313 (1986).

214. Dobbs LG, Wright JR, Hawgood S, Gonzales R, Venstrom K, Nellenbogen J. Pulmonary surfactant and its components inhibit secretion of phosphatidylcholine from cultured rat alveolar type II cells. Proc Natl Acad Sci USA 84:1010–1014 (1987).

215. Doering EB, Hanson CW, Reily DJ, Marshall C, Marshall BE. Improvement in oxygenation by phenylephrine and nitric oxide in patients with adult respiratory distress syndrome. Anesthesiology 87:18–25 (1997).

216. Donnelly SC, Haslett C, Reid PT, Grant IS, Wallace WAH, Metz CN, Bruce LJ. Regulatory role for macrophage migration inhibitory factor in acute respiratory distress syndrome. Nature Med 3:320–323 (1997).

217. Donnelly SC, Strieter RM, Kunkel SL. IL-8 and development of adult respiratory distress syndrome in at-risk patients. Lancet 341:643–647 (1993).

218. Dormans JA, Vanbree L. Function and response of type II cells to inhaled toxicants. Inhal Tox 7:319–342 (1995).

219. Douzinas EE, Tsidemiadou PD, Pitaridis MT, Andrianakis I, Bobota-Chloraki A, Katsouyanni K, Sfyras D, Malagari K, Roussos C. The regional production of cytokines and lactate in sepsis-related multiple organ failure. Am J Respir Crit Care Med 155:53–59 (1997).

220. Doyle RL, Szaflarski N, Modin GW, Wiener-Kronish JP, Matthay MA. Identification of patients with acute lung injury: Predictors of mortality. Am J Respir Crit Care Med 152:1818–1824 (1995).

221. Ducker TB, Simmonds RL. Increased intracranial pressure and pulmonary edema. II. The hemodynamic response of dogs and monkeys to increase intracranial pressure. J Neurosurg 28:118–123 (1968).

222. Dunn MS, Shennan AT, Possmayer F. Single vs multi-dose surfactant replacement therapy in neonates 30 to 36 weeks gestation with respiratory distress syndrome. Pediatrics 86:564–571 (1990).

223. Dunn MS, Shennan AT, Zayack D, Possmayer F. Bovine surfactant replacement in neonates of less than 30 weeks gestation: A randomized controlled trial of prophylaxis vs. treatment. Pediatrics 87:377–386 (1991).

224. Durand DJ, Clyman RI, Heymann MA, Clements JA, Mauray F, Kitterman J, Ballard P. Effects of protein-free, synthetic surfactant on survival and pulmonary function in preterm lambs. J Pediatr 107:775–780 (1985).

225. Edberg KE, Ekstrom-Jodal B, Hallman M, Hjalmarson O, Sandberg K, Silberberg A. Immediate effect on lung function of instilled human surfactant in mechanically ventilated newborn infants with IRDS. Acta Paediatr Scand 79:750–755 (1990).

226. Efrati H, Hawgood S, Williams MC, Hong K, Benson BJ. Divalent cation and hydrogen ion effects on the structure and surface activity of pulmonary surfactant. Biochemistry 26:7986–7993 (1987).

227. Egan EA, Notter RH, Kwong MS, Shapiro DL. Natural and artificial lung surfactant replacement therapy in premature lambs. J Appl Physiol 55:875–883 (1983).

228. Egberts J, de Winter JP, Sedin G, de Kleine MJ, Broberger U, van Bel F, Curstedt T, Robertson B. Comparison of prophylaxis and rescue treatment with Curosurf in babies less than 30 weeks gestation. A randomized trial. Pediatrics 92:768–774 (1993).

229. Ehrly A. The effect of pentoxifylline on the deformability of erythrocytes and the muscular oxygen pressure in patients with chronic arterial disease. J Med 10:331–336 (1979).

230. Eijking EP, van Daal GJ, Tenbrinck R, Luyenduijk A, Sluiters JF, Hannappel E, Lachmann B. Effect of surfactant replacement on *Pneumocystis carinii* pneumonia in rats. Intensive Care Med 17:475–478 (1990).

231. Ellefson RD, Caraway WT. Cholesterol. In: Tietz NW, Ed. Fundamentals of Clinical Chemistry. Philadelphia: WB Saunders, pp. 506–509 (1976).

232. Ellyett KM, Broadbent RS, Fawcett ER, Campbell AJ. Surfactant aerosol treatment of respiratory distress syndrome in the spontaneously breathing rabbit. Pediatr Res 39:953–957 (1996).

233. Emrie PA, Shannon JM, Mason RJ, Fisher JH. cDNA and deduced amino acid sequence for the rat hydrophobic pulmonary surfactant–associated protein, SP-B. Biochim Biophys Acta 994:215–221 (1989).

234. Emrie PA, Shannon JM, Mason RJ, Fisher JH. The coding sequence for the human 18,000 dalton hydrophobic surfactant protein is located on chromosome 2 and identifies a restriction fragment length polymorphism. Somat Cell Mol Genet 14:105–110 (1988).

235. Engstrom PC, Holm BA, Matalon S. Surfactant replacement attenuates the increase in alveolar permeability in hyperoxia. J Appl Physiol 67:688–693 (1989).

236. Enhorning G. Pulsating bubble technique for evaluation of pulmonary surfactant. J Appl Physiol 43:198–203 (1977).

237. Enhorning G, Duffy LC, Welliver RC. Pulmonary surfactant maintains patency of conducting airways in the rat. Am J Respir Crit Care Med 151:554–556 (1995).

238. Enhorning G, Grossman G, Robertson B. Pharyngeal deposition of surfactant in the premature rabbit fetus. Biol Neonate 22:126–132 (1973).

239. Enhorning G, Grossman G, Robertson B. Tracheal deposition of surfactant before the first breath. Am Rev Respir Dis 107:921–927 (1973).

240. Enhorning G, Hill D, Sherwood G, Cutz E, Robertson B, Bryan C. Improved ventilation of prematurely derived primates following tracheal deposition of surfactant. Am J Obstet Gynecol 132:529–536 (1978).

241. Enhorning G, Holm BA. Disruption of pulmonary surfactant's ability to maintain openness of a narrow tube. J Appl Physiol 74:2922–2927 (1993).

242. Enhorning G, Robertson B. Lung expansion in the premature rabbit fetus after tracheal deposition of surfactant. Pediatrics 50:58–66 (1972).

243. Enhorning G, Shennan A, Possmayer F, Dunn M, Chen CP, Milligan J. Prevention of neonatal respiratory distress syndrome by tracheal instillation of surfactant: A randomized clinical trial. Pediatrics 76:145–153 (1985).

244. Enhorning G, Shumel B, Keicher L, Sokolowski J, Holm BA. Phospholipases

introduced into the hypophase affect the surfactant film outlining a bubble. J Appl Physiol 73:941–945 (1992).

245. Enhorning G, Yarussi A, Rao P, Vargas I. Increased airway resistance due to surfactant dysfunction can be alleviated with aerosol surfactant. Can J Physiol Pharmacol 74:687–691 (1996).

246. Espinosa FF, Shapiro AH, Fredberg JJ, Kamm RD. Spreading of exogenous surfactant in an airway. J Appl Physiol 75:2028–2039 (1993).

247. Evans DA, Wilmott RW, Whitsett JA. Surfactant replacement therapy for adult respiratory distress syndrome in children. Pediatr Pulmonol 21:328–336 (1996).

248. Exosurf Neonatal Official Package Circular. Burroughs-Wellcome., Research Triangle Park, NC (August 1990).

249. Fan BR, Bruni R, Taeusch HW, Findlay R, Waring A. Antibodies against synthetic amphipathic helical sequences of surfactant protein SP-B detect a conformational change in the native protein. FEBS Lett 282:220–224 (1991).

250. Farrell PM, Ed. Lung Development: Biological and Clinical Perspectives. Vol. I, II. New York: Academic Press (1982).

251. Farrukh IS, Gurtner GH, Michael JR. Pharmacologic modification of pulmonary vascular injury: Possible role of cAMP. J Appl Physiol 62:47–54 (1987).

252. Fauci AS, Braunwald E, Isselbacher KJ, Martin JB, Kasper DL, Hauser SL, Longo DL, Eds. Harrison's Principles of Internal Medicine. 14th ed. New York: McGraw-Hill (1998).

253. Fields CG, Loyd DH, McDonald RL, Ottenson KM, Nobel RL. HBTU activation for automated Fmoc solid-phase peptide synthesis. Peptide Res 4:95–101 (1991).

254. Fierobe L, Brunet F, Dhainaut J-F, Monchi M, Belghith M, Mira J-P, Dallava-Santucci J, Dinh-Xuan A. Effect of inhaled nitric oxide on right ventricular function in adult respiratory distress syndrome. Am J Respir Crit Care Med 151:1414–1419 (1995).

255. Findlay EJ, Barton PG. Phase behavior of synthetic phosphatidylglycerols and binary mixtures with phosphatidylcholines in the presence and absence of calcium ions. Biochemistry 17:2400–2405 (1978).

256. Findlay RD, Taeusch HW, Walther FJ. Surfactant replacement therapy for meconium aspiration syndrome. Pediatrics 97:48–52 (1996).

257. Fine JB, Sprecher H. Unidimensional thin layer chromatography of phospholipids on boric acid–impregnated plates. J Lipid Res 23:660–663 (1982).

258. Finkelstein JN. Physiologic and toxicologic response of alveolar type II cells. Toxicology 60:41–52 (1990).

259. Finkelstein JN, Mavis RD. Biochemical evidence for internal proteolytic damage during isolation of type II alveolar epithelial cells. Lung 156:243–254 (1979).

260. Finkelstein JN, Shapiro DL. Isolation of type II alveolar epithelial cells using low protease concentrations. Lung 160:85–98 (1982).

261. Fisher AB, Furia L. Isolation and metabolism of granular pneumocytes from rat lungs. Lung 154:155–165 (1977).

262. Fisher CJ, Dhainaut JFA, Opal SM, Pribble JP, Balk RA, Slotman GJ, Iberti TJ, Rackow EC, Shapiro MJ, Greenman RL. Recombinant human IL-1 receptor antagonist in the treatment of patients with sepsis syndrome. Results from a

randomized, double blind, placebo-controlled trial. Phase III rhIL-1ra sepsis syndrome study group. JAMA 271:1836–1843 (1994).

263. Fisher CJ, Slotman GJ, Opal SM, Pribble JP, Bone RC, Emmanuel G, Ng D, Bloedow DC, Catalano MA. Initial evaluation of human recombinant interleukin-1 receptor antagonist in the treatment of sepsis syndrome: A randomized open-label, placebo-controlled multicenter study. Crit Care Med 22:12–21 (1994).

264. Fisher E, Marano MA, Van Zee KJ, Rock CS, Hawes AS, Thompson WA, DeForge L, Kenney JS, Remick DG, Bloedow DC, Thompson RC, Lowry SF, Moldawer LL. IL-1 receptor blockade improves survival and hemodynamic performance in *E. coli* septic shock, but fails to alter host responses to sublethal endotoxemia. J Clin Invest 89:1551–1557 (1992).

265. Fisher JH, Emrie PA, Drabkin HA, Kushnik T, Gerber M, Hofman T, Jones C. The gene encoding the hydrophobic surfactant protein SP-C is located on 8p and identifies an *Eco*RI restriction fragment length polymorphism. Am J Hum Genet 43:436–441 (1988).

266. Fisher JH, Kao FT, Jones C, White RT, Benson BJ, Mason RJ. The coding sequence for the 32,000 dalton pulmonary surfactant–associated protein A is located on chromosome 10 and identifies two separate restriction fragment length polymorphisms. Am J Hum Genet 40:503–511 (1987).

267. Fisher JH, Shannon JM, Hofmann T, Mason RJ. Nucleotide and deduced amino acid sequence of the hydrophobic surfactant protein SP-C from rat: Expression in alveolar type II cells and homology with SP-C from other species. Biochim Biophys Acta 995:225–230 (1989).

268. Fisher MH, Wilson MF, Weber KC. Determination of alveolar surface area and tension from in situ pressure-volume data. Respir Physiol 10:159–170 (1970).

269. Flecher EJ. A colorimetric method for estimating serum triglycerols. Clin Chim Acta 22:393–397 (1968).

270. Fleming BD, Keough KMW. Surface respreading after collapse of monolayers containing major lipids of pulmonary surfactant. Chem Phys Lipids 49:81–86 (1988).

271. Fletcher MA, McKenna TM, Owens EH, Nadkarni VM. Effects of in vivo pentoxifylline treatment on survival and ex vivo vascular contractility in a rat lipopolysaccharide shock model. Circ Shock 36:74–80 (1992).

272. Floros J, Steinbrink R, Jacobs K, Phelps D, Kriz R, Rency M, Sultzman L, Jones S, Taeusch HW, Frank HA, Fritsch EF. Isolation and characterization of cDNA clones for the 35 kDa pulmonary surfactant associated protein. J Biol Chem 261:9029–9033 (1986).

273. Folch J, Lees M, Slaone-Stanely GH. A simple method for the isolation and purification of total lipids from animal tissues. J Biol Chem 226:497–509 (1957).

274. Folkesson HF, Matthay MA, Hebert C, Broaddus CV. Acid aspiration-induced lung injury in rabbits is mediated by IL-8 dependent mechanisms. J Clin Invest 96:107–116 (1995).

275. Fracica PJ, Knapp MJ, Crapo JD. Patterns of progression and markers of lung injury in rodents and subhuman primates exposed to hyperoxia. Exp Lung Res 14:869–885 (1988).

276. Fracica PJ, Knapp MJ, Piantadosi CA, Takeda K, Fulkerson WJ, Coleman RE,

Wolfe WG, Crapo JD. Responses of baboons to prolonged hyperoxia: Physiology and qualitative pathology. J Appl Physiol 71:2352–2362 (1991).

277. Fratacci MD, Frostell C, Chen TY, Wain JC, Robinson DR, Zapol WM. Inhaled nitric oxide: A selective pulmonary vasocilator of heparin-protamine vasoconstriction in sheep. Anesthesiology 75:990–999 (1991).

278. Fredberg JJ, Stamenovic D. On the imperfect elasticity of lung tissue. J Appl Physiol 67:2408–2419 (1989).

279. Fries K, Gaspari A, Blieden T, Looney RJ, Phipps RP. CD40 expression by human fibroblasts. Clin Immunopathol 154:162–170 (1995).

280. Froh D, Ballard PL, Williams MC, Gonzales J, Goerke J, Odom MW, Gonzales LW. Lamellar bodies of cultured human fetal lung: Content of surfactant protein A (SP-A), surface film formation and structural transformation in vitro. Biochim Biophys Acta 1052:78–89 (1990).

281. Frosolonmo MF, Charms BL, Pawlowski R, Slivka S. Isolation, characterization, and surface chemistry of a surface-active fraction from dog lung. J Lipid Res 11:439–457 (1970).

282. Fuchimukai T, Fujiwara T, Takahashi A, Enhorning G. Artificial pulmonary surfactant inhibited by proteins. J Appl Physiol 62:429–437 (1987).

283. Fujita Y, Kogishi K, Suzuki Y. Pulmonary damage induced in mice by a monoclonal antibody to proteins associated with pig pulmonary surfactant. Exp Lung Res 14:247–260 (1988).

284. Fujiwara T. Surfactant replacement in neonatal RDS. In: Robertson B, van Golde L, Batenburg J, Eds. Pulmonary Surfactant. Amsterdam: Elsevier, pp. 479–504 (1984).

285. Fujiwara T. Tracheal instillation of artificial surfactant for treatment of hyaline membrane disease. Ann Nestle 48:24–29 (1981).

286. Fujiwara T, Adams FH, El-Salawy A, Sipos S. "Alveolar" and whole lung phospholipids of newborn lambs. Proc Soc Exp Biol Med 127:962–969 (1968).

287. Fujiwara T, Adams FH, Seto K. Lipids and surface tension of extracts of normal and oxygen-treated guinea pig lungs. J Pediatr 65:47–52 (1964).

288. Fujiwara T, Konishi M, Chida S, Okuyama Y, Takeuchi Y, Nishida H, Kito H, Fujiwara M, Nakamura H. Surfactant replacement therapy with a single post-ventilatory dose of a reconstituted bovine surfactant in preterm neonates with respiratory distress syndrome: Final analysis of a multicenter, double-blind, randomized trial and comparison with similar trials. Pediatrics 86:753–764 (1990).

289. Fujiwara T, Maeta H, Chida S, Morita T, Watabe Y, Abe T. Artificial surfactant therapy in hyaline membrane disease. Lancet 1:55–59 (1980).

290. Fujiwara T, Robertson B. Pharmacology of exogenous surfactants. In: Robertson B, van Golde LMG, Batenburg JJ, Eds. Pulmonary Surfactant: From Molecular Biology to Clinical Practice. Amsterdam: Elsevier Science Publishers, pp. 561–592 (1992).

291. Fujiwara T, Tanaka Y, Takei T. Surface properties of artifical surfactant in comparison with natural and synthetic surfactant lipids. Int Res Comm Syst (IRCS) Med Sci 7:311 (1979).

292. Fulkerson WJ, Macintyre N, Stamler J, Crapo JD. Pathogenesis and treatment of the adult respiratory distress syndrome. Arch Intern Med 156:29–38 (1996).

293. Fung YC. Does the surface tension make the lung inherently unstable? Circ Res 37:497–502 (1975).

294. Fung YC. Stress, deformation, and atelectasis of the lungs. Circ Res 37:481–496 (1975).

295. Gaines G. Insoluble Monolayers at Liquid-Gas Interfaces. New York: Interscience Publishers (1966).

296. Galdston M, Shah DO, Shinowara GY. Isolation and characterization of a lung lipoprotein surfactant. J Colloid Interface Sci 29:319–334 (1969).

297. Gaver DP, Samsel RW, Solway J. Effects of surface tension and viscosity on airway reopening. J Appl Physiol 75:1323–1333 (1993).

298. Ge Z, Brown CW, Turcotte JG, Wang Z, Notter RH. FTIR studies of calcium-dependent molecular order in lung surfactant and surfactant extract dispersions. J. Colloid Interface Sci. 173:471–477 (1995).

299. Gehr P, Geiser M, Im Hof V, Schurch S, Waber U, Baumann M. Surfactant and inhaled particles in the conducting airways: Structural, stereological, and biophysical aspects. Microsc Res Tech 26:423–436 (1993).

300. Gelbart WM, Ben-Shaul A. The "new" science of complex fluids. J Phys Chem 100:13169–13189 (1996).

301. Gennis RB. Biomembranes. Molecular Structure and Function. New York: Springer-Verlag (1989).

302. Gerlach H, Rossaint R, Pappert D, Falke KJ. Time-course and dose-response of nitric oxide inhalation for systemic oxygenation and pulmonary hypertension in patients with adult respiratory distress syndrome. Eur J Clin Invest 23:499–502 (1993).

303. Gil J. Alveolar surface, intra-alveolar fluid pools, and respiratory volume changes. J Appl Physiol 54:321–324 (1983).

304. Gil J, Bachofen H, Gehr P, Weibel ER. Alveolar volume-surface area relation in air- and saline-filled lungs fixed by vascular perfusion. J Appl Physiol 47:990–1001 (1979).

305. Gil J, Reiss OK. Isolation and characterization of a lamellar body fraction from rat lung homogenates. J Cell Biol 58:152–171 (1973).

306. Gil J, Weibel ER. Improvements in demonstration of lining layer of lung alveoli by electron microscopy. Respir Physiol 8:13–36 (1969).

307. Gil J, Weibel ER. Morphological study of pressure-volume hysteresis in rat lungs fixed by vascular perfusion. Respir Physiol 15:190–213 (1972).

308. Gilfillan AM, Rooney SA. Purinoceptor agonists stimulate phosphatidylcholine secretion in primary culture of adult rat type II pneumocytes. Biochim Biophys Acta 917:18–23 (1987).

309. Gillart T, Bazin JE, Cosserant B, Guelon D, Aigouy L, Mansoor O, Schoeffler P. Combined nitric oxide, prone positioning and almitrine infusion improve oxygenation in servere ARDS. Can J Anesth 45:402–409 (1998).

310. Gilliard N, Heldt GP, Loredo J, Gasser H, Redl H, Merritt TA, Spragg RG. Exposure of the hydrophobic components of porcine lung surfactant to oxidant stress alters surface tension properties. J Clin Invest 93:2608–2615 (1994).

311. Gillissen A, Nowak D. Characterization of *N*-acetylcysteine and ambroxol in anti-oxidant therapy. Respir Med 92:609–623 (1998).

312. Girod S, Fuchey C, Galabert C, Lebonvallet S, Bonnet N, Ploton D, Puchelle E. Identification of phospholipids in secretory granules of human submucosal gland respiratory cells. J Histochem Cytochem 39:193–198 (1991).

313. Girod S, Galabert C, Pierrot D, Boissonnade M, Zahm JM, Baszkin A, Puchelle E. Role of phospholipid lining on respiratory mucus clearance by cough. J Appl Physiol 71:2262–2266 (1991).

314. Gitlin JD, Soll RF, Parad RB, Horbar JD, Feldmann HA, Lucey JF, Taeusch HW. Randomized controlled trial of exogenous surfactant for the treatment of hyaline membrane disease. Pediatrics 79:31–37 (1987).

315. Glasser SW, Korfhagen TR, Perme CM, Pilot-Matias TJ, Kister SE, Whitsett JA. Two SP-C gene encoding human pulmonary surfactant proteolipid. J Biol Chem 263:10326–10331 (1988).

316. Glasser SW, Korfhagen TR, Weaver TE, Clark JC, Pilot-Matias T, Meuth J, Fox JL, Whitsett JA. cDNA, deduced polypeptide structure and chromosomal assignment of human pulmonary surfactant proteolipid SPL (pVAL). J Biol Chem 263:9–12 (1988).

317. Glasser SW, Korfhagen TR, Weaver TE, Pilot-Matias T, Fox JL, Whitsett JA. cDNA and deduced amino acid sequence of human pulmonary surfactant–associated proteolipid SPL (Phe). Proc Natl Acad Sci USA 84:4007–4011 (1987).

318. Gluck L, Kulovich MV. Lecithin-sphingomyelin ratios in amniotic fluid in normal and abnormal pregnancies. Am J Obstet Gynecol 115:539–546 (1973).

319. Goerke J. Lung surfactant. Biochim Biophys Acta 344:241–261 (1974).

320. Goerke J, Clements JA. Alveolar surface tension and lung surfactant. In: Fishman AP, Mackelroy PT, Mead J, Geiger GR, Eds. Handbook of Physiology. The Respiratory System. Mechanics of Breathing, Vol. III. Bethesda, MD: American Physiological Society Press, pp. 247–261 (1986).

321. Goerke J, Gonzales J. Temperature dependence of dipalmitoyl phosphatidylcholine monolayer stability. J Appl Physiol 51:1108–1114 (1981).

322. Goldenberg VE, Buckingham S, Sommers SC. Pulmonary alveolar lesions in vagotomized rats. Lab Invest 16:693–705 (1967).

323. Goldman AP, Tasker RC, Haworth SG, Sigston PE, Macrae DJ. Four patterns of response to inhaled nitric oxide for persistent pulmonary hypertension of the newborn. Pediatrics 98:706–713 (1996).

324. Goldman AP, Tasker RC, Hosiasson S, Henrichsen T, Macrae DJ. Early response to inhaled nitric oxide and its relationship to outcome in children with severe hypoxemic respiratory failure. Chest 112:752–758 (1997).

325. Goldman G, Welbourn R, Kobzik L, Valeri CR, Shepro D, Hechtman HB. Tumor necrosis factor-alpha mediates acid aspiration-induced systemic organ injury. Ann Surg 212:513–520 (1990).

326. Goldman S, Ellis R, Dhar V, Cairo MS. Rationale and potential use of cytokines in the prevention and treatment of neonatal sepsis. Clin Perinatol 25:699–709 (1998).

327. Gommers D, Eijking EP, van't Veen A, Lachmann B. Bronchoalveolar lavage with a diluted surfactant suspension prior to surfactant instillation improves the effectiveness of surfactant therapy in experimental acute respiratory distress syndrome (ARDS). Intensive Care Med 24:494–500 (1998).

328. Gommers D, Hartog A, van't Veen A, Lachmann B. Improved oxygenation by nitric oxide is enhanced by prior lung reaeration with surfactant, rather than positive end-expiratory pressure, in lung-lavaged rabbits. Crit Care Med 25:1868–1873 (1997).

329. Gommers D, van't Veen A, Verbrugge SJC, Lachmann B. Comparison of eight different surfactant preparations on improvement of blood gases in lung-lavaged rats. Appl Cardiopulm Pathophysiol 7:95–102 (1998).

330. Gonzalez PK, Zhuang J, Doctrow SR, Malfroy B, Benson PF, Menconi MJ, Fink MP. Role of oxidant stress in the adult respiratory distress syndrome: Evaluation of a novel antioxidant strategy in a porcine model of endotoxin-induced acute lung injury. Shock 6:S23–S26 (1996).

331. Gonzalez PK, Zhuang J, Doctrow SR, Malfroy B, Benson PF, Menconi MJ, Fink MP. EUK-8, a synthetic superoxide dismutase and catalase mimetic, ameliorates acute lung injury in endotoxemic mice. J Pharmacol Exp Ther 275:798–806 (1995).

332. Goodman ER, Kleinstein E, Fusco AM, Quinlan DP, Lavery R, Livingstone DH, Deitch EA, Hauser CJ. Role of interleukin 8 in the genesis of acute respiratory distress syndrome through an effect on neutrophil apoptosis. Arch Surg 13:1234–1239 (1998).

333. Goodman RB, Strieter RM, Martin DP, Steinberg KP, Milberg JA, Maunder RJ, Kunkel SL, Walz A, Hudson LD, Martin TR. Inflammatory cytokines in patients with persistence of the acute respiratory distress syndrome. Am J Respir Crit Care Med 154:602–611 (1996).

334. Gordon LM, Horvath S, Longo ML, Zasadzinski JAN, Taeusch HW, Faull K, Leung C, Waring AJ. Conformation and molecular topography of the N-terminal segment of surfactant protein B in structure-promoting environments. Protein Sci 5:1662–1675 (1996).

335. Gortner L, Bartmann P, Pohlandt F, Bernsau U, Porz F, Hellwege HH, Seitz RC, Hieronimi G, Bremer C, Jorch G, Hentschel R, Reiter HL, Wolf H, Ball F. Early treatment of respiratory distress syndrome with bovine surfactant in very preterm infants: A multicenter controlled clinical trial. Pediatr Pulmonol 14:4–9 (1992).

336. Gortner L, Bernsau U, Hellwege HH, Heironimi G, Jorch G, Reiter HL. A multicenter randomized controlled clinical trial of bovine surfactant for prevention of respiratory distress syndrome. Lung (Suppl) 168:864–869 (1990).

337. Gortner L, Bernsau U, Hellwege HH, Hieronimi G, Jorch G, Reiter HL. Does prophylactic use of bovine surfactant change drug utilization in very premature infants during neontal period? Dev Pharmacol Ther 16:1–6 (1991).

338. Gortner L, Pohlandt F, Disse B, Weller E. Effects of bovine surfactant in premature lambs after intratracheal application. Eur J Pediatr 149:280–283 (1990).

339. Graybeal JD. Molecular Spectroscopy. revised 1st ed. New York: McGraw-Hill (1988).

340. Greaves IA, Hildebrandt J, Hoppin FC. Micromechanics of the lung. In: Handbook of Physiology: Section 3. The Respiratory System. Bethesda: American Physiological Society, pp. 217–232 (1986).

341. Greenman RL, Schein RMH, Martin MA, Wenzel RP, MacIntyre NR, Emmanuel G, Chmel H, Kohler RB, McCarthy M, Plouffe J, Russell JA, and the XOMA

Sepsis Study Group. A controlled clinical trial of E5 murine monoclonal IgM antibody to endotoxin in the treatment of gram negative sepsis. JAMA 266:1097–1102 (1991).

342. Gregoriadis G. The carrier potential of liposomes in biology and medicine. N Engl J Med 295:704–710 (1976).

343. Gregory GA, Ketterman JA, Phibbs RH, Tooley WH, Hamilton WK. Treatment of idiopathic respiratory distress syndrome with continuous positive airway pressure. N Engl J Med 284:1332–1340 (1971).

344. Gregory TJ, Longmore WJ, Moxley MA, Whitsett JA, Reed CR, Fowler AA, Hudson LD, Maunder RJ, Crim C, Hyers TM. Surfactant chemical composition and biophysical activity in acute respiratory distress syndrome. J Clin Invest 88:1976–1981 (1991).

345. Gregory TJ, Steinberg KP, Spragg R, Gadek JE, Hyers TM, Longmere WJ, Moxley MA, Guang-Zuan CAI, Hite RD, Smith RM, Hudson LD, Crim C, Newton P, Mitchell BR, Gold AJ. Bovine surfactant therapy for patients with acute respiratory distress syndrome. Am J Respir Crit Care Med 155:109–131 (1997).

346. Griese M. Pulmonary surfactant in health and human lung diseases: State of the art. Eur Respir J 13:1455–1476 (1999).

347. Gross I. Regulation of fetal lung maturation. Am J Physiol 259:L337–L344 (1990).

348. Gross NJ. Inhibition of surfactant subtype convertase in radiation model of adult respiratory distress syndrome. Am J Physiol 4:L311–L317 (1991).

349. Gross NJ. Surfactant subtypes in experimental lung damage: Radiaiton pneumonitis. Am J Physiol 4:L302–L310 (1991).

350. Gross NJ, Narine KR. Surfactant subtypes in mice: Characterization and quantitation. J Appl Physiol 66:342–349 (1989).

351. Gross NJ, Schultz RM. Requirements for extracellular metabolism of pulmonary surfactant: Tentative identification of serine protease. Am J Physiol 262:L446–L453 (1992).

352. Gross NJ, Schultz RM. Serine protease requirement of the extra-cellular metabolism of rabbit alveolar surfactant subfractions. Biochim Biophys Acta 1044:222–230 (1990).

353. Grossman G, Larsson I, Nilsson R, Robertson B, Rydhag L, Stenius P. Lung expansion in premature newborn rabbits treated with emulsified synthetic surfactant; principles for experimental evaluation of synthetic substitutes for pulmonary surfactant. Respiration 45:327–338 (1984).

354. Grotberg J. Pulmonary flow and transport phenomena. Annu Rev Fluid Mech 26:529–571 (1994).

355. Grotberg JB, Halpern D, Jensen OE. Interaction of exogenous and endogenous surfactant: Spreading-rate effects. J Appl Physiol 78:750–756 (1995).

356. Gruenwald P. The mechanism of abnormal expansion of the lungs of mature and premature newborn infants. Bull Margaret Hague Maternity Hosp 8:100–106 (1955).

357. Gruenwald P. Surface tension as a factor in the resistance of neonatal lungs to aeration. Am J Obstet Gynecol 53:996–1007 (1947).

358. Gruenwald P, Johnson RP, Hustead RF, Clements JA. Correlation of mechanical

properties of infant lungs with surface activity of extracts. Proc Soc Exp Biol Med 109:369–371 (1962).

359. Guinard N, Beloucif S, Gatecel C, Maeto J, Payen D. Interest of a therapeutic optimization strategy in severe ARDS. Chest 111:1000–1007 (1997).

360. Günther A, Bleyl H, Seeger W. Apoprotein-based synthetic surfactants inhibit plasmic cleavage of fibrinogen in vitro. Am J Physiol 265:L186–L192 (1993).

361. Günther A, Siebert C, Schmidt R, Ziegle S, Grimminger F, Yabut M, Temmesfeld B, Walmrath D, Morr H, Seeger W. Surfactant alterations in severe pneumonia, acute respiratory distress syndrome, and cardiogenic lung edema. Am J Respir Crit Care Med 153:176–184 (1996).

362. Haagsman HP. Surfactant protein A and D. Biochem Soc Trans 22:100–106 (1994).

363. Haagsman HP, Hawgood S, Sargeant T, Buckley D, White RT, Drickamer K, Benson BJ. The major lung surfactant protein, SP 28-36, is a calcium-dependent, carbohydrate-binding protein. J Biol Chem 262:13877–13800 (1987).

364. Haagsman HP, Sargeant T, Hanschka PV, Benson BJ, Hawgood S. Binding of calcium to SP-A, a surfactant-associated protein. Biochemistry 29:8894–8900 (1990).

365. Haagsman HP, van Golde LMG. Synthesis and assembly of lung surfactant. Annu Rev Physiol 53:441–464 (1991).

366. Haagsman HP, White RT, Schilling J, Lau K, Benson BJ, Golden J, Hawgood S, Clements JA. Studies on the structure of lung surfactant protein SP-A. Am J Physiol 257:L421–L429 (1989).

367. Haas C, Voss T, Engel J. Assembly and disulfide rearrangement of recombinant surfactant protein A in vitro. Eur J Biochem 197:799–803 (1991).

368. Hack CE, Aarden LA, Thijs LG. Role of cytokines in sepsis. Adv Immunol 66:101–195 (1997).

369. Hack M, Fanaroff A. Outcomes of extremely low-birthweight infants between 1982 and 1988. N Engl J Med 321:1642–1647 (1989).

370. Hack M, Horbar JD, Malloy MH, Tyson JE, Wright E, Wright L. Very low birth weight outcomes of the National Insititute of Child Health and Human Development neonatal network. Pediatrics 87:587–597 (1991).

371. Haddad IY, Crow JP, Hu P, Ye Y, Beckman J, Matalon S. Concurrent generation of nitric oxide and superoxide damages surfactant protein A. Am J Physiol 267:L242–L249 (1994).

372. Haddad IY, Ischiropoulos H, Holm BA, Beckman JS, Baker JR, Matalon S. Mechanisms of peroxynitrite-induced injury to pulmonary surfactants. Am J Physiol 265:L555–L564 (1993).

373. Haddad IY, Nieves-Cruz G, Matalon S. Inhibition of surfactant function by copper-zinc superoxide dismutase (CuZn-SOD). J Appl Physiol 83:1545–1550 (1997).

374. Hafner D, Germann P-G, Hauschke D. Comparison of rSP-C surfactant with natural and synthetic surfactants after late treatment in a rat model of the acute respiratory distress syndrome. Br J Pharmacol 124:1083–1090 (1998).

375. Hafner D, Germann P-G, Hauschke D. Effects of rSP-C surfactant on oxygenation and histology in a rat-lung-lavage model of acute lung injury. Am J Respir Crit Care Med 158:270–278 (1998).

376. Haies DM, Gil J, Weibel ER. Morphometric study of rat lung cells. I. Numeric and dimensional characteristics of parenchymal cell populations. Am Rev Respir Dis 123:533–541 (1981).

377. Hall JB, Schmidt GA, Wood LDH. Principles of Critical Care. 2nd ed. New York: McGraw-Hill (1998).

378. Hall SB, Bermel MS, Ko YT, Palmer HJ, Enhorning GA, Notter RH. Approximations in the measurement of surface tension with the oscillating bubble surfactometer. J Appl Physiol 75:468–477 (1993).

379. Hall SB, Hyde RW, Notter RH. Changes in subphase surfactant aggregates in rabbits injured by free fatty acid. Am J Respir Crit Care Med 149:1099–1106 (1994).

380. Hall SB, Lu ZR, Venkitaraman AR, Hyde RW, Notter RH. Inhibition of pulmonary surfactant by oleic acid: Mechanisms and characteristics. J Appl Physiol 72:1708–1716 (1992).

381. Hall SB, Notter RH, Smith RJ, Hyde RW. Altered function of pulmonary surfactant in fatty acid lung injury. J Appl Physiol 69:1143–1149 (1990).

382. Hall SB, Venkitaraman AR, Whitsett JA, Holm BA, Notter RH. Importance of hydrophobic apoproteins as constituents of clinical exogenous surfactants. Am Rev Respir Dis 145:24–30 (1992).

383. Hall SB, Wang Z, Notter RH. Separation of subfractions of the hydrophobic components of calf lung surfactant. J Lipid Res 35:1386–1394 (1994).

384. Halliday HL. Overview of clinical trials comparing natural and synthetic surfactants. Biol Neonate 67(Suppl):32–47 (1995).

385. Halliday HL, McClure G, Reid MM, Lappin TR, Meban C, Thomas PS. Controlled trial of artifical surfactant to prevent respiratory distress syndrome. Lancet 1:476–478 (1984).

386. Hallman M, Bry K. Nitric oxide and lung surfactant. Semin Perinatol 20:173–185 (1996).

387. Hallman M, Bry K, Lappalainen U. A mechanism of nitric oxide–induced surfactant dysfunction. J Appl Physiol 80:2035–2043 (1996).

388. Hallman M, Enhorning G, Possmayer F. Composition and surface activity of normal and phosphatidylglycerol deficient lung surfactant. Pediatr Res 19:286–292 (1985).

389. Hallman M, Epstein BL, Gluck L. Analysis of labeling and clearance of lung surfactant phospholipids in rabbit. Evidence of bidirectional surfactant flux between lamellar bodies and alveolar lavage. J Clin Invest 68:742–751 (1981).

390. Hallman M, Feldman BH, Kirkpatrick E, Gluck L. Absence of phosphatidylglycerol (PG) in respiratory distress in the newborn. Pediatr Res 11:714–720 (1977).

391. Hallman M, Gluck L. Phosphatidylglycerol in lung surfactant. III. Possible modifier of surfactant function. J Lipid Res 17:257–262 (1976).

392. Hallman M, Kulovich MV, Kirkpatrick E, Sugerman RG, Gluck L. Phosphatidylinositol and phosphatidylglycerol in amniotic fluid: Indices of lung maturity. Am J Obstet Gynecol 125:613–617 (1976).

393. Hallman M, Merritt TA, Bry K. The fate of exogenous surfactant in neonates with respiratory distress syndrome. Clin Pharmacokinet 26:215–232 (1994).

394. Hallman M, Merritt TA, Jarvenpaa AL, Boynton B, Mannino F, Gluck L, Moore T, Edwards D. Exogenous human surfactant for treatment of severe respiratory distress syndrome: A randomized, prospective clinical trial. J Pediatr 106:963–969 (1985).

395. Hallman M, Merritt TA, Pohjavuori M, Gluck L. Effect of surfactant substitution on lung effluent phospholipids in respiratory distress syndrome: Evaluation of surfactant phospholipid turnover, pool size, and the relationship to severity of respiratory failure. Pediatr Res 20:1228–1235 (1986).

396. Hallman M, Merritt TA, Schneider H, Epstein BL, Mannino F, Edwards DK, Gluck L. Isolation of human surfactant from amniotic fluid and a pilot study of its efficacy in respiratory distress syndrome. Pediatrics 71:473–482 (1983).

397. Hallman M, Miyai K, Wagner RM. Isolated lamellar bodies from rat lung. Correlated ultrastructural and biochemical studies. Lab Invest 35:79–86 (1976).

398. Hallman M, Spragg R, Harrell JH, Moser KM, Gluck L. Evidence of lung surfactant abnormality in respiratory failure. J Clin Invest 70:673–683 (1982).

399. Hallman M, Waffarn F, Bry K, Turbow R, Kleinman MT, Mautz WJ, Rasmussen RE, Bhalla DK, Phalen RF. Surfactant dysfunction after inhalation of nitric oxide. J Appl Physiol 80:2026–2034 (1996).

400. Halpern D, Grotberg JB. Surfactant effects on fluid-elastic instabilities of liquid-lined flexible tubes: A model of airway closure. J Biomech Eng 115:271–277 (1993).

401. Hamers RJ. Scanned probe microscopies in chemistry. J Phys Chem 100:13103–13120 (1996).

402. Hamm H, Kroegel C, Hohlfeld J. Surfactant: A review of its functions and relevance in adult respiratory disorders. Respir Med 90:251–270 (1996).

403. Hamvas A, Cole FS, deMello DE, Moxley M, Whitsett JA, Colten HR, Nogee LM. Surfactant protein B deficiency: Antenatal diagnosis and prospective treatment with surfactant replacement. J Pediatr 125:356–361 (1994).

404. Hamvas A, Nogee LM, deMello DE, Cole FS. Pathophysiology and treatment of surfactant protein-B deficiency. Biol Neonate 67(Suppl 1):18–31 (1995).

405. Hamvas A, Nogee LM, Mallory GB, Spray TL, Huddleston CB, August A, Dehner LP, deMello DE, Moxley M, Nelson R, Cole FS, Colten HR. Lung transplantation for treatment of infants with surfactant protein B deficiency. J Pediatr 130:231–239 (1997).

406. Hamvas A, Wise PH, Yang RK, Wampler NS, Noguchi A, Maurer MM, Walentik CA, Schramm WF, Cole FS. The influence of the wider use of surfactant therapy on neonatal mortality among blacks and whites. N Engl J Med 334:1635–1640 (1996).

407. Harkins WD, Kirkwood JG. Note on surface viscosimetry (letter). J Chem Phys 6:298 (1938).

408. Harkins WD, Kirkwood JG. The viscosity of monolayers: Theory of the surface slit viscosimeter (letter). J Chem Phys 6:53 (1938).

409. Harris JD, Jackson F, Moxley MA, Longmore WJ. Effect of exogenous surfactant instillation on experimental acute lung injury. J Appl Physiol 66:1846–1851 (1989).

410. Hartog A, Gommers D, van't Veen A, Erdmann W, Lachmann B. Exogenous

surfactant and nitric oxide have a synergistic effect in improving gas exchange in experimental ARDS. Adv Exp Med Biol 428:277–279 (1997).

411. Hartshorn KL, Crouch EC, White MR, Eggleton P, Tauber AI, Chang D, Sastry K. Evidence for a protective role of pulmonary surfactant protein D (SP-D) against influenza A viruses. J Clin Invest 94:311–319 (1994).

412. Haslam PL, Hughes DA, MacNaughton PD, Baker CS, Evans TW. Surfactant replacement therapy in late-stage adult respiratory distress syndrome. Lancet 343:1009–1011 (1994).

413. Hassett RJ, Engleman W, Kuhn C. Extramembranous particles in tubular myelin from rat lung. J Ultrastruct Res 71:60–67 (1980).

414. Hauser H, Pascher I, Pearson RH, Sundell S. Preferred conformation and molecular packing of phosphatidylethanolamine and phosphatidylcholine. Biochim Biophys Acta 650:21–51 (1981).

415. Hawco MW, Coolbear KP, Davis PJ, Keough KMW. Exclusion of fluid during compression of monolayers of mixtures of dipalmitoylphosphatidylcholine with some other phosphatidylcholines. Biochim Biophys Acta 646:185–187 (1981).

416. Hawco MW, Davis PJ, Keough KMW. Lipid fluidity in lung surfactant: Monolayers of saturated and unsaturated lecithins. J Appl Physiol 51:509–515 (1981).

417. Hawgood S. The hydrophilic surfactant protein SP-A: Molecular biology, structure, and function. In: Robertson B, van Golde L, Batenburg JE, Eds. Pulmonary Surfactant: From Molecular Biology to Clinical Practice. Amsterdam: Elsevier Science Publishers, pp. 33–45 (1992).

418. Hawgood S. Pulmonary surfactant apoproteins—A review of protein and genomic structure. Am J Physiol 257:L13–L22 (1989).

419. Hawgood S. Surfactant: Composition, structure, and metabolism. In: Crystal RG, West JB, Weibel ER, Barnes PJ, Eds. The Lung: Scientific Foundations. 2nd ed. Philadelphia: Lippincott-Raven, pp. 557–571 (1997).

420. Hawgood S, Benson BJ. The molecular biology of the surfactant apoproteins. In: Massaro D, Ed. Lung Cell Biology. New York: Marcel Dekker, pp. 701–734 (1989).

421. Hawgood S, Benson BJ, Hamilton RJ. Effects of a surfactant-associated protein and calcium ions on the structure and surface activity of lung surfactant lipids. Biochemistry 24:184–190 (1985).

422. Hawgood S, Benson BJ, Schilling J, Damm D, Clements JA, White RT. Nucleotide and amino acid sequences of pulmonary surfactant protein SP 18 and evidence for cooperation between SP 18 and SP 28-36 in surfactant lipid adsorption. Proc Natl Acad Sci USA 84:66–70 (1987).

423. Hawgood S, Clements JA. Pulmonary surfactant and its apoproteins. J Clin Invest 86:1–6 (1990).

424. Hawgood S, Latham D, Borchelt J, Damm D, White T, Benson B, Wright JR. Cell-specific posttranslational processing of the surfactant-associated protein SP-B. Am J Physiol 264:L290–L299 (1993).

425. Hawgood S, Ogawa A, Benson B. Lung function in premature rabbits treated with recombinant human surfactant protein-C. Am J Respir Crit Care Med 154:484 (1996).

426. Hawgood S, Schiffer K. Structures and properties of the surfactant-associated proteins. Annu Rev Physiol 53:375–394 (1991).
427. Headley AS, Tolley E, Meduri GU. Infections and the inflammatory response in acure respiratory distress syndrome. Chest 111:1306–1321 (1997).
428. Heffner JE, Brown LK, Barbieri CA, Harpel KS, Deleo J. Prospective validation of an acute respiratory distress syndrome predictive score. Am J Respir Crit Care Med 152:1518–1526 (1995).
429. Heffner JE, Repine JE. Pulmonary strategies of antioxidant defense. Am Rev Respir Dis 140:531–554 (1989).
430. Hennes HM, Lee MB, Rimm AA, Shapiro DL. Surfactant replacement therapy in respiratory distress syndrome. Am J Dis Child 145:102–104 (1991).
431. Henon S, Meunier J. Microscope at the Brewster-angle: Direct observation of first-order phase transitions in monolayers. Rev Sci Instrum 62:936–939 (1991).
432. Hildebran JN, Goerke J, Clements JA. Pulmonary surface film stability and composition. J Appl Physiol 47:604–611 (1979).
433. Hills BA. What is the true role of surfactant in the lung? Thorax 36:1–4 (1981).
434. Hills BA, Bryan-Brown CW. Role of surfactant in the lung and other organs. Crit Care Med 11:951–956 (1983).
435. Hinshaw LB. Sepsis/septic shock: Participation of the microcirculation: An abbreviated review. Crit Care Med 24:1072–1078 (1996).
436. Hirs CHW, Timasheff SN, Eds. Enzyme Structure (Part I). Methods in Enzymology. Vol. 91. New York: Academic Press (1983).
437. Hirschl RB, Merz SI, Montoya JP, Parent A, Wolfson MR, Shaffer TH, Bartlett RH. Development and application of a simplified liquid ventilator. Crit Care Med 23:157–163 (1995).
438. Hirschl RB, Parent A, Tooley R, McCracken M, Johnson K, Shaffer TH, Wolfson MR, Bartlett RH. Liquid ventilation improves pulmonary function, gas exchange, and lung injury in a model of respiratory failure. Ann Surg 221:79–88 (1995).
439. Hirschl RB, Pranikoff T, Gauger P, Schreiner RJ, Dechert R, Bartlett RH. Liquid ventilation in adults, children, and full-term neonates. Lancet 346:1201–1202 (1995).
440. Hirschl RB, Pranikoff T, Wise C, Overbeck MC, Gauger P, Schreiner RJ, Dechert R, Bartlett RH. Initial clinical experience with partial liquid ventilation in adult patients with the acute respiratory distress syndrome. JAMA 275:383–389 (1996).
441. Hirschl RB, Tooley R, Parent A, Johnson K, Bartlett RH. Evaluation of gas exchange, pulmonary compliance, and lung injury during total and partial liquid ventilation in the acute respiratory distress syndrome. Crit Care Med 24:1001–1008 (1996).
442. Hirschl RB, Tooley R, Parent A, Johnson K, Bartlett RH. Improvement of gas exchange, pulmonary function, and lung injury with partial liquid ventilation: A study model in a setting of severe respiratory failure. Chest 108:500–508 (1995).
443. Hoekstra RE, Jackson JC, Myers TF, Frantz ID, Stern ME, Powers WF, Maurer M, Raye JR, Carrier ST, Gunkel JH, Gold AJ. Improved neonatal survival following multiple doses of bovine surfactant in very premature infants at risk for respiratory distress syndrome. Pediatrics 88:10–18 (1991).
444. Hoffman GM, Ross RA, Day SE, Rice TB, Nelin LD. Inhaled nitric oxide reduces

the utilization of extracorporeal membrane oxygenation in persistent pulmonary hypertension of the newborn. Crit Care Med 25:352–359 (1997).

445. Hoffmann H, Hatherill JR, Crowley J, Harada H, Yonemaru M, Zheng H, Ishizaka A, Raffin TA. Early post-treatment with pentoxifylline or dibutyril cAMP attenuates *Escherichia coli*–induced acute lung injury in guinea pigs. Am Rev Respir Dis 143:289–293 (1991).

446. Hohlfeld J, Fabel H, Hamm H. The role of pulmonary surfactant in obstructive airway disease. Eur Respir J 10:482–491 (1997).

447. Holm BA. Surfactant inactivation in ARDS. In: Robertson B, van Golde LMG, Batenburg JJ, Eds. Pulmonary Surfactant: From Molecular Biology to Clinical Practice. Amsterdam: Elsevier, pp. 665–684 (1992).

448. Holm BA, Enhorning G, Notter RH. A biophysical mechanism by which plasma proteins inhibit lung surfactant activity. Chem Phys Lipids 49:49–55 (1988).

449. Holm BA, Keicher L, Liu M, Sokolowski J, Enhorning G. Inhibition of pulmonary surfactant function by phospholipases. J Appl Physiol 71:1–5 (1991).

450. Holm BA, Matalon S, Finkelstein JN, Notter RH. Type II pneumocyte changes during hyperoxic lung injury and recovery. J Appl Physiol 65:2672–2678 (1988).

451. Holm BA, Matalon S, Notter RH. Pulmonary surfactant effects and replacement in oxygen toxicity and other ARDS-type injuries. In: Lachmann B, Ed. Surfactant Replacement Therapy in Neonatal and Adult Respiratory Distress Syndrome. Berlin: Springer-Verlag, pp. 223–244 (1988).

452. Holm BA, Notter RH. Animal models for testing surfactants. In: Jobe A, Taeusch HW, Eds. Surfactant Treatment of Lung Diseases. Columbus: Ross Laboratories Press, pp. 53–60 (1988).

453. Holm BA, Notter RH. Effects of hemoglobin and cell membrane lipids on pulmonary surfactant activity. J Appl Physiol 63:1434–1442 (1987).

454. Holm BA, Notter RH. Pulmonary surfactant effects in sublethal hyperoxic lung injury. In: Taylor AE, Matalon S, Eds. Physiology of Oxygen Radicals. Bethesda, MD: American Physiological Society Press (1986).

455. Holm BA, Notter RH. Surfactant therapy in adult respiratory distress syndrome and lung injury. In: Shapiro DL, Notter RH, Eds. Surfactant Replacement Therapy. New York: AR Liss, pp. 273–304 (1989).

456. Holm BA, Notter RH, Finkelstein JH. Surface property changes from interactions of albumin with natural lung surfactant and extracted lung lipids. Chem Phys Lipids 38:287–298 (1985).

457. Holm BA, Notter RH, Leary JF, Matalon S. Alveolar epithelial changes in rabbits following a 21 day exposure to 60% O_2. J Appl Physiol 62:2230–2236 (1987).

458. Holm BA, Notter RH, Siegle J, Matalon S. Pulmonary physiological and surfactant changes during injury and recovery from hyperoxia. J Appl Physiol 59:1402–1409 (1985).

459. Holm BA, Venkitaraman AR, Enhorning G, Notter RH. Biophysical inhibition of synthetic lung surfactants. Chem Phys Lipids 52:243–250 (1990).

460. Holm BA, Wang Z, Egan EA, Notter RH. Content of dipalmitoyl phosphatidylcholine in lung surfactant: Ramifications for surface activity. Pediatr Res 39:805–811 (1996).

461. Holm BA, Wang Z, Notter RH. Multiple mechanisms of lung surfactant inhibition. Pediatr Res 46:85–93 (1999).

462. Holm BA, Waring AJ. Designer surfactants: The next generation in surfactant replacement. Clin Perinatol 20:813–829 (1993).

463. Honda Y, Kuroki Y, Shijubo N, Fujishima T, Takahashi H, Hosoda K, Akino T, Abe S. Aberrant appearance of lung surfactant protein A in sera of patients with idiopathic pulmonary fibrosis and its clinical significance. Respiration 62:64–69 (1995).

464. Honig D, Mobius D. Reflectometry at the Brewster angle and Brewster angle microscopy at the air-water interface. Thin Solid Films 210/211:64–68 (1992).

465. Hooper RG, Kearl RA. Established adult respiratory distress syndrome successfully treated with corticosteroids. South Med J 89:359–364 (1996).

466. Hope MJ, Bally MB, Mayer LD, Janoff AS, Cullis PR. Generation of multilamellar and unilamellar phospholipid vesicles. Chem Phys Lipids 40:89–107 (1986).

467. Hopewell PC, Murray J. The adult respiratory distress syndrome. In: Shibel EM, Moser KM, Eds. Resp. Emergencies. St. Louis: CV Mosby, pp. 101–128 (1977).

468. Hoppin FG, Stothert JC, Greaves IA, Lai Y, Hildebrandt J. Lung recoil: Elastic and rheological properties. In: Handbook of Physiology: The Respiratory System. Mechanics of Breathing. Bethesda, MD: American Physiological Society, pp. 195–216 (1986).

469. Horbar J, Soll R, Schachinder H, Kewitz G, Versmold HT, Lindner W, Duc G, Mieth D, Linderkamp O, Zilow EP. A European multicenter randomized trial of single dose surfactant therapy for idiopathic respiratory distress syndrome. Eur J Pediatr 149:416–423 (1990).

470. Horbar JD, Soll R, Sutherland JM, Kotagal U, Philip AG, Kessler DL, Little GA, Edwards WH, Vidyasagar D, Raju TN. A multicenter randomized, placebo-controlled trial of surfactant therapy for respiratory distress syndrome. N Engl J Med 320:959–965 (1989).

471. Horbar JD, Wright LL, Soll RF, Wright EC, Fanaroff AA, Korones SB, Shankaran S, Oh W, Fletcher BD, Bauer CR (NIH NICHHD Neonatal Research Network). A multicenter randomized trial comparing two surfactants for the treatment of neonatal respiratory distress syndrome. J Pediatr 123:757–766 (1993).

472. Horie T, Hildebrandt J. Dynamic compliance, limit cycles, and static equilibria of excised cat lung. J Appl Phsyiol 31:423–430 (1971).

473. Horn LW, Davis SH. Apparent surface tension hysteresis of a dynamical system. J Colloid Interface Sci 51:459–476 (1975).

474. Horowitz AD. Exclusion of SP-C, but not SP-B, by gel phase palmitoyl lipids. Chem Phys Lipids 76:27–39 (1995).

475. Horowitz AD, Baatz JE, Whitsett JA. Lipid effects on aggregation of pulmonary surfactant protein SP-C studied by fluorescence energy transfer. Biochemistry 32:9513–9523 (1993).

476. Horowitz AD, Elledge B, Whitsett JA, Baatz JE. Effects of lung surfactant proteolipid SP-C on the organization of model membrane lipids: A fluorescence study. Biochim Biophys Acta 107:44–54 (1992).

477. Houmes R-JM, Hartog A, Verbrugge SJ, Bohm S, Lachmann B. Combining partial

liquid ventilation with nitric oxide to improve gas exchange in acute lung injury. Intensive Care Med 23:163–169 (1997).

478. Huang YC, Caminiti SP, Fawcett TA, Moon RE, Fracica PJ, Miller FJ, Young SL, Piantadosi CA. Natural surfactant and hyperoxic lung injury in primates. I. Physiology and biochemistry. J Appl Physiol 76:991–1001 (1994).

479. Hubbard AE. The Handbook of Surface Imaging and Visualization. Boca Raton, FL: CRC Press (1995).

480. Hudak ML, Farrell EE, Rosenberg AA, Jung AL, Auten RL, Durand DJ, Horgan MJ, Buckwald S, Belcastro MR, Donohue PK, Carrion V, Maniscalco WM, Balsan MJ, Torres BA, Miller RR, et al. A multicenter randomized masked comparison of natural vs synthetic surfactant for the treatment of respiratory distress syndrome. J Pediatr 128:396–406 (1996).

481. Hudak ML, Martin DJ, Egan EA, Matteson EJ, Cummings J, Jung AL, Kimberlin LV, Auten RL, Rosenberg AA, Asselin JM, Belcastro MR, Donahue PK, Hamm CR, Jansen RD, Brody AS, et al. A multicenter randomized masked comparison trial of synthetic surfactant versus calf lung surfactant extract in the prevention of neonatal respiratory distress syndrome. Pediatrics 100:39–50 (1997).

482. Hudson LD. New therapies for ARDS. Chest 108(Suppl):79S-91S (1995).

483. Hudson LD, Milberg JA, Anardi D, Maunder RJ. Clinical risks for development of the acute respiratory distress syndrome. Am J Respir Crit Care Med 151:293–301 (1995).

484. Hulsey TC, Alexander GR, Robillard PY, Annibale DJ, Kennan A. Hyaline membrane disease: The role of ethnicity and maternal risk characteristics. Am J Obstet Gynecol 168:572–576 (1993).

485. Hunninghake GW, Gadek JE, Kawanami O, Ferrans VJ, Crystal RG. Inflammatory and immune processes in human lung in health and disease: Evaluation by bronchoalveolar lavage. Am J Pathol 97:146–206 (1979).

486. Hunt AN, Kelly FJ, Postle AD. Developmental variation in whole human lung phosphatidylcholine molecular species: A comparison with guinea pig and rat. Early Hum Dev 25:157–171 (1991).

487. Hwang J, Tamm LK, Bohm C, Ramalingam TS, Betzig E, Edidin M. Nanoscale complexity of phospholipid monolayers investigated by near-field scanning optical microscopy. Science 270:610–614 (1995).

488. Hyers TM. Prediction of survival and mortality in patients with the adult respiratory distress syndrome. New Horizons 1:466–470 (1993).

489. Ignarro LJ, Buga GM, Wood KS, Byrns RE, Chaudhuri G. Endothelium-derived relaxing factor produced and released from artery and vein is nitric oxide. Proc Natl Acad Sci USA 84:9265–9269 (1987).

490. Ikegami M, Adams FH, Towers B, Osher AB. The quantity of natural surfactant necessary to prevent the respiratory distress syndrome in premature lambs. Pediatr Res 14:1082–1085 (1980).

491. Ikegami M, Agata Y, Elkady T, Hallman M, Berry D, Jobe A. Comparison of four surfactants: In vitro surface properties and responses of preterm lambs to treatment at birth. Pediatrics 79:38–46 (1987).

492. Ikegami M, Hesterberg T, Nozaki M, Adams FH. Restoration of lung pressure-

volume characteristics with surfactant: Comparison of nebulization versus instillation and natural versus synthetic surfactant. Pediatr Res 11:178–182 (1977).

493. Ikegami M, Horowitz AD, Whitsett JA, Jobe AH. Clearance of SP-C and recombinant SP-C in vivo and in vitro. Am J Physiol 274:L933–L939 (1998).

494. Ikegami M, Jobe A, Jacobs H, Jones SJ. Sequential treatments of premature lambs with an artificial surfactant and natural surfactant. J Clin Invest 68:491–496 (1981).

495. Ikegami M, Jobe A, Jacobs H, Lam R. A protein from airways of premature lambs that inhibits surfactant function. J Appl Physiol 57:1134–1142 (1984).

496. Ikegami M, Jobe A, Yamada T, Priestly A, Ruffini L, Rider E, Seidner S. Surfactant metabolism in surfactant-treated preterm ventilated lambs. J Appl Physiol 67:429–437 (1989).

497. Ikegami M, Jobe AH. Surfactant protein-C in ventilated premature lamb lung. Pediatr Res 44:860–864 (1998).

498. Ikegami M, Jobe AH, Glatz T. Surface activity following natural surfactant treatment of premature lambs. J Appl Physiol 51:306–312 (1981).

499. Ikegami M, Jobe AH, Nathanieisz PW. The labeling of pulmonary surfactant phosphatidylcholine in newborn and adult sheep. Exp Lung Res 2:197–206 (1981).

500. Ikegami M, Silverman J, Adams FH. Restoration of lung pressure-volume characteristics with various phospholipids. Pediatr Res 13:777–780 (1979).

501. Ikegami M, Ueda T, Absolom D, Baxter C, Rider E, Jobe AH. Changes in exogenous surfactant in ventilated preterm lamb lungs. Am Rev Respir Dis 148:837–844 (1993).

502. Ikegami M, Wada K, Emerson GA, Rebello CM, Hernandez RE, Jobe AH. Effects of ventilation style on surfactant metabolism and treatment response in preterm lambs. Am J Respir Crit Care Med 157:638–644 (1998).

503. Ingram RH, Pedley TJ. Pressure-flow relationships in the lungs. In: Handbook of Physiology. Section 3. The Respiratory System. Bethesda, MD: American Physiological Society, pp. 277–293 (1986).

504. Inoue T, Matsuura E, Nagata A, Ogasawara Y, Hattori A, Kuroki Y, Fujimoto S, Akino T. Enzyme-linked immunosorbent assay for human pulmonary surfactant protein D. J Immunol Methods 173:157–164 (1994).

505. Ishizaka A, Wu ZH, Stephens KE, Horada H, Hogue RS, O'Hanley PT, Raffin TA. Attenuation of acute lung injury in septic guinea pigs by pentoxifylline. Am Rev Respir Dis 138:376–384 (1988).

506. Itaya K, Ui M. Colorimetric determination of free fatty acids in biological fluids. J Lipid Res 6:16–20 (1965).

507. Ito Y, Gofin J, Veldhuizen R, Joseph M, Fjarneson J, McCaig L, Yao L-J, Marcou J, Lewis J. Timing of exogenous surfactant administration in a rabbit model of acute lung injury. J Appl Physiol 80:1357–1364 (1996).

508. Jacobs H, Jobe A, Ikegami M, Miller D, Jones S. Reutilization of phosphatidylcholine analogues by the pulmonary surfactant system: the lack of specificity. Biochim Biophys Acta 793:300–309 (1984).

509. Jacobs H, Jobe AH, Ikegami M, Conaway D. The significance of reutilization of surfactant phosphatidylcholine in 3-day-old rabbits. J Biol Chem 258:4159–4165 (1983).

510. Jacobs HC, Ikegami M, Jobe AH, Berry DD, Jones S. Reutilization of surfactant phosphatidylcholine in adult rabbits. Biochim Biophys Acta 837:77–84 (1985).

511. Jacobs HC, Jobe A, Ikegami M, Glatz T, Jones SJ, Barajas L. Premature lambs rescued with natural surfactant: Clinical and biophysical correlates. Pediatr Res 16:424–429 (1982).

512. Jacobs HC, Jobe AH, Ikegami M, Jones S. Surfactant phosphatidylcholine source, fluxes, and turnover times in 3-day old, 10-day old, and adult rabbits. J Biol Chem 257:1805–1810 (1982).

513. Jacobs KA, Phelps DS, Steinbrink R, Fisch J, Kriz R, Mitsock L, Dougherty JP, Taeusch HW, Floros J. Isolation of a cDNA clone encoding a high molecular weight precursor to a 6 kDa pulmonary surfactant–associated protein. J Biol Chem 262:9808–9811 (1987).

514. Jepsen S, Herlevsen P, Knudsen P, Bud MI, Klausen N-O. Antioxidant treatment with N-acetylcysteine during adult respiratory distress syndrome: A prospective, randomized, placebo-controlled study. Crit Care Med 20:918–923 (1992).

515. Jobe A. Metabolism of endogenous surfactant and exogenous surfactants for replacement therapy. Semin Perinatol 12:231–244 (1988).

516. Jobe A, Ikegami M, Jacobs H, Jones S. Surfactant and pulmonary blood flow distributions following treatment of premature lambs with natural surfactant. J Clin Invest 73:848–856 (1984).

517. Jobe AH. Lung development. In: Fanaroff A, Martin RJE, Eds. Neonatal-Perinatal Medicine. St. Louis: Mosby–Year Book, pp. 991–1009 (1997).

518. Jobe AH. Pulmonary surfactant therapy. N Engl J Med 328:861–868 (1993).

519. Jobe AH, Gluck K. The labeling of lung surfactant phosphatidylcholine in premature rabbits. Pediatr Res 13:635–640 (1979).

520. Jobe AH, Ikegami M. Surfactant metabolism. Clin Perinatol 20:683–696 (1993).

521. Jobe AH, Ikegami M, Glatz T, Yoshida Y, Diakomanolis E. Saturated phosphatidylcholine secretion and the effect of natural surfactant on premature and term lambs ventilated for 2 days. Exp Lung Res 4:259–267 (1983).

522. Jobe AH, Ikegami M, Glatz T, Yoshida Y, Diakomanolis E, Padbury J. Duration and characteristics of treatment of premature lambs with natural surfactant. J Clin Invest 67:372–391 (1981).

523. Jobe AH, Ikegami M, Sarton-Miller I, Barajas L. Surfactant metabolism of newborn lamb lungs in vivo. J Appl Physiol 49:1091–1098 (1980).

524. Jobe AH, Ikegami M, Seidner SR, Pettenazzo A, Ruffini L. Surfactant phosphatidylcholine metabolism and surfactant function in preterm, ventilated lambs. Am Rev Respir Dis 139:352–359 (1989).

525. Jobe AH, Kirkpatrick E, Gluck L. Lecithin appearance and apparent biologic half life in term newborn rabbit lung. Pediatr Res 12:669–675 (1978).

526. Jobe AH, Mitchell BR, Gunkel JH. Beneficial effects of the combined use of prenatal corticosteroids and postnatal surfactant on preterm lambs. Am J Obstet Gynceol 168:508–513 (1993).

527. Johannsen EC, Chung JB, Chang CH, Franses EI. Lipid transport to air-water interfaces. Colloids Surf 53:117–134 (1991).

528. Johansson J, Curstedt T, Jornvall H. Surfactant protein B: Disulfide bridges, structural properties, and kringle similarities. Biochemistry 30:6917–6921 (1991).

529. Johansson J, Curstedt T, Robertson B. The proteins of the surfactant system. Eur Respir J 7:372–391 (1994).

530. Johansson J, Curstedt T, Robertson B. Synthetic protein analogues in artificial surfactants. Acta Paediatr 85:642–646 (1996).

531. Johansson J, Curstedt T, Robertson B, Jornvall H. Size and structure of the hydrophobic low moleculalr weight surfactant-associated polypeptide. Biochemistry 27:3544–3547 (1988).

532. Johansson J, Gustafsson M, Zaltash S, Robertson B, Curstedt T. Synthetic surfactant protein analogs. Biol Neonate 74(Suppl):9–14 (1998).

533. Johansson J, Jornvall H, Curstedt T. Human surfactant polypeptide SP-B: Disulfide bridges, C-terminal end, and peptide analysis of the airway form. FEBS Lett 301:165–167 (1992).

534. Johansson J, Jörnvall H, Eklund A, Christensen N, Robertson B, Curstedt T. Hydrophobic 3.7 kDa surfactant polypeptide: Structural characterization of the human and bovine forms. FEBS Lett 232:61–64 (1988).

535. Johansson J, Nilsson G, Strömberg R, Robertson B, Jörnvall H, Curstedt T. Secondary structure and biophysical activity of synthetic analogues of the pulmonary surfactant polypeptide SP-C. Biochem J 307:535–541 (1995).

536. Johansson J, Szyperski T, Curstedt T, Wuthrich K. The NMR structure of the pulmonary surfactant–associated polypeptide SP-C in an apolar solvent contains a valyl-rich alpha-helix. Biochemistry 33:6015–6023 (1994).

537. Johansson J, Szyperski T, Wuthrich K. Pulmonary surfactant–associated polypeptide SP-C in lipid micelles: CD studies of intact SP-C and NMR secondary structure determination of depalmitoyl-SP-C(1–17). FEBS Lett 362:261–265 (1995).

538. Jolliet P, Bulpa P, Ritz M, Ricou B, Lopez J, Chevrolet J-C. Additive beneficial effects of the prone position, nitric oxide, and almitrine bismesylate on gas exchange and oxygen transport in acute respiratory distress syndrome. Crit Care Med 25:786–794 (1997).

539. Joos P. Cholesterol as a liquifier in phospholipid membranes studied by surface viscosity measurements of mixed monolayers. Chem Phys Lipids 4:162–168 (1970).

540. Jorens PG, Sibille Y, Goulding NJ, van Overveld FJ, Herman AG, Bossaert L, DeBacker WA, Lauwerys R, Flower RJ, Bernard A. Potential role of Clara cell protein, and endogenous phospholipase A_2 inhibitor, in acute lung injury. Eur Respir J 8:1643–1653 (1995).

541. Kahn MC, Anderson GJ, Anyan WR, Hall SB. Phosphatidylcholine molecular species of calf lung surfactant. Am J Physiol 13:L567–L573 (1995).

542. Kahovkova J, Odavic R. A simple method for the quantitative analysis of phopholipids separated by thin layer chromatography. J Chromatogr 40:90–96 (1969).

543. Kalina M, Mason RJ, Shannon JM. Surfactant protein C is expressed in alveolar type II cells but not in Clara cells of rat lung. Am J Respir Cell Mol Biol 6:594–600 (1992).

544. Kamm RD, Schroter RC. Is airway closure caused by a liquid film instability? Respir Physiol 75:141–156 (1989).

545. Kaplin RS, Pedersen PL. Sensitive protein assay in the presence of high levels of lipid. Anal Biochem 150:97–104 (1989).

546. Karamanoukian HL, Glick PL, Wilcox DL, Rossman JE, Morin FC, Holm BA. Pathophysiology of congenital diaphragmatic hernia VII: Inhaled nitric oxide requires exogenous surfactant therapy in the lamb model of CDH. J Pediatr Surg 30:1–4 (1995).

547. Kari MA, Hallman M, Eronen M, Teramo K, Virtanen M, Koivisto M, Ikonen RS. Prenatal dexamethasone treatment in conjunction with rescue therapy of human surfactant: A randomized placebo-controlled multicenter study. Pediatrics 93:730–736 (1994).

548. Karima R, Matsumoto S, Higashi H, Matsushima K. The molecular pathogenesis of endotoxin shock and organ failure. Mol Med Today 5:123–132 (1999).

549. Kates M. Techniques of lipidology: Isolation, analysis, and identification of lipids. In: Work TS, Work E, Eds. Laboratory Techniques in Biochemistry and Molecular Biology. New York: Elsevier, pp. 275–610 (1972).

550. Kattwinkel J. Surfactant: Evolving issues. Clin Perinatol 25:17–32 (1998).

551. Kattwinkel J, Bloom B, Delmore P, Davis C, Farrell E, Friss H, Jung A, King K, Mueller D. Prophylactic administration of calf lung surfactant extract (CLSE) is more effective than early treatment of RDS in 29–32 weeks gestation babies. Pediatrics 92:90–98 (1993).

552. Keel JB, Hauser M, Stocker R, Bauman PC, Speich R. Established acute respiratory distress syndrome: Benefit of corticosteroid rescue therapy. Respiration 65:258–264 (1998).

553. Kendig JW, Notter RH, Cox C, Aschner JL, Benn S, Bernstein RM, Hendricks-Munoz K, Maniscalco WM, Metlay LA, Phelps DL, Sinkin RA, Wood BP, Shapiro DL. Surfactant replacement therapy at birth: Final analysis of a clinical trial and comparisons with similar trials. Pediatrics 82:756–762 (1988).

554. Kendig JW, Notter RH, Cox C, Reubens LJ, Davis JM, Maniscalco WM, Sinkin RA, Bartoletti A, Dweck HS, Horgan MJ, Risemberg H, Phelps DL, Shapiro DL. A comparison of surfactant as immediate prophylaxis and as rescue therapy in newborns of less than 30 weeks gestation. N Engl J Med 324:865–871 (1991).

555. Kendig JW, Notter RH, Maniscalco WM, Davis JM, Shapiro DL. Clinical experience with calf lung surfactant extract. In: Shapiro DL, Notter RH, Eds. Surfactant Replacement Therapy. New York: AR Liss, pp. 257–271 (1989).

556. Kendig JW, Ryan RM, Sinkin RA, Maniscalco WM, Notter RH, Guillet R, Cox C, Dweck H, Horgan MJ, Reubens LJ, Risemberg H, Phelps DL. Comparison of two strategies for surfactant prophylaxis in very premature infants: A multicenter randomized trial. Pediatrics 101:1006–1012 (1998).

557. Keough K, Taeusch HW. Surface balance and differential scanning calorimetric studies on aqueous dispersions of mixtures of dipalmitoyl phosphatidylcholine and short-chain saturated phosphatidylcholines. J Colloid Interface Sci 109:365–374 (1986).

558. Keough KMW. Modifications of lipid structure and their influence on mesomorphism in model membranes: The influence of hydrocarbon chains. Biochem Cell Biol 64:44–49 (1986).

559. Keough KMW. Physical chemical properties of some mixtures of lipids and their potential for use in exogenous surfactants. Prog Respir Res 18:257–262 (1984).

560. Keough KMW. Physical chemistry of pulmonary surfactant in the terminal air

spaces. In: Robertson B, van Golde LMG, Batenburg JJ, Eds. Pulmonary Surfactant: From Molecular Biology to Clinical Practice. Amsterdam: Elsevier, pp. 109–164 (1992).

561. Keough KMW. Physicochemical properties of surfactant lipids. Biochem Soc Trans 13:1081–1084 (1985).

562. Keough KWM, Parsons CS, Tweeddale MG. Interactions between plasma proteins and pulmonary surfactant: Pulsating bubble studies. Can J Physiol Pharmacol 67:663–668 (1989).

563. Khammash H, Perlman M, Wojtulewicz J, Dunn M. Surfactant therapy in full-term neonates with severe respiratory failure. Pediatrics 92:135–139 (1993).

564. Kikkawa Y, Yoneda K. The type II epithelial cell of the lung. I. Method of isolation. Lab Invest 30:76–84 (1974).

565. Kikkawa Y, Yoneda K, Smith F, Packard B, Suzuki K. The type II epithelial cells of the lung. II. Chemical composition and phospholipid synthesis. Lab Invest 32:295–302 (1975).

566. Kim JT, Mattai J, Shipley GG. Gel phase polymorphism in ether-linked dihexadecyl PC bilayers. Biochemistry 26:6599–6603 (1987).

567. Kimmel E, Budiansky B. Surface tension and the dodecahedron model for lung elasticity. J Biomech Eng 112:160–167 (1990).

568. King RJ. Isolation and chemical composition of pulmonary surfactant. In: Robertson B, van Golde LMG, Batenburg JJ, Eds. Pulmonary Surfactant. Amstersdam: Elsevier, pp. 1–15 (1984).

569. King RJ. The surfactant system of the lung. Fed Proc 33:2238–2247 (1974).

570. King RJ, Clements JA. Surface active materials from dog lung. I. Method of isolation. Am J Physiol 223:707–714 (1972).

571. King RJ, Clements JA. Surface active materials from dog lung. II. Composition and physiological correlations. Am J Physiol 223:715–726 (1972).

572. King RJ, Clements JA. Surface active materials from dog lung. III. Thermal analysis. Am J Physiol 223:727–733 (1972).

573. King RJ, Coalson JJ, Seidenfeld JJ, Anzueto AR, Smith DB, Peters JI. O_2- and pneumonia-induced lung injury. II. Properties of pulmonary surfactant. J Appl Physiol 67:357–365 (1989).

574. King RJ, Klass DJ, Gikas EG, Clements JA. Isolation and apoproteins from canine surface-active material. Am J Physiol 224:788–795 (1973).

575. King RJ, MacBeth MC. Interactions of the lipid and protein components of pulmonary surfactant: Role of phosphatidylglyerol and calcium. Biochim Biophys Acta 647:159–168 (1981).

576. King RJ, Macbeth MC. Physiochemical properties of dipalmitoylphosphatidylcholine after interaction with an apolipoprotein of pulmonary surfactant. Biochim Biophys Acta 577:86–101 (1979).

577. King RJ, Simon D, Horowitz PM. Aspects of secondary and quaternary structure of surfactant protein A from canine lung. Biochim Biophys Acta 1001:294–301 (1989).

578. Kinsella JP, Abman SH. Controversies in the use of inhaled nitric oxide therapy in the newborn. Clin Perinatol 25:203–217 (1998).

579. Kinsella JP, Neish SR, Shaffer E, Abman SH. Low dose inhalational nitric oxide in persistent pulmonary hypertension of the newborn. Lancet 340:819–820 (1992).
580. Kinsella JP, Truog WE, Walsh WF, Goldberg RN, Bancalari E, Clark RH, Mayock DE, Redding GJ, deLemos RA, Sardesai S, McCurnin DC, Yoder BA, Moreland SG, Cutter GR, Abamn SH. Randomized, multicenter trial of inhaled nitric oxide and high frequency oscillatory ventilation in severe persistent pulmonary hypertension of the newborn. J Pediatr 131:55–62 (1997).
581. Klaus MH, Clements JA, Havel HJ. Composition of surface-active material isolated from beef lung. Proc Natl Acad Sci USA 47:1858–1859 (1961).
582. Knaus WA, Harrell FE, Fisher CJ, Wagner DP, Opal SM, Sadoff JC, Draper EA, Walawander CA, Conboy K, Grasela TM. The clinical evaluation of new drugs for sepsis: A prospective study design based on survival analysis. JAMA 270:1233–1241 (1993).
583. Knight DP, Knight JA. Pulmonary alveolar proteinosis in the newborn. Arch Pathol Lab Med 109:529–531 (1985).
584. Knobler CM, Desal RC. Phase transitions in monolayers. Annu Rev Phys Chem 43:207–236 (1992).
585. Kobayashi H, Tanaka N, Winkler M, Zapol WM. Combined effects of NO inhalation and intravenous $PGF_{2\alpha}$ on pulmonary circulation and gas exchange in an ovine ARDS model. Intensive Care Med 22:656–663 (1996).
586. Kobayashi T, Ganzuka M, Taniguchi J, Nitta K, Murakami S. Lung lavage and surfactant replacement for hydrochloric acid aspiration in rabbits. Acta Anaesthesiol Scand 34:216–221 (1990).
587. Kobayashi T, Grossman G, Robertson B, Ueda T. Effects of artificial and natural surfactant supplementation in immature newborn rabbits. J Jpn Med Soc Biol Interface 15:125–131 (1984).
588. Kobayashi T, Kataoka H, Murakami S. A case of isiopathic respiratory distress syndrome treated by newly-developed surfactant (Surfactant CK). J Jpn Med Soc Biol Interface 12:1–6 (1981).
589. Kobayashi T, Kataoka H, Ueda T, Murakami S, Takada Y, Kobuko M. Effect of surfactant supplementation and end expiratory pressure in lung-lavaged rabbits. J Appl Physiol 57:995–1001 (1984).
590. Kobayashi T, Nitta K, Takahashi K, Kurashima B, Robertson B, Suzuki Y. Activity of pulmonary surfactant after blocking the associated protein SP-A and SP-B. J Appl Physiol 71:530–536 (1991).
591. Kobayashi T, Robertson B, Grossmann G, Nitta K, Curstedt T, Suzuki Y. Exogenous porcine surfactant (Curosurf) is inactivated by monoclonal antibody to the surfactant-associated hydrophobic protein SP-B. Acta Paediatr 81:665–671 (1992).
592. Kobayashi Y, Nitta K, Ganzuka M, Inui S, Grossman G, Robertson B. Inactivation of exogenous surfactant by pulmonary edema fluid. Pediatr Res 29:353–356 (1991).
593. Kollef MH. Rescue therapy for the acute respiratory distress syndrome (editorial). Chest 111:845–846 (1997).
594. Kollef MH, Schuster DP. The acute respiratory distress syndrome. N Engl J Med 332:27–37 (1995).
595. Konishi M, Fujiwara T, Chida S, Maeta H, Shimada S, Kasai T, Fujii Y, Murakami

Y. A prospective, randomized trial of early vs late administration of a single dose of Surfactant-TA. Early Hum Dev 29:275–282 (1992).

596. Konishi M, Fujiwara T, Naito T, Takeuchi Y, Ogawa Y, Inukai K, Fujiwara M, Nakamura H, Hashimoto T. Surfactant replacement therapy in neonatal respiratory distress syndrome—A multicenter, randomized clinical trial: Comparison of high versus low dose of Surfactant-TA. Eur J Pediatr 147:20–25 (1988).

597. Korfhagen TR, Bruno MD, Ross GF, Huelsman KM, Ikegami M, Jobe AH, Wert SB, Stripp BR, Morris RE, Glasser SW, Bachurski CJ, Iwamoto HS, Whitsett JA. Altered surfactant function and structure in SP-A gene targeted mice. Proc Natl Acad Sci USA 93:9594–9599 (1996).

598. Kott AT, Gardner JW, Schechter RS, DeGroot W. The elasticity of pulmonary surfactant. J Colloid Interface Sci 47:265–266 (1974).

599. Krafft P, Fridrich P, Pernerstorfer T, Fitzgerald RD, Koc D, Schneider B, Hammerle AF, Steltzer H. The acute respiratory distress syndrome; definitions, severity, and clinical outcome. An analysis of 101 clinical investigations. Intensive Care Med 22:519–529 (1996).

600. Kramer HJ, Schmidt R, Gunther A, Becker G, Suzuki Y, Seeger W. ELISA technique for quantification of surfactant protein B (SP-B) in bronchoalveolar lavage fluid. Am J Respir Crit Care Med 152:1540–1544 (1995).

601. Krause MF, Lienhart H-G, Haberstroh J, Hoehn T, Shulte-Monting J, Leititis JU. Effect of inhaled nitric oxide on intrapulmonary right-to-left shunting in two rabbit models of saline lavage induced surfactant deficiency and meconium instillation. Eur J Pediatr 157:410–415 (1998).

602. Krueger P, Schalke M, Wang Z, Notter RH, Dluhy RA, Loesche M. Effect of hydrophobic surfactant peptides SP-B and SP-C on binary phospholipid monolayers. I. Fluorescence and dark-field microscopy. Biophys J 77:903–914 (1999).

603. Kuan S, Persson A, Parghi D, Crouch E. Lectin-mediated interactions of surfactant protein D with alveolar macrophages. Am J Respir Cell Mol Biol 10:430–436 (1994).

604. Kuan SF, Rust K, Crouch E. Interactions of surfactant protein D with bacterial lipopolysaccharides. J Clin Invest 90:97–106 (1992).

605. Kuint J, Reichman B, Neumann L, Shinwell ES. Prognostic value of the immediate response to surfactant. Arch Dis Child 71:F170–F173 (1994).

606. Kumar V, Cotran RS, Robbins SL. Basic Pathology. 6th ed. Philadelphia: WB Saunders (1997).

607. Kunc L, Kuncova R, Holusa R, Soldan F. Physical properties and biochemistry of lung surfactant following vagotomy. Respiration 35:192–197 (1978).

608. Kurdowska A, Carr FK, Stevens MD, Paughman RP, Martin TR. Studies on the interaction of IL-8 with human α_2-macroglobulin: Evidence of the presence of IL-8 complexed to α_2-macroglobulin in lung fluids of patients with adult respiratory distress syndrome. J Immunol 158:1930–1940 (1997).

609. Kurdowska A, Miller EJ, Noble JM, Baughman RP, Matthay MA, Brelsford WG, Cohen AB. Anti-IL-8 antibodies in alveolar fluid from patients with the adult respiratory distress syndrome. J Immunol 157:2699–2706 (1996).

610. Kuroki Y. Surfactant protein D. In: Robertson B, van Golde LMG, Batenburg JJE,

Eds. Pulmonary Surfactant: From Molecular Biology to Clinical Practice. Amsterdam: Elsevier Science Publishers, pp. 77–85 (1992).

611. Kuroki Y, Mason RJ, Voelker DR. Pulmonary surfactant apoprotein A structure and modulation of surfactant secretion by rat alveolar type II cells. J Biol Chem 263:3388–3394 (1988).

612. Kuroki Y, Shiratori M, Murata Y, Akino T. Surfactant protein D (SP-D) counteracts the inhibitory effect of surfactant protein A (SP-A) on phospholipid secretion from alveolar type II cells: Interaction of native SP-D with SP-A. Biochemistry 279:115–119 (1991).

613. Kuroki Y, Shiratori M, Ogasawara T, Tsuzuki A, Akino T. Characterization of pulmonary surfactant protein D: Its copurification with lipids. Biochim Biophys Acta 1086:185–190 (1991).

614. Kuroki Y, Voelker DR. Pulmonary surfactant proteins. J Biol Chem 269:25943–25946 (1994).

615. Kwong MS, Egan EA, Notter RH, Shapiro DL. A double blind clinical trial of calf lung surfactant extract for the prevention of hyaline membrane disease in extremely premature infants. Pediatrics 76:585–592 (1985).

616. Lachmann B. Animal models and clinical pilot studies of surfactant replacement in adult respiratory distress syndrome. Eur Respir J 2(Suppl):98S-103S (1989).

617. Lachmann B, Fujiwara T, Chida S, Morita T, Konishi M, Nakamura K, Maeta H. Surfactant replacement therapy in experimental adult respiratory distress syndrome (ARDS). In: Cosmi EV, Scarpelli EM, Eds. Pulmonary Surfactant System. Amsterdam: Elsevier, pp. 221–235 (1983).

618. Lachmann B, Hallman M, Bergman K-C. Respiratory failure following anti-lung serum: Study on mechanisms associated with surfactant system damage. Exp Lung Res 12:163–180 (1987).

619. Lachmann B, Robertson B, Vogel J. In vivo lung lavage as an experimental model of the respiratory distress syndrome. Acta Anaesthesiol Scand 24:231–236 (1980).

620. Lachmann B, van Daal G-J. Adult respiratory distress syndrome: Animal models. In: Robertson B, van Golde LMG, Batenburg JJ, Eds. Pulmonary Surfactant: From Molecular Biology to Clinical Practice. Amsterdam: Elsevier Science Publishers, pp. 635–663 (1992).

621. Lamm WJ, Albert RK. Surfactant replacement improves lung recoil in rabbit lungs after acid aspiration. Am Rev Respir Dis 142:1279–1283 (1990).

622. Lang MJ, Hall RT, Reddy NS, Kurth CG, Merritt TA. A controlled trial of human surfactant replacement therapy for severe respiratory distress syndrome in very low birth weight infants. J Pediatr 116:295–300 (1990).

623. Langley SC, Kelly FJ. N-Acetylcysteine ameliorates hyperoxic lung injury in the preterm guinea pig. Biochem Pharmacol 45:841–846 (1993).

624. Langmuir I. The constitution and fundamental properties of solids and liquids. II. Liquids. J Am Chem Soc 39:1848–1906 (1917).

625. Laurent T, Markert M, Feihl F, Schaller M-D, Perret C. Oxidant-antioxidant balance in granulocytes during ARDS. Chest 109:163–166 (1996).

626. Lauterbach R, Zembala M. Pentoxifylline reduces plasma tumour necrosis factor-alpha concentration in premature infants with sepsis. Eur J Pediatr 155:404–409 (1996).

627. Law WR, Nadkarni VM, Fletcher MA, Nevola JJ, Eckstein JM, Quance J, McKenna TM, Lee CH, Williams TJ. Pentoxifylline treatment of sepsis in conscious Yucatan minipigs. Circ Shock 37:291–300 (1992).

628. Leach CL, Fuhrman BP, Morin FC 3d, Rath MG. Perfluorocarbon-assisted gas exchange (partial liquid ventilation) in respiratory distress syndrome: A prospective, randomized, controlled study. Crit Care Med 21:1270–1278 (1993).

629. Leach CL, Greenspan JS, Rubenstein SD, Shaffer TH, Wolfson MR, Jackson JC, DeLemos R, Fuhrman BP. Partial liquid ventilation with perflubron in premature infants with severe respiratory distress syndrome. N Engl J Med 335:761–767 (1996).

630. Leach CL, Holm BA, Morin FC, Fuhrman BP, Papo MC, Steinhorn D, Hernan LJ. Partial liquid ventilation in premature lambs with respiratory distress syndrome: Efficacy and compatibility with exogenous surfactant. J Pediatr 126:412–420 (1995).

631. Leary JF, Finkelstein JN, Notter RH, Shapiro DL. Isolation of type II pneumocytes by laser flow cytometry. Am Rev Respir Dis 125:326–330 (1982).

632. Leary JF, Finkelstein JN, Notter RH, Shapiro DL. A quantitative study of the development of type II pneumocytes in fetal lung. Cytometry 7:431–438 (1986).

633. Lee Y, Chan SJ. Effect of lysolecithin on the structure and permeability of lecithin bilayer vesicles. Biochemistry 16:1303–1309 (1977).

634. Leff JA, Wilke CP, Hybertson BM, Shanley PF, Beehler CJ, Repine JE. Postinsult treatment with N-acetyl-L-cysteine decreases IL-1-induced neutrophil influx and lung leak in rats. Am J Physiol 265:L501–L506 (1993).

635. LeVine AM, Lotze A, Stanley S, Stroud C, O'Donnell R, Whitsett J, Pollack MM. Surfactant content in children with inflammatory lung disease. Crit Care Med 24:1062–1067 (1996).

636. Levy PC, Utell MC, Sickel JZ, Apostolakos MJ. The acute respiratory distress syndrome: Current trends in pathogenesis and management. Compr Ther 21:438–444 (1995).

637. Lewis J, Ikegami M, Higuchi R, Jobe A, Absolom D. Nebulized vs. instilled exogenous surfactant in an adult lung injury model. J Appl Physiol 71:1270–1276 (1991).

638. Lewis J, McCaig L, Hafner D, Spragg R, Veldhuizen R, Kerr C. Dosing and delivery of a recombinant surfactant in lung-injured sheep. Am J Respir Crit Care Med 159:741–747 (1999).

639. Lewis JF, Goffin J, Yue P, McCaig LA, Bjarneson D, Veldhuizen RAW. Evaluation of exogenous surfactant treatment strategies in an adult model of acute lung injury. J Appl Physiol 80:1156–1164 (1996).

640. Lewis JF, Ikegami M, Jobe AH. Altered surfactant function and metabolism in rabbits with acute lung injury. J Appl Physiol 69:2303–2310 (1990).

641. Lewis JF, Ikegami M, Jobe AH. Metabolism of exogenously administered surfactant in the acutely injured lungs of adult rabbits. Am Rev Respir Dis 145:19–23 (1992).

642. Lewis JF, Ikegami M, Jobe AH, Tabor H. Aerosolized surfactant treatment of preterm lambs. J Appl Physiol 70:869–876 (1991).

643. Lewis JF, Jobe AH. Surfactant and the adult respiratory distress syndrome. Am Rev Respir Dis 147:218–233 (1993).

644. Lewis JF, Veldhuizen RAW. Factors influencing efficacy of exogenous srufactant in acute lung injury. Biol Neonate 67(Suppl):48–60 (1995).

645. Li XY, Donaldson K, Brown D, MacNee W. The role of tumor necrosis factor in increased airspace epithelial permeability in acute lung inflammation. Am J Respir Cell Mol Biol 13:185–195 (1995).

646. Li Z, Daniel EE, Lane CG, Arnaout MA, O'Bryne PM. Effect of an anti-Mo1 on ozone-induced airway inflammation and airway hyperresponsiveness in dogs. Am J Physiol 263:L723–L726 (1992).

647. Liechty EA, Donovan E, Purohit D, Gilhooly J, Feldman B, Noguchi A, Denson SE, Sehgal SS, Gross I, Steven D, Ikegami M, Zachman RD, Carrier ST, Gunkel JH, Gold AJ. Reduction of neonatal mortality after multiple doses of bovine surfactant in low birth weight neonates with respiratory distress syndrome. Pediatrics 88:19–28 (1991).

648. Liggins CG, Howie RN. A controlled trial of antepartum glucocorticoid treatment for prevention of the respiratory distress syndrome in premature infants. Pediatrics 50:515–525 (1972).

649. Liggins GC. Premature delivery of fetal lambs infused with glucocorticoids. J Endocrinol 45:515–523 (1969).

650. Liley HG, White RT, Warr RG, Benson BJ, Hawgood S, Ballard PL. Regulation of messenger RNAs for the hydrophobic surfactant proteins in human lung. J Clin Invest 83:1191–1197 (1989).

651. Lin WH, Cramer SG, Turcotte JG, Thrall RS. A diether phosphonolipid surfactant analog, DEPN-8, is resistant to phospholipase-C cleavage. Respiration 64:96–101 (1997).

652. Lindsey HJ, Kisala JM, Ayala A, Lehman D, Hedron CD, Chaudry IH. Pentoxifylline attentuates oxygen-induced lung injury. J Surg Res 56:543–548 (1994).

653. Lipp MM, Lee KYC, Waring A, Zasadinski JA. Fluorescence, polarized fluorescence, and Brewster-angle microscopy of palmitic acid and lung surfactant protein B monolayers. Biophys J 72:2783–2804 (1997).

654. Lipp MM, Lee KYC, Zasadzinski JA, Waring A. Protein and lipid interactions in lung surfactant monolayers. Prog Colloid Polym Sci 103:268–279 (1997).

655. Lipp MM, Lee KYC, Zasadzinski JA, Waring AJ. Design and performance of an integrated fluorescence, polarized fluorescence, and Brewster angle microscope/Langmuir trough assembly for the study of lung surfactant monolayers. Rev Sci Instrum 68:2574–2583 (1997).

656. Lipp MM, Lee KYC, Zasadzinski JA, Waring AJ. Phase and morphology changes in lipid monolayers induced by SP-B protein and its amino-terminal peptide. Science 273:1196–1199 (1996).

657. Liu H, Lu RZ, Turcotte JG, Notter RH. Dynamic interfacial properties of surface-excess films of phospholipids and phosphonolipid analogs. I. Effects of pH. J Colloid Interface Sci 167:378–390 (1994).

658. Liu H, Turcotte JG, Notter RH. Dynamic interfacial properties of surface-excess films of phospholipid and phosphonolipid analogs: II. Effects of chain linkage and headgroup structure. J. Colloid Interface Sci 167:391–400 (1994).

659. Liu H, Turcotte JG, Notter RH. Thermotropic behavior of structurally-related phospholipids and phosphonolipid analogs of lung surfactant glycerophospholipids. Langmuir 11:101–107 (1995).

660. Liu M, Wang L, Li E, Enhorning G. Pulmonary surfactant will secure free airflow through a narrow tube. J Appl Physiol 71:742–748 (1991).

661. Loewen GM, Holm BA, Milanowski L, Wild LM, Notter RH, Matalon S. Alveolar hyperoxic injury in rabbits receiving exogenous surfactant. J. Appl. Physiol 66:1987–1992 (1989).

662. Long W, Corbet A, Cotton R, Courtney S, McGuiness G, Walter D, Watts J, Smyth J, Bard H, Chernick V. A controlled trial of synthetic surfactant in infants weighing 1,250 g or more with respiratory distress syndrome. N Engl J Med 325:1696–1703 (1991).

663. Long W, Thompson R, Sundell H, Schumacher R, Volberg F, Guthrie R. Effects of two rescue doses of a synthetic surfactant on mortality rate and survival without bronchopulmonary dysplasia in 700 to 1350 gram infants with respiratory distress syndrome. J Pediatr 11:595–605 (1991).

664. Long W, Zucker J, Kraybill E. Symposium on synthetic surfactant II: Perspective and commentary. J Pediatr 126:S1–S4 (1995).

665. Longmuri KJ, Resele-Tiden C, Rossi ME. Fatty acids of pulmonary surfactant phosphatidylcholine from fetal rabbit lung tissue in culture. J Lipid Res 29:1065–1077 (1988).

666. Longo ML, Bisagno AM, Zasadzinski JAN, Bruni R, Waring A. A function of lung surfactant protein SP-B. Science 261:453–456 (1993).

667. Longo ML, Waring A, Zasadzinski JAN. Lipid bilayer surface association of lung surfactant protein SP-B amphiphatic segment detected by flow immunofluorescence. Biophys J 63:760–773 (1992).

668. Lotze A, Knight GR, Martin GR, Bulas DI, Hull WM, O'Donnell RM, Whitsett JA, Short BL. Improved pulmonary outcome after exogenous surfactant therapy for respiratory failure in term infants requiring extracorporeal membrane oxygenation. J Pediatr 122:261–268 (1993).

669. Lowry OH, Rosebrough NJ, Farr AL, Randall RJ. Protein measurement with the Folin phenol reagent. J Biol Chem 132:265–275 (1951).

670. Lu J, Wiedemann H, Holmskov U, Thiel S, Timpl R, Reid KBM. Structural similarity between lung surfactant protein D and conglutinin. Two distinct C-type lectins containing collagen-like sequences. Eur J Biochem 215:793–799 (1993).

671. Lu J, Willis AC, Reid KB. Purification, characterization and cDNA cloning of human lung surfactant protein D. Biochem J 284:795–802 (1992).

672. Lu Q, Mourgeon E, Law-Koune J, Roche S, Vezinet C, Abdennour L, Vicaut E, Puybasset L, Diaby M, Coriat P, Rouby J. Dose-response curves of inhaled nitric oxide with and without intravenous almitrine in nitric oxide–responding patients with acute respiratory distress syndrome. Anesthesiology 83:929–943 (1995).

673. Lu RZ, Turcotte JG, Lin WH, Steim JM, Notter RH. Differential scanning calorimetry studies of phosphonolipid analogs of lung surfactant phospholipids. J Colloid Interface Sci 154:24–34 (1992).

674. Lucas R, Lou J, Morel DR, Ricou B, Suter PM, Grau GE. TNF receptors in the

microvascular pathology of acute respiratory distress syndrome and cerebral malaria. J Leukoc Biol 61:551–558 (1997).

675. Luce JM. Acute lung injury and the acute respiratory distress syndrome. Crit Care Med 26:369–376 (1998).

676. Luce JM, Montgomery AB, Marks JD, Turner J, Metz CA, Murray JF. Ineffectiveness of high-dose methylprednisone in preventing parenchymal lung injury and improving mortality in patients with septic shock. Am Rev Respir Dis 138:62–68 (1988).

677. Lutz C, Carney D, Finck C, Picone A, Gatto L, Paskanik A, Langenbeck E, Nieman G. Aerosolized surfactant improves pulmonary function in endotoxin-induced lung injury. Am J Respir Crit Care Med 158:840–845 (1998).

678. Lutz CJ, Picone A, Gatto LA, Paskanik A, Landas S, Nieman G. Exogenous surfactant and positive end-expiratory pressure in the treatment of endotoxin-induced lung injury. Crit Care Med 26:1379–1389 (1998).

679. Ma J, Koppennol S, Yu H, Zografti G. Effects of a cationic and hydrophobic peptide KL4 on model lung surfactant lipid monolayers. Biophys J 74:1899–1907 (1998).

680. Mabrey S, Sturtevant JM. Investigation of phase transition of lipids and lipid mixtures by high sensitivity differential scanning calorimetry. Proc Natl Acad Sci USA 73:3862–3866 (1976).

681. Macklem PT, Proctor DF, Hogg JC. The stability of peripheral airways. Respir Physiol 8:191–203 (1970).

682. Macnaughton PD, Evans TW. The effect of exogenous surfactant therapy on lung function following cardiopulmonary bypass. Chest 105:421–425 (1994).

683. Maeta H, Vidyasagar D, Raju T, Bhat R, Matsuda H. Early and late surfactant treatment in baboon model of hyaline membrane disease. Pediatrics 81:277–283 (1988).

684. Maeta H, Vidyasagar D, Raju T, Bhat R, Matsuda H. Response to bovine surfactant (Surfactant TA) in two different HMD models (lambs and baboons). Eur J Pediatr 147:162–167 (1988).

685. Magoon MW, Wright JR, Baritussio A, Williams MC, Goerke J, Benson BJ, Hamilton RL, Clements JA. Subfractionation of lung surfactant: Implications for metabolism and surface activity. Biochim Biophys Acta 750:18–31 (1983).

686. Malcolm JD, Elliott CD. Interfacial tension from height and diameter of a single profile drop of captive bubble. Can J Chem Eng 58:151–153 (1980).

687. Malik AB. Pulmonary edema after pancreatitis: Role of humoral factors. Circ Shock 10:71–80 (1983).

688. Mandell GL. ARDS, neutrophils and pentoxifylline. Am Rev Respir Dis 138:1103–1105 (1988).

689. Manz-Keinke H, Plattner H, Schlepper-Schafer J. Lung surfactant protein A (SP-A) enhances serum-independent phagocytosis of bacteria by alveolar macrophages. Eur J Cell Biol 57:95–100 (1992).

690. Marklund S, Karlsson K. Extracellular superoxide dismutase, distribution in the body and therapeutic applications. In: Advances in Experimental Medicine and Biology. Orlando: Plenum Press, pp. 1–4 (1990).

691. Marklund SL. Extracellular superoxide dismutase and other superoxide isoenzymes in tissues from nine mammalian species. Biochem J 222:649–655 (1984).

692. Marks LB, Notter RH, Oberdoerster G, McBride JT. Ultrasonic and jet aerosolization of phospholipids and the effects on surface activity. Pediatr Res 17:742–747 (1984).

693. Marks LB, Oberdoerster G, Notter RH. Generation and characterization of aerosols of dispersed surface active phospholipids by ultrasonic and jet nebulization. J Aerosol Sci 14:683–694 (1983).

694. Martin TR. Cytokines and the acute respiratory distress syndrome (ARDS): A question of balance. Nature Med 3:272–273 (1997).

695. Mason RJ, Dobbs LG, Greenleaf RD, Williams MC. Alveolar type II cells. Fed Proc 36:2697–2702 (1977).

696. Mason RJ, Greene K, Voelker DR. Surfactant protein A and surfactant protein D in health and disease. Am J Physiol 275:L1–L13 (1998).

697. Mason RJ, Nellenbogen J, Clements JA. Isolation of disaturated phosphatidylcholine with osmium tetroxide. J Lipid Res 17:281–284 (1976).

698. Mason RJ, Williams MC. Alveolar type II cells. In: Crystal RG, West JB, Weibel ER, Barnes PJ, Eds. The Lung: Scientific Foundations. 2nd ed. Philadelphia: Lipincott-Raven, pp. 235–246 (1997).

699. Matalon S, DeMarco V, Haddad IY, Myles C, Skimming JW, Schurch S, Cheng S, Cassin S. Inhaled nitric oxide injures the pulmonary surfactant system of lambs in vivo. Am J Physiol 270:L273–L280 (1996).

700. Matalon S, Egan EA. Effects of 100% O_2 breathing on permeability of alveolar epithelium to solute. J Appl Physiol 50:859–863 (1981).

701. Matalon S, Egan EA. Interstitial fluid volumes and albumin spaces in pulmonary oxygen toxicity. J Appl Physiol 57:1767–1722 (1984).

702. Matalon S, Holm BA, Baker RR, Whitfield K, Freeman BA. Characterization of antioxidant activities of pulmonary surfactant mixtures. Biochim Biophys Acta 1035:121–127 (1990).

703. Matalon S, Holm BA, Loewen GM, Baker RR, Notter RH. Sublethal hyperoxic injury to the alveolar epithelium and the pulmonary surfactant system. Exp Lung Res 14:1021–1033 (1988).

704. Matalon S, Holm BA, Notter RH. Mitigation of pulmonary hyperoxic injury by administration of exogenous surfactant. J Appl Physiol 62:756–761 (1987).

705. Matthay MA. The acute respiratory distress syndrome (editorial). N Engl J Med 334:1469–1470 (1996).

706. Matthay MA, Pittet JF, Jayr C. Just say NO to inhaled nitric oxide for the acute respiratory distress syndrome. Crit Care Med 26:1–2 (1998).

707. Matute-Bello G, Liles WC, Radella F. Neutrophil apoptosis in the acute respiratory distress syndrome. Am J Respir Crit Care Med 156:1969–1977 (1997).

708. McConnell HM. Structures and transitions in lipid monolayers at the air-water interface. Ann Rev Phys Chem 42:171–195 (1991).

709. McCormack FX, Calvert HM, Watson PA, Smith DL, Mason RJ, Voelker DR. The structure and function of surfactant protein A. J Biol Chem 269:5833–5841 (1994).

710. McIver D, Possmayer S, Schurch S. A synthetic emulsion reproduces the functional properties of pulmonary surfactant. Biochim Biophys Acta 751:74–80 (1985).
711. McLean LR, Lewis JE. Biomimetic pulmonary surfactant. Life Sci 56:363–378 (1995).
712. McLean LR, Lewis JE, Hagaman KA, Owen TJ, Jackson RL. Amphipathic alpha-helical peptides based on surfactant apoprotein SP-A. Biochim Biophys Acta 1166:31–38 (1993).
713. McLean LR, Lewis JE, Krstenansky JL, Hagaman KA, Cope AS, Olsen KF, Matthews ER, Uhrhammer DC, Owen TJ, Payne MH. An amphipathic α-helical decapeptide and phosphatidylcholine is an effective synthetic lung surfactant. Am Rev Respir Dis 147:462–465 (1993).
714. Mead J, Whittenberger JL, Radford EP. Surface tension as a factor in pulmonary volume-pressure hysteresis. J Appl Physiol 10:191–196 (1957).
715. Meban C. Surface dilatational viscosity of pulmonary surfactant films. Biorheology 15:251–259 (1978).
716. Meduri GU. The role of host defense response in the progression and outcome of ARDS: Pathophysiological correlations and response to glucocorticoid treatment. Eur Respir J 9:2650–2670 (1996).
717. Meduri GU, Headley AS, Golden E, Carson SJ, Umberger RA, Kelso T, Tolley EA. Effect of prolonged methylprednisone therapy in unresolving acute respiratory distress syndrome: A randomized controlled trial. JAMA 280:159–165 (1998).
718. Meduri GU, Headley S, Tolley E, Shelby M, Stentz F, Postlewaite A. Plasma and BAL cytokine response to corticosteroid rescue treatment in late ARDS. Chest 108:1315–1325 (1995).
719. Meduri GU, Healey S, Kohler G, Stentz F, Tolley E, Umberger R, Leeper K. Persistent elevation of inflammatory cytokines predicts poor outcome in ARDS. Plasma IL-1β and IL-6 levels are consistent and efficient predictors of outcome over time. Chest 107:1062–1073 (1995).
720. Meduri GU, Kohler G, Tolley E, Headley AS, Stentz F, Postlethwaite A. Inflammatory cytokines in the BAL of patients with ARDS: Persistent elevation over time predicts poor outcome. Chest 108:1303–1314 (1995).
721. Members of the American College of Chest Physicians Society of Critical Care Medicine Consensus Conference Committee. Definitions for sepsis and organ failure and guidelines for the use of innovative therapies for sepsis. Crit Care Med 20:864–874 (1992).
722. Mendelson CR, Alcorn JL, Gao E. The pulmonary surfactant protein genes and their regulation in fetal lung. Semin Perinatol 17:223–232 (1993).
723. Mendenhall RM. Surface spreading of lung alveolar surfactant. Respir Physiol 16:175–178 (1972).
724. Mendenhall RM, Sun CN, Mendenhall AL. Lung alveolar surfactant and the Thompson-Marangoni effect. Respir Physiol 2:360–374 (1967).
725. Mercier CE, Soll RF. Clinical trials of natural surfactant extract in respiratory distress syndrome. Clin Perinatol 20:711–735 (1993).
726. Merrifield B. Solid phase synthesis. Science 232:341–347 (1986).
727. Merrifield RB. Solid phase peptide synthesis. I. The synthesis of a tetrapeptide. J Am Chem Soc 85:2149–2154 (1963).

728. Merritt TA, Hallman M, Berry C, Pohjavuori M, Edwards DK, Jaaskelainen J, Grafe MR, Vaucher Y, Wozniak P, Heldt G, Rapola J. Randomized, placebo-controlled trial of human surfactant given at birth versus rescue administration in very low birth weight infants with lung immaturity. J Pediatr 118:581–594 (1991).

729. Merritt TA, Hallman M, Bloom BT, Berry C, Benirschke K, Sahn D, Key T, Edwards D, Jarvenpaa AJ, Pohjavouri M. Prophylactic treatment of very premature infants with human surfactant. N Engl J Med 315:785–790 (1986).

730. Merritt TA, Kheiter A, Cochrane CG. Positive end-expiratory pressure during KL4 surfactant instillation enhances intrapulmonary distribution in a simian model of respiratory distress syndrome. Pediatr Res 38:211–217 (1995).

731. Mescher EJ, Dobbs LG, Mason RJ. Cholera toxin stimulates secretion of saturated phosphatidylcholine and increases cAMP in isolated rat alveolar type II epithelial cells. Exp Lung Res 5:173–182 (1983).

732. Metcalfe IL, Enhorning G, Possmayer F. Pulmonary surfactant–associated proteins: Their role in expression of surface activity. J Appl Physiol 49:34–41 (1980).

733. Michael JR, Barton RG, Saffle JR, Mone M, Markewitz BA, Hillier K, Elstad MR, Campbell EJ, Troyer BE, Whatley RE, Liou TG, Samuelson WM, Carveth HJ, Hinson DM, Morris SE, et al. Inhaled nitric oxide versus conventional therapy: Effect on oxygenation in ARDS. Am J Respir Crit Care Med 157:1372–1380 (1998).

734. Michie HR, Mangue DR, Spriggs A, Revhaug S, O'Dwyer CA, Dinarello, Carem A, Wolf SM, Wilmore DW. Detection of circulating tumor necrosis factor after endotoxin administration. N Engl J Med 318:1481–1486 (1988).

735. Milberg JA, Davis DR, Steinberg KP, Hudson LD. Improved survival of patients with acute respiratory distress syndrome. JAMA 273:306–309 (1995).

736. Miller EJ, Cohen AB, Matthay MA. Increased interleukin-8 concentrations in the pulmonary edema fluid in patients with acute respiratory distress syndrome from sepsis. Crit Care Med 24:1448–1454 (1996).

737. Miller EJ, Cohen AB, Nagao S, Griffith DG, Maunder RJ, Martin TR, Weiner-Kronish JP, Sticherling M, Christophers E, Matthay MA. Elevated levels of NAP-1/interleukin-8 are present in the airspaces of patients with adult respiratory distress syndrome and are associated with increased mortality. Am Rev Respir Dis 146:427–432 (1992).

738. Milner AD, Vyas H, Hopkin IE. Effects of artifical surfactant on lung function and blood gases in idopathic respiratory distress syndrome. Arch Dis Child 58:458–460 (1983).

739. Miyazawa N, Suzuki S, Akahori T, Okubo T. Effects of surfactant on the in vivo alveolar surface-to-volume ratio. J Appl Physiol 80:86–90 (1996).

740. Mizuno K, Ikegami M, Chen C-M, Ueda T, Jobe AH. Surfactant protein-B supplementation improves in vivo function of a modified natural surfactant. Pediatr Res 37:271–276 (1995).

741. Moen A, Yu X-Q, Almaas R, Curstedt T, Saugstad OD. Acute effects on systemic circulation after intratracheal instillation of Curosurf or Survanta in surfactant-depleted newborn piglets. Acta Paediatr 87:297–303 (1998).

742. Mohr MS. The New Games Treasury. Boston: Houghton-Mifflen (1997).

743. Montravers P, Fagon JY, Gilbert C, Blanchet F, Novara A, Chastre J. Pilot study

of cardiopulmonary risk from pentoxifylline in adult repiratory distress syndrome. Chest 103:1017–1022 (1993).

744. Morgan TE, Finley TN, Fialkow H. Comparison of the composition and surface activity of "alveolar" and whole lung lipids in the dog. Biochim Biophys Acta 106:403–413 (1965).

745. Morgenroth K, Bolz J. Morphological features of the interaction between mucus and surfactant on the bronchial mucosa. Respiration 47:225–231 (1985).

746. Morley CJ. Clinical experience with artificial surfactants. In: Robertson B, van Golde LMG, Batenburg JJ, Eds. Pulmonary Surfactant: From Molecular Biology to Clinical Practice. Amsterdam: Elsevier Science Publishers, pp. 605–633 (1992).

747. Morley CJ, Bagham AD, Miller N, Davis JA. Dry artificial lung surfactant and its effect on very premature babies. Lancet 1:64–68 (1981).

748. Morris PE, Bernard GR. Significance of glutathione in lung disease and implications for therapy. Am J Med Sci 307:119–127 (1994).

749. Morrow MR, Pérez-Gil J, Simatos GA, Boland C, Stewart J, Absolom D, Sarin V, Keough KMW. Pulmonary surfactant–associated protein SP-B has little effect on acyl chains in dipalmitoylphosphatidylcholine dispersions. Biochemistry 32:4397–4402 (1993).

750. Moses D, Holm BA, Spitale P, Liu M, Enhorning G. Inhibition of pulmonary surfactant function by meconium. Am J Obstet Gynecol 164:477–481 (1991).

751. Moses R, Schleiffenblaum B, Groscurth P. Interleukin-1 and tumour necrosis factor stimulate human endothelial cells to promote transendothelial neutrophil passage. J Clin Invest 83:444–455 (1989).

752. Moulton SL, Krous HF, Merritt TA, Odell RM, Gangitano E, Cornish JD. Congenital pulmonary alveolar proteinosis: Failure of treatment with extracorporeal life support. J Pediatr 120:297–302 (1992).

753. Mueller P, Chien TF, Rudy B. Formation and properties of cell-size lipid bilayer vesicles. Biophys J 44:375–381 (1983).

754. Munden JW, Blois DW, Swarbrick J. Surface pressure relaxation and hysteresis in stearic acid monolayers at the air-water interface. J Pharm Sci 58:1308–1312 (1969).

755. Munden JW, Swarbrick J. Time-dependent surface behavior of dipalmitoyl lecithin and lung alveolar surfactant molecules. Biochim Biophys Acta 291:244–250 (1973).

756. Murray JF, Matthay MA, Luce JM, Flick MR. An expanded definition of the adult respiratory distress syndrome. Am Rev Respir Dis 138:720–723 (1988).

757. Mustard RA, Fusher J, Hayman S, Matlow A, Mullen JBM, Odumeru J, Roomi MW, Schouten BD, Swanson HT. Cardiopulmonary responses to *Pseudomonas* septicemia in swine: An improved model of the adult respiratory distress syndrome. Lab Anim Sci 39:37–43 (1989).

758. Nag K, Boland C, Rich NH, Keough KMW. Epifluorescence microscopic observation of monolayers of dipalmitoyl phosphatidylcholine: Dependence of domain size on compression rates. Biochim Biophys Acta 1068:157–160 (1991).

759. Nag K, Keough KMW. Epifluorescence microscopic studies of monolayers containing mixtures of dioleoyl- and dipalmitoylphosphatidylcholines. Biophys J 65:1019–1026 (1993).

760. Nag K, Perez-Gil J, Cruz A, Rich NH, Keough KMW. Spontaneous formation of interfacial lipid-protein monolayers during adsorption from vesicles. Biophys J 71:1356–1363 (1996).

761. Nag K, Perez-Gil J, Ruano M, Worthman LD, Stewart J, Casals C, Keough KMW. Phase transitions in films of lung surfactant at the air-water interface. Biophys J 74:2983–2995 (1998).

762. Nag K, Taneva SG, Perez-Gil J, Cruz A, Keough KMW. Combinations of fluorescently labeled pulmonary surfactant proteins SP-B and SP-C in phospholipid films. Biophys J 72:2638–2650 (1997).

763. Nakagawa TA, Morris A, Gomez RJ, Johnston SJ, Sharkey PT, Zaritsky AL. Dose response to inhaled nitric oxide in pediatric patients with pulmonary hypertension and acute respiratory distress syndrome. J Pediatr 131:63–69 (1997).

764. Nakos G, Kitsiouli EI, Tsangaris I, Lekka ME. Bronchoalveolar lavage fluid characteristics of early intermediate and late phases of ARDS. Intensive Care Med 24:296–303 (1998).

765. National Center for Health Statistics, MacDorman MF, Rosenberg HM. Trends in infant mortality by cause of death and other characteristics, 1960–1988. Vital and health statistics. (DHHS publication PHS 93-1857):1–51 (1993).

766. National Research Council of the United States of America. International Critical Tables. New York: McGraw-Hill (1928).

767. Neiman GF, Bredenberg CE. High surface tension pulmonary edema induced by detergent aerosol. J Appl Physiol 58:129–136 (1985).

768. Nelin LD, Hoffman GM. The use of inhaled nitric oxide in a wide variety of clinical problems. Pediatr Clin North Am 45:531–548 (1998).

769. Neonatal Inhaled Nitric Oxide Study Group. Inhaled nitric oxide in full-term and nearly full-term infants with hypoxic respiratory failure. N Engl J Med 336:597–604 (1997).

770. Nesti FD, Fuhrman BP, Steinhorn DM, Papo MC, Hernan LJ, Duffy LC, Fisher JE, Leach CL, Paczan PR, Burak BA. Perfluorocarbon-associated gas exchange in gastric aspiration. Crit Care Med 22:1445–1452 (1994).

771. Nicholas TE, Barr HA. The release of surfactant in the rat lung by brief periods of hyperventilation. Respir Physiol 52:69–83 (1983).

772. Nicholas TE, Doyle IR, Bersten AD. Surfactant replacement therapy in ARDS: White knight or noise in the system? Thorax 52:195–197 (1997).

773. Nieman G, Gatto L, Paskanik A, Yang B, Fluck R, Picone A. Surfactant replacement in the treatment of sepsis-induced adult respiratory distress syndrome in pigs. Crit Care Med 24:1025–1033 (1996).

774. Niewoehner D, Rice K, Sinha A, Wangensteen D. Injurious effects of lysophosphatidylcholine on barrier properties of alveolar epithelium. J Appl Physiol 63:1979–1986 (1987).

775. Nilsson G, Gustafsson M, Vandenbussche G, Veldhuizen E, Griffiths W, Sjovall J, Haagsman H, Ruysschaert J-M, Robertson B, Curstedt T, Johansson J. Synthetic peptide–containing surfactants: Evaluation of transmembrane versus amphipathic helices and surfactant protein C poly-valyl to poly-leucyl substitution. Eur J Biochem 255:116–124 (1998).

776. Nitta K, Kobayashi T. Impairment of surfactant activity and ventilation by proteins in lung edema fluid. Respir Physiol 95:43–51 (1994).

777. Noack G, Berggren P, Curstedt T, Grossman G, Herin P, Mortensson W, Nilsson R, Robertson B. Severe neonatal respiratory distress syndrome treated with the isolated phospholipid fraction of natural surfactant. Acta Pediatr Scand 76:697–705 (1987).

778. Nogee LM, deMello DE, Dehner LP, Colten HR. Brief report: Deficiency of pulmonary surfactant protein B in congenital alveolar proteinosis. N Engl J Med 328:406–410 (1993).

779. Nogee LM, Garnier G, Dietz HC, Singer L, Murphy AM, deMello DE, Colten HR. A mutation in the surfactant protein B gene responsible for fatal neonatal respiratory disease in multiple kindreds. J Clin Invest 93:1860–1863 (1994).

780. Nohara K, Muramatsu K, Oda T. Six cases of RDS treated with Surfactant CK. J Jpn Med Soc Biol Interface 14:61–66 (1983).

781. Northway WHJ, Rosan RC, Porter DY. Pulmonary disease following respirator therapy of hyaline membrane disease. N Engl J Med 276:357–368 (1967).

782. Nosaka S, Sakai T, Yonekura K, Yoshikawa K. Surfactant for adults with respiratory distress failure. Lancet 336:947–948 (1990).

783. Notter R, Taubold R, Finkelstein J. Comparative adsorption of natural lung surfactant, extracted phospholipids, and synthetic phospholipid mixtures. Chem Phys Lipids 33:67–80 (1983).

784. Notter RH. Physical chemistry and physiological activity of pulmonary surfactant. In: Shapiro DL, Notter RH, Eds. Surfactant Replacement Therapy. New York: AR Liss, pp. 19–70 (1989).

785. Notter RH. Surface chemistry of pulmonary surfactant: the role of individual components. In: Roberson B, van Golde LMG, Batenburg JJ, Eds. Pulmonary Surfactant. Amsterdam: Elsevier Science Publishers, pp. 17–53 (1984).

786. Notter RH. Biophysical behavior of lung lung surfactant: Implications for respiratory physiology and pathophysiology. Semin Perinatol 12:180–212 (1988).

787. Notter RH, Egan EA, Kwong MS, Holm BA, Shapiro DL. Lung surfactant replacement in premature lambs with extracted lipids from bovine lung lavage: Effects of dose, dispersion technique, and gestational age. Pediatr Res 19:569–577 (1985).

788. Notter RH, Finkelstein JN. Pulmonary surfactant: An interdisciplinary approach. J Appl Physiol 57:1613–1624 (1984).

789. Notter RH, Holcomb S, Mavis RD. Dynamic surface properties of phosphatidylglycerol-dipalmitoyl phosphatidylcholine mixed films. Chem Phys Lipids 27:305–319 (1980).

790. Notter RH, Morrow PE. Pulmonary surfactant: A surface chemistry viewpoint. Ann Biomed Eng 3:119–159 (1975).

791. Notter RH, Penney DP, Finkelstein JN, Shapiro DL. Adsorption of natural lung surfactant and phospholipid extracts related to tubular myelin formation. Pediatr Res 20:97–101 (1986).

792. Notter RH, Shapiro DL. Lung surfactants for replacement therapy: Biochemical, biophysical, and clinical aspects. Clin Perinatol 14:433–479 (1987).

793. Notter RH, Shapiro DL, Ohning B, Whitsett JA. Biophysical activity of synthetic

phospholipids combined with purified lung surfactant 6000 dalton apoproteins. Chem Phys Lipids 44:1–17 (1987).

794. Notter RH, Smith S, Taubold RD, Finkelstein JN. Path dependence of adsorption behavior of mixtures containing dipalmitoyl phosphatidylcholine. Pediatr Res 16:515–519 (1982).

795. Notter RH, Tabak SA, Holcomb S, Mavis RD. Post-collapse dynamic surface pressure relaxation in binary surface films containing dipalmitoyl phosphatidylcholine. J Colloid Interface Sci 74:370–377 (1980).

796. Notter RH, Tabak SA, Mavis RD. Surface properties of binary mixtures of some pulmonary surfactant components. J Lipid Res 21:10–22 (1980).

797. Notter RH, Taubold R, Mavis RD. Hysteresis in saturated phospholipid films and its potential relevance for lung surfactant function in vivo. Exp Lung Res 3:109–127 (1982).

798. Notter RH, Wang Z. Pulmonary surfactant: Physical chemistry, physiology and replacement. Rev Chem Eng 13:1–118 (1997).

799. Novotny WE, Hudak BB, Matalon S, Holm BA. Hyperoxic lung injury reduces exogenous surfactant clearance in vitro. Am J Respir Crit Care Med 151:1843–1847 (1995).

800. Ohlsson K, Bjork P, Bergenfeldt M, Hageman R, Thompson RC. Interleukin-1 receptor antagonist reduces mortality from endotoxin shock. Nature 348:550–552 (1990).

801. Okamoto K, Hamaguchi M, Kukita I, Kikuta K, Sato T. Efficacy of inhaled nitric oxide in children with ARDS. Chest 114:827–833 (1998).

802. Oldham KT, Guice KS, Stetson PS, Wolfe RR. Bacteremia-induced suppression of alveolar surfactant production. J Surg Res 47:397–402 (1989).

803. Olivieri D. Corticosteroids in late adult respiratory distress syndrome—Towards a better use. Respiration 65:256–257 (1998).

804. Oosterlaken-Dijksterhuis MA, Haagsman HP, van Golde LM, Demel RA. Characterization of lipid insertion into monomolecular layers mediated by lung surfactant proteins SP-B and SP-C. Biochemistry 30:10965–10971 (1991).

805. Oosterlaken-Dijksterhuis MA, Haagsman HP, van Golde LM, Demel RA. Interaction of lipid vesicles with monomolecular layers containing lung surfactant proteins SP-B or SP-C. Biochemistry 30:8276–8281 (1991).

806. Oosterlaken-Dijksterhuis MA, van Eijk M, van Buel BLM, van Golde LMG, Haagsman HP. Surfactant protein composition of lamellar bodies isolated from rat lung. Biochem J 274:115–119 (1991).

807. Oosterlaken-Dijksterhuis MA, van Eijk M, van Golde LMG, Haagsman HP. Lipid mixing is mediated by the hydrophobic surfactant protein SP-B but not by SP-C. Biochim Biophys Acta 1110:45–50 (1992).

808. O'Reilly MA, Nogee L, Whitsett JA. Requirement of the collagenous domain for carbohydrate processing and secretion of a surfactant protein, SP-A. Biochim Biophys Acta 969:176–184 (1988).

809. O'Reilly MA, Weaver TE, Pilot-Matias TJ, Sarin VK, Gazdar AF, Whitsett JA. In vitro translation, post-translational processing and secretion of pulmonary surfactant protein B precursors. Biochim Biophys Acta 1011:140–148 (1989).

810. OSIRIS Collaborative Group. Early versus delayed neonatal administration of a synthetic surfactant—The judgement of OSIRIS. Lancet 340:1364–1369 (1992).

811. Otis DR, Ingenito EP, Kamm RD, Johnson M. Dynamic surface tension of TA surfactant: Experiments and theory. J Appl Physiol 77:2681–2688 (1994).

812. Otis DR Jr, Johnson M, Pedley TJ, Kamm RD. Role of pulmonary surfactant in airway closure: A computational study. J Appl Physiol 75:1323–1333 (1993).

813. Oury T, Chang L, Marklund S, Day B, Crapo J. Immunocytochemical localization of extracellular superoxide dismutase in human lung. Lab Invest 70:889–898 (1994).

814. Oury TD, Ho YS, Piantadosi CA, Crapo JD. Extracellular superoxide dismutase, nitric oxide, and central venous system oxygen toxicity. Proc Natl Acad Sci USA 89:9715–9719 (1991).

815. Oyarzun MJ, Clements JA. Control of lung surfactant by ventilation, adrenergic mediators, and prostaglandin in the rabbit. Am Rev Respir Dis 117:879–891 (1978).

816. Pacht E, Timerman A, Lykens M, Merola A. Deficiency of alveolar fluid glutathione in patients with sepsis and the adult respiratory distress syndrome. Chest 100:1397–1403 (1991).

817. Padmanabhan RV, Gudapaty R, Liener IE, Schwartz BA, Hoidal JR. Protection against pulmonary oxygen toxicity in rats by the intratracheal administration of liposomes-encapsulated superoxide dismutase or catalase. Am Rev Respir Dis 132:164–167 (1985).

818. Palmer R, Ashton D, Moncada S. Vascular endothelial cells synthesize nitric oxide from L-arginine. Nature 333:664–666 (1988).

819. Palmer R, Ferrige A, Moncada S. Nitric oxide release accounts for the biologic activity of endothelium-derived relaxing factor. Nature 327:524–526 (1987).

820. Papazian L, Bregeon F, Gaillat F, Kaphan E, Thiroon X, Saux P, Badier M, Gregoire R, Gouin F, Jammes Y, Auffray J-P. Does norepinephrine modify the effects of inhaled nitric oxide in septic patients with acute respiratory distress syndrome? Anesthesiology 89:1089–1098 (1998).

821. Papazian L, Bregeon F, Gaillat F, Thirion X, Gainnier M, Gregoire R, Saux P, Gouin F, Jammes Y, Auffray J-P. Respective and combined effect of prone position and inhaled nitric oxide in patients with acute respiratory distress syndrome. Am J Resp Crit Care Med 157:580–585 (1998).

822. Pappert D, Busch T, Gerlach H, Lewandowski K, Radermacher P, Rossaint R. Aerosolized prostacyclin versus inhaled nitric oxide in children with severe acute respiratory distress syndrome. Anesthesiology 82:1507–1511 (1995).

823. Paranka MS, Walsh WF, Stancombe BB. Surfactant lavage in a piglet model of meconium aspiration syndrome. Pediatr Res 31:625–628 (1992).

824. Parker MM. Surfactant replacement in pediatric respiratory failure: Promising new therapy? Crit Care Med 24:1281–1282 (1996).

825. Parsons P, Gillesis M, Moore E, Moore F, Worthen G. Neutrophil response to endotoxin in the adult respiratory distress syndrome: Role of CD14. Am J Respir Cell Mol Biol 13:152–160 (1995).

826. Parsons P, Moore F, Ikle D, Henson P, Worthen G. Studies on the role of

tumor necrosis factor in adult respiratory distress syndrome. Am Rev Respir Dis 146:694–700 (1992).

827. Pastrana B, Mautone AJ, Mendelsohn R. Fourier transform infrared studies of secondary structure and orientation of pulmonary surfactant SP-C and its effect on the dynamic surface properties of phospholipids. Biochemistry 30:10058–10064 (1991).

828. Pastrana-Rios B, Flach CR, Brauner JW, Mautone AJ, Mendelsohn R. A direct test of the "squeeze-out" hypothesis of lung surfactant function: External reflection FT-IR at the air/water interface. Biochemistry 33:5121–5127 (1994).

829. Pattle R. Surface lining of lung alveoli. Physiol Rev 45:48–75 (1965).

830. Pattle RE. Properties, function and origin of the alveolar lining layer. Proc R Soc (Lond) Ser B 148:217–240 (1958).

831. Pattle RE. Properties, function, and origin of the alveolar lining layer. Nature 175:1125–1126 (1955).

832. Patton HD, Fuchs AF, Hille B, Scher AM, Steiner R. Textbook of Physiology. 21st ed. Philadelphia: WB Saunders (1989).

833. Paulson T, Spear R, Peterson B. New concepts in the treatment of children with acute respiratory failure. J Pediatr 127:163–175 (1995).

834. Pawlowski R, Frosolono M, Charms B, Przybylski R. Intra- and extracellular compartmentalization of the surface active fraction in dog lung. J Lipid Res 12:538–544 (1971).

835. Payen D, Muret J, Beloucif S, Gatecel C, Kermarrec N, Guinard N, Mateo J. Inhaled nitric oxide, almitrine infusion, or their coadministration as a treatment of severe hypoxemic focal lung lesions. Anesthesiology 89:1158–1165 (1998).

836. Pearson DJ, Mentnech MS, Gamble M, Taylor G, Green FHY. Acute pulmonary injury induced by immune complexes. Exp Lung Res 1:323–334 (1980).

837. Pedley TJ, Kamm RD. Dynamics of gas-flow and pressure-flow relationships. In: Crystal RG, West JB, Weibel ER, Barnes PJ, Eds. The Lung: Scientific Foundations. 2nd ed. Philadelphia: Lippincott-Raven, pp. 1365–1381 (1997).

838. Pérez-Gil J, Casals C, Marsh D. Interactions of hydrophobic lung surfactant proteins SP-B and SP-C with dipalmitoylphosphatidylcholine and dipalmitoyl phosphatidylglycerol bilayers studies by electron spin resonance spectroscopy. Biochemistry 34:3964–3791 (1995).

839. Pérez-Gil J, Cruz A, Casals C. Solubility of hydrophobic surfactant proteins in organic solvent/water mixtures. Structural studies on SP-B and SP-C in aqueous organic solvents and lipids. Biochim Biophys Acta 1168:261–270 (1993).

840. Pérez-Gil J, Keough KMW. Structural similarities between myelin and hydro-phobic surfactant associated proteins: Protein motifs for interacting with bilayers. J Theor Biol 169:221–229 (1994).

841. Pérez-Gil J, Lopez-Lacomba JL, Cruz A, Beldarrain A, Casals C. Deacylated pulmonary surfactant protein SP-C has different effects on the thermotropic behaviour of bilayers of dipalmitoylphosphatidyl-glycerol (DPPG) than the native acylated protein. Biochem Soc Trans 22:372S (1994).

842. Pérez-Gil J, Nag K, Taneva S, Keough KMW. Pulmonary surfactant protein SP-C causes packing rearrangements of dipalmitoylphosphatidylcholine in spread mono-layers. Biophys J 63:197–204 (1992).

843. Perkins W, Dause R, Parente R, Minchey S, Neuman K, Gruner S, Taraschi T, Janoff A. Role of lipid polymorphism in pulmonary surfactant. Science 273:330–332 (1996).

844. Persson A, Chang D, Crouch E. Surfactant protein D is a divalent cation-dependent carbohydrate-binding protein. J Biol Chem 265:5755–5760 (1990).

845. Persson A, Chang D, Rust K, Moxley M, Longmore W, Crouch E. Purification and biochemical characterization of CP4 (SP-D), a collagenous surfactant-associated protein. Biochemistry 28:6361–6367 (1989).

846. Persson A, Rust K, Chang D, Moxley M, Longmore W, Crouch E. CP4: A pneumocyte-derived collagenous surfactant-associated protein. Evidence for heterogeneity of collagenous surfactant proteins. Biochemistry 27:8576–8584 (1988).

847. Petty T, Reiss O, Paul G, Silvers G, Elkins N. Characteristics of pulmonary surfactant in adult respiratory distress syndrome associated with trauma and shock. Am Rev Respir Dis 115:531–536 (1977).

848. Petty TL, Ashbaugh DG. The adult repiratory distress syndrome. Clinical features, factors influencing prognosis and principles of management. Chest 60:233–239 (1971).

849. Pfleger RC. Type II epithelial cells from the lung of Syrian hamsters: Isolation and metabolism. Exp Mol Pathol 27:152–166 (1977).

850. Phelps DS, Floros J. Localization of surfactant protein synthesis in human lung by in situ hybridization. Am Rev Respir Dis 137:939–942 (1988).

851. Phibbs RH, Ballard RA, Clements JA, Heilbron DC, Phibbs CS, Schlueter MA, Sniderman SH, Tooley WH, Wakeley A. Initial clinical trial of Exosurf, a protein-free synthetic surfactant, for the prophylaxis and early treatment of hyaline membrane disease. Pediatrics 88:1–9 (1991).

852. Phillips MC, Hauser H. Spreading of solid glycerides and phospholipids at the air-water interface. J Colloid Interface Sci 49:31–39 (1974).

853. Phizackerely PJR, Town M-H, Newman GE. Hydrophobic proteins of lamellated osmiophilic bodies isolated from pig lung. Biochemistry 183:731–736 (1979).

854. Pian MS, Dobbs LG, Duzgunes N. Positive correlation between cytosolic free calcium and surfactant secretion in cultured rat alveolar type II cells. Biochim Biophys Acta 960:43–53 (1988).

855. Pilot-Matias TJ, Kister SE, Fox JL, Kropp K, Glasser SW, Whitsett JA. Structure and organization of the gene encoding human pulmonary surfactant proteoplipid SP-B. DNA 8:75–86 (1989).

856. Pinto RA, Wright JR, Lesikar D, Benson BJ, Clements JA. Uptake of pulmonary surfactant protein C into adult rat lung lamellar bodies. J Appl Physiol 74:1005–1011 (1993).

857. Pison U, Max M, Neuendank A, Weissbach S, Pietschmann S. Host defence capacities of pulmonary surfactant: Evidence for 'non-surfactant' functions of the surfactant system. Eur J Clin Invest 24:586–599 (1994).

858. Pison U, Seeger W, Buchhorn R, Joka T, Brand M, Obertacke U, Neuhof H, Schmit-Neuerberg K. Surfactant abnormalities in patients with respiratory failure after multiple trauma. Am Rev Respir Dis 140:1033–1039 (1989).

859. Pison U, Tam EK, Caughey GH, Hawgood S. Proteolytic inactivation of dog

lung surfactant–associated proteins by neutrophil elastase. Biochim Biophys Acta 992:251–257 (1989).

860. Pittet J-F, Wiener-Kronish JP, Serokiv V, Matthay MA. Resistance of the alveolar epithelium to injury from septic shock in sheep. Am J Respir Crit Care Med 151:1093–1100 (1995).

861. Possmayer F. A proposed nomenclature for pulmonary surfactant–associated proteins. Am Rev Respir Dis 138:990–998 (1988).

862. Possmayer F, Yu S-H, Weber JM, Harding PGR. Pulmonary surfactant. Can J Biochem Cell Biol 62:1121–1133 (1984).

863. Post M, Battenburg JJ, Schuurmans EAJM, Lyons CB, van Golde LMG. Lamellar bodies isolated from adult human lung tissue. Exp Lung Res 3:17–28 (1982).

864. Post M, Smith BT. Hormonal control of surfactant metabolism. In: Robertson B, van Golde LMG, Batenburg J, Eds. Pulmonary Surfactant: From Molecular Biology to Clinical Practice. Amsterdam: Elsevier Science Publishers, pp. 379–424 (1992).

865. Post M, van Golde LMG. Metabolic and developmental aspects of the pulmonary surfactant system. Biochim Biophys Acta 947:249–286 (1988).

866. Poulain FR, Clements JA. Pulmonary surfactant therapy. West J Med 162:43–50 (1995).

867. Preamanik A, Holtzman R, Merritt T. Surfactant replacement therapy for pulmonary diseases. Pediatr Clin North Am 40:913–936 (1993).

868. Pugin J, Ricou B, Steinberg KP, Suter PM, Martin TR. Proinflammatory activity in bronchoalveolar lavage fluids from patients with ARDS, a prominent role for interleukin-1. Am J Respir Crit Care Med 153:1850–1856 (1996).

869. Putensen C, Hormann C, Kleinsasser A, Putensen-Himmer G. Cardiopulmonary effects of aerosolized prostaglandin E_1 and nitric oxide inhalation in patients with acute respiratory distress syndrome. Am J Resp Crit Care Med 157:1743–1747 (1998).

870. Putensen C, Rasanen J, Lopez F, Downs J. Continuous positive pressure modulates effect of inhaled nitric oxide on the ventilation-perfusion distributions in canine lung injury. Chest 106:1563–1569 (1994).

871. Putman E, Boere AJ, van Bree L, van Golde LMG, Haagsman HP. Pulmonary surfactant subtype metabolism is altered after short-term ozone exposure. Toxicol Appl Pharmacol 134:132–138 (1995).

872. Putman E, Creuwels LAJM, van Golde LMG, Haagsman HP. Surface properties, morphology and protein composition of pulmonary surfactant subtypes. Biochem J 320:599–605 (1996).

873. Putz G, Goerke J, Clements JA. Surface activity of rabbit pulmonary surfactant subfractions at different concentrations in a captive bubble. J Appl Physiol 77:597–605 (1994).

874. Putz G, Goerke J, Schurch S, Clements JA. Evaluation of pressure-driven captive bubble surfactometer. J Appl Physiol 76:1417–1424 (1994).

875. Putz G, Goerke J, Taeusch HW, Clements JA. Comparison of captive and pulsating bubble surfactometers with use of lung surfactants. J Appl Physiol 76:1425–1431 (1994).

876. Putz G, Hormann C, Koller W, Schon G. Surfactant replacement therapy in acute

respiratory distress syndrome from viral pneumonia. Intensive Care Med 22:588–590 (1996).

877. Putz G, Walch M, van Eijk M, Haagsman H. A spreading technique for forming film in captive bubble. Biophys J 75:2229–2239 (1998).

878. Qanbar R, Possmayer F. On the surface activity of surfactant-associated protein C (SP-C): Effects of palmitoylation and pH. Biochim Biophys Acta 1255:251–259 (1995).

879. Rabinovitch W, Robertson RF, Mason SG. Relaxation of surface pressure and collapse of unimolecular films of stearic acid. Can J Chem 38:1881–1890 (1960).

880. Radford EP Jr. Method for estimating respiratory surface area of mammalian lungs from their physical characteristics. Proc Soc Exp Biol Med 87:58–61 (1954).

881. Radford EP Jr. Recent studies of mechanical properties of mammalian lungs. In: Remington JW, Ed. Tissue Elasticity. Bethesda, MD: American Physiological Society, pp. 177–190 (1957).

882. Rais-Bahrami K, Rivera O, Seale W, Short B. Effect of nitric oxide in meconium aspiration syndrome after treatment with surfactant. Crit Care Med 25:1744–1747 (1997).

883. Raju TNK, Bhat R, McCulloch KM, Maeta H, Vidyasagar D, Sobel D, Anderson M, Levy P, Furner S. Double-blind controlled trial of single dose treatment with bovine surfactant in severe hyaline membrane disease. Lancet 1:651–656 (1987).

884. Rana FR, Mautone AJ, Dluhy RA. Surface chemistry of binary mixtures of phospholipids in monolayers: Infrared studies of surface composition at varying surface pressures in a pulmonary surfactant model system. Biochemistry 32:3169–3177 (1993).

885. Rannels DE, Rannels S. Influence of the extracellular matrix on type 2 cell differentiation. Chest 96:165–173 (1989).

886. Raponi G, Antonelli M, Gaeta A, Bufi M, De Blasi RA, Conti G, D'Errico RR, Mancini C, Filadoro F, Gasparetto A. Tumor necrosis factor in serum and broncho-alveolar lavage of patients at risk for the adult respiratory distress syndrome. J Crit Care 7:183–188 (1992).

887. Rauvala H, Hallman M. Glycolipid accumulation in bronchoalveolar space in adult respiratory distress syndrome. J Lipid Res 25:1257–1262 (1984).

888. Rayleigh L. Measurements of the amount of oil necessary in order to check the motions of camphor upon water. Proc R Soc (Lond) 47:364–367 (1890).

889. Redenti E, Peveri T, Ventura P, Zanol M, Selva A. Characterization of phospholipidic components of the natural pulmonary surfactant Curosurf. Il Farmaco 49:285–289 (1994).

890. Reifenrath R. The significance of alveolar geometry and surface tension in the respiratory mechanics of the lung. Respir Physiol 24:115–137 (1975).

891. Reilly KE, Mautone AJ, Mendelsohn R. Fourier tranform infrared spectroscopy studies of lipid/protein interactions in pulmonary surfactant. Biochemistry 28:7368–7373 (1989).

892. Revak S, Merritt T, Cochrane C, Heldt G, Alberts M, Anderson D, Kheiter A. Efficacy of synthetic peptide–containing surfactant in the treatment of respiratory

distress syndrome in preterm infant rhesus monkeys. Pediatr Res 39:715–724 (1996).

893. Revak SD, Merritt TA, Degryse E, Stefani L, Courtney M, Hallman M, Cochrane CG. The use of human low molecular weight (LMW) apoproteins in the reconstitution of surfactant biological activity. J Clin Invest 81:826–833 (1988).

894. Revak SD, Merritt TA, Hallman M, Cochrane CG. Reconstitution of surfactant activity using purified human apoprotein and phospholipids measured in vitro and in vivo. Am Rev Respir Dis 134:1258–1265 (1986).

895. Revak SD, Merritt TA, Hallman M, Heldt G, La Polla RJ, Hoey K, Houghten RA, Cohrane CG. The use of synthetic peptides in the formation of biophysically and biologically active surfactants. Pediatr Res 29:460–465 (1991).

896. Rice WR, Ross GF, Singleton FM, Dingle S, Whitsett JA. Surfactant-associated protein inhibits phospholipid secretion from type II cells. J Appl Physiol 63:692–698 (1987).

897. Rice WR, Sarin VK, Fox JL, Baatz J, Wert S, Whitsett JA. Surfactant peptides stimulate uptake of phosphatidylcholine by isolated cells. Biochim Biophys Acta 1006:237–245 (1989).

898. Rider E, Ikegami M, Whitsett J, Hull W, Absolom D, Jobe A. Treatment responses to surfactants containing natural surfactant proteins in preterm rabbits. Am Rev Respir Dis 147:669–676 (1993).

899. Rider ED, Ikegami M, Jobe AH. Intrapulmonary catabolism of surfactant saturated phosphatidylcholine in rabbits. J Appl Physiol 69:1856–1862 (1990).

900. Rider ED, Ikegami M, Jobe AH. Localization of alveolar surfactant clearance in rabbit lung cells. Am J Physiol 263:L201–L209 (1992).

901. Rider ED. Stable ridges in a collapsing monolayer. Nature 281:287–289 (1979).

902. Ries HE, Matsumoto M, Vyeda N, Suito E. Electron micrographs of cholesterol monolayers. J Colloid Interface Sci 57:396–398 (1976).

903. Ries HE, Swift H. Electron microscope and π-A studies on a gramicidin and its binary mixtures with cerebronic acid, cholesterol and valinomycin. Colloids Surf 40:145–165 (1989).

904. Ries HE, Swift H. Twisted double-layer ribbons and the mechanism for monolayer collapse. Langmuir 3:853–855 (1987).

905. Ries HE, Walker DC. Films of mixed horizontally and vertically oriented compounds. J Colloid Interface Sci 16:361–374 (1961).

906. Rinaldo JE, Rogers RM. Adult respiratory distress syndrome, changing concepts of lung injury and repair. N Engl J Med 15:900–909 (1982).

907. Ring J, Stidham G. Novel therapies for acute respiratory failure. Pediatr Clin North Am 41:1325–1363 (1994).

908. Robbins C, Horowitz S, Merritt TA, Kheiter A, Tierney J, Narula P, Davis JM. Recombinant human superoxide dismutase reduces lung injury caused by inhaled nitric oxide and hyperoxia. Am J Physiol 272:L903–L907 (1997).

909. Robbins CG, Davis JM, Merritt TA, Amirkhanian JD, Sahgal N, Morin FC, Horowitz S. Combined effects of nitric oxide and hyperoxia on surfactant function and pulmonary inflammation. Am J Physiol 269:L545–L550 (1995).

910. Roberts JD, Fineman JR, Morin FC, Shaul PW, Rimar S, Schreiber MD, Polin RA, Zwass MS, Zayek MM, Gross I, Heymann MA, Zapol WM. Inhaled nitric oxide

and persistent pulmonary hypertension of the newborn. N Engl J Med 336:605–610 (1997).

911. Roberts RJ, Frank L. Developmental consequences of oxygen toxicity. In: Kacew S, Reasor MJ, Eds. Toxicology of the Newborn. New York: Elsevier, pp. 141–171 (1984).

912. Robertson B. Surfactant inactivation and surfactant replacement in experimental models of ARDS. Acta Anaesthesiol 35(Suppl):22–28 (1991).

913. Robertson B. Surfactant substitution: Experimental models and clinical applications. Lung 158:57–68 (1980).

914. Robertson B, Curstedt T, Grossmann G, Kobayashi T, Kokubo M, Suzuki Y. Prolonged ventilation of the premature newborn rabbit after treatment with natural or apoprotein-based artificial surfactant. Eur J Pediatr 147:168–173 (1988).

915. Robertson B, Curstedt T, Johansson J, Jornvall H, Kobayashi T. Structural and functional characterization of porcine surfactant isolated by liquid-gel chromatography. Prog Respir Res 25:237–246 (1990).

916. Robertson B, Enhorning G. The alveolar lining of the premature newborn rabbit after pharyngeal deposition of surfactant. Lab Invest 31:54–59 (1974).

917. Robertson B, Kobayashi T, Ganzuka M, Grossmann G, Li WZ, Suzuki Y. Experimental neonatal respiratory failure induced by a monoclonal antibody to the hydrophobic surfactant-associated protein SP-B. Pediatr Res 30:239–243 (1991).

918. Robertson B, van Golde LMG, Batenburg JJ, Eds. Pulmonary Surfactant. Amsterdam: Elsevier (1984).

919. Robertson B, van Golde LMG, Batenburg JJ, Eds. Pulmonary Surfactant: from Molecular Biology to Clinical Practice. Amsterdam: Elsevier (1992).

920. Robillard E, Alarie Y, Dagenais-Perusse P, Baril E, Guilbeault A. Microaerosol administration of synthetic β,γ-dipalmitoyl-L-α-lecithin in the respiratory distress syndrome: A preliminary report. Can Med Assoc J 90:55–57 (1964).

921. Rollins M, Jenkins J, Tubman R, Corkey C, Wilson D. Comparison of clinical responses to natural and synthetic surfactants. J Perinat Med 21:341–347 (1993).

922. Rooney SA. The surfactant system and lung phospholipid biochemistry. Am Rev Respir Dis 131:439–460 (1985).

923. Rooney SA, Canavan PM, Motoyama EK. The identification of phosphatidylglycerol in the rat, rabbit, monkey and human lung. Biochim Biophys Acta 360:56–67 (1974).

924. Rooney SA, Wai-Lee TS, Gobran L, Motoyama EK. Phospholipid content, composition and biosynthesis during fetal lung development in the rabbit. Biochim Biophys Acta 431:447–458 (1976).

925. Rooney SA, Young SL, Mendelson CR. Molecular and cellular processing of lung surfactant. FASEB J 8:957–967 (1994).

926. Rosenfeld WN, Davis JM, Parton L, Richter SE, Price A, Flaster E, Kassem N. Safety and pharmacokinetics of recombinant human superoxide dismutase administered intratracheally to premature neonates with respiratory distress syndrome. Pediatrics 97:811–817 (1996).

927. Ross GF, Notter RH, Meuth J, Whitsett JA. Phospholipid binding and biophysical activity of pulmonary surfactant–associated protein SAP-35 and its non-collagenous C-terminal domains. J Biol Chem 261:14283–14291 (1985).

928. Rossaint R, Falke KJ, Lopez F, Slama K, Pison U, Zapol W. Inhaled nitric oxide for the acute respiratory distress syndrome. N Engl J Med 328:399–405 (1993).

929. Rossaint R, Gerlach H, Schmidt-Ruhnke H, Pappert D, Lewandowski K, Steudel W, Falke K. Efficacy of inhaled nitric oxide in ARDS. Chest 107:1107–1115 (1995).

930. Ruano M, Miguel E, Perez-Gil J, Casals C. Comparison of lipid aggregation and self-aggregation activities of pulmonary surfactant–associated protein A. Biochem J 313:683–689 (1996).

931. Ruano M, Nag K, Worthman L-A, Casals C, Perez-Gil J, Keough K. Differential partitioning of pulmonary surfactant protein SP-A into regions of monolayers of dipalmitoylphosphatidylcholine and dipalmitoylphosphatidylcholine/dipalmitoyl-phosphatidylglycerol. Biophys J 74:1101–1109 (1998).

932. Ruano M, Perez-Gil J, Casals C. Effect of acidic pH on the structure and lipid binding properties of porcine surfactant protein A. J Biol Chem 273:15183–15191 (1998).

933. Rubin BK, Ramirez O, King M. The role of mucus rheology and transport in neonatal respiratory distress syndrome and the effect of surfactant therapy. Chest 101:1080–1085 (1992).

934. Ryan RM, Morris RE, Rice WR, Ciraolo G, Whitsett JA. Binding and uptake of pulmonary surfactant protein (SP)-A by pulmonary type II epithelial cells. J Histochem Cytochem 37:429–440 (1989).

935. Ryan SF, Barrett CR, Liau DF. Nitrosourethane induced lung injury. In: Cantor JO, Ed. Handbook of Animal Models of Pulmonary Disease. Boca Raton, FL: CRC Press, pp. 67–106 (1989).

936. Sanders RL. The composition of pulmonary surfactant. In: Farrell PM, Ed. Lung Development: Biological and Clinical Perspectives. New York: Academic Press, pp. 193–219 (1982).

937. Sanderson R, Paul G, Vatter A, Filley G. Morphological and physical basis for lung surfactant action. Respir Physiol 27:379–392 (1976).

938. Sano K, Voelker DR, Mason R. Involvement of protein kinase C in pulmonary surfactant secretion from alveolar type II cells. J Biol Chem 260:12725–12729 (1985).

939. Sarin VK, Gupta S, Leung TK, Taylor VE, Ohning BL, Whitsett JA, Fox JL. Biophysical and biological activity of a synthetic 8.7 kDa hydrophobic pulmonary surfactant protein SP-B. Proc Natl Acad Sci USA 87:2633–2637 (1990).

940. Sastre J, Asensi M, Rodrigo F, Pallardo F, Vento M, Vina J. Antioxidant adminis-tration to the mother prevents oxidative stress associated with birth in the neonatal rat. Life Sci 54:2055–2059 (1994).

941. Scarpelli EM. The alveolar surface network: A new anatomy and its physiological significance. Anat Res 251:491–527 (1998).

942. Scarpelli EM. Pulmonary Surfactant. Philadelphia: Lea & Febiger (1968).

943. Schafer KP, Voss T, Melchers K, Eistetter H. Lung surfactant: A biotechnological challenge. Lung 168:851S–859S (1990).

944. Schagger H, von Jagow G. Tricine-SDS-polyacrylamide gel electrophoresis for the separation of proteins in the range from 1 to 100 kDa. Anal Biochem 166:368–379 (1987).

945. Scheeren T, Radermacher P. Prostacyclin (PGI$_2$): New aspects of an old substance in the treatment of critically ill patients. Intensive Care Med 23:146–158 (1997).

946. Scherzer H, Ward PA. Lung injury produced by immune complexes of varying composition. J Immunol 121:947–952 (1978).

947. Schilling JWJ, White RT, Cordell BI, inventors. Recombinant alveolar surfactant protein. USA patent 4,659,805. 1987 April 21.

948. Schlag G, Strohmaier W. Experimental aspiration trauma: Comparison of steroid treatment versus exogenous natural surfactant. Exp Lung Res 19:397–405 (1993).

949. Schlame M, Casals C, Rustow B, Rabe H, Kunze D. Molecular species of phosphatidylcholine and phosphatidylglycerol in rat lung surfactant and different pools of pneumocytes type II. Biochem J 253:209–215 (1988).

950. Schneider H, Hallman M, Benirschke K, Gluck L. Human surfactant: A therapeutic trial in premature rabbits. J Pediatr 100:619–622 (1982).

951. Schoel WM, Schurch S, Goerke J. The captive bubble method for the evaluation of pulmonary surfactant: surface tension, area, and volume calculations. Biochim Biophys Acta 1200:281–290 (1994).

952. Schurch S. Surface tension at low lung volumes and dependence on time and alveolar size. Respir Physiol 48:339–355 (1982).

953. Schurch S, Bachofen H. Biophysical aspects in the design of a therapeutic surfactant. In: Robertson B, Taeusch HWE, Eds. Surfactant Therapy for Lung Disease. New York: Marcel Dekker, pp. 3–32 (1995).

954. Schurch S, Bachofen H, Goerke J, Green F. Surface properties of rat pulmonary surfactant studied with the captive bubble method: Adsorption, hysteresis, stability. Biochim Biophys Acta 1103:127–136 (1992).

955. Schurch S, Bachofen H, Goerke J, Possmayer F. A captive bubble method reproduces the in situ behavior of lung surfactant monolayers. J Appl Physiol 67:2389–2396 (1989).

956. Schurch S, Bachofen H, Weibel ER. Alveolar surface tension in excised rabbit lungs: Effects of temperature. Respir Physiol 62:31–45 (1985).

957. Schurch S, Gehr P, Im Hof V, Geiser M, Green FHY. Surfactant displaces particles toward the epithelium of airways and alveoli. Respir Physiol 80:17–32 (1990).

958. Schurch S, Goerke J, Clements J. Direct determination of volume and time-dependence of alveolar surface tension in excised lungs. Proc Natl Acad Sci USA 75:3417–3421 (1978).

959. Schurch S, Goerke J, Clements JA. Direct determination of surface tension in the lung. Proc Natl Acad Sci USA 73:7720–7726 (1976).

960. Schurch S, Possmayer F, Cheng S, Cockshutt AM. Pulmonary SP-A enhances adsorption and appears to induce surface sorting of lipid extract surfactant. Am J Physiol 263:L210–L218 (1992).

961. Schurch S, Schurch D, Curstedt T, Robertson B. Surface activity of lipid extract surfactant in relation to film area compression and collapse. J Appl Physiol 77:974–986 (1994).

962. Schwartz RM, Kellogg RJ, Scanlon JW, Zhao Q. Permanent presence of managed care and outcome of critically-ill infants. Pediatr Clin North Am 45:635–650 (1998).

963. Schwartz RM, Luby AM, Scanlon JW, Kellogg RJ. Effect of surfactant on

morbidity, mortality, and resource use in newborn infants weighing 500 to 1,500 g. N Engl J Med 330:1476–1480 (1994).

964. Scopes R. Protein Purification. Principles and Practice. 2nd ed. New York: Springer-Verlag (1987).

965. Scriven L. Dynamics of a fluid interface. Equation of motion for Newtonion surface fluids. Chem Eng Sci 12:98–108 (1960).

966. Searcy RL, Bergquist LM. A new color reaction for the quantitation of serum cholesterol. Clin Chim Acta 5:192–199 (1960).

967. Seeger W, Grube C, Günther A, Schmidt R. Surfactant inhibition by plasma proteins: Differential sensitivity of various surfactant preparations. Eur Respir J 6:971–977 (1993).

968. Seeger W, Günther A, Thede C. Differential sensitivity to fibrinogen inhibition of SP-C- vs. SP-B-based surfactants. Am J Physiol 261:L286–L291 (1992).

969. Seeger W, Günther A, Walmrath HD, Grimminger F, Lasch HG. Alveolar surfactant and adult respiratory distress syndrome. Pathogenic role and therapeutic prospects. Clin Invest 71:177–190 (1993).

970. Seeger W, Lepper H, Hellmut RD, Neuhof H. Alteration of alveolar surfactant function after exposure to oxidant stress and to oxygenated and native arachadonic acid in vitro. Biochim Biophys Acta 835:58–67 (1985).

971. Seeger W, Pison U, Buchhorn R, Obestacke U, Joka T. Surfactant abnormalities and adult respiratory failure. Lung 168(Suppl):891–902 (1990).

972. Seeger W, Stohr G, Wolf HRD, Neuhof H. Alteration of surfactant function due to protein leakage: Special interaction with fibrin monomer. J Appl Physiol 58:326–338 (1985).

973. Seeger W, Thede C, Günther A, Grube C. Surface properties and sensitivity to protein-inhibition of a recombinant apoprotein C–based phospholipid mixture in vitro: Comparison to natural surfactant. Biochim Biophys Acta 1081:45–52 (1991).

974. Segerer H, Stevens P, Schadow B, Maier R, Kattner E, Schwarz H, Curstedt T, Robertson B, Obladen M. Surfactant substitution in ventilated very low birth weight infants: Factors related to response types. Pediatr Res 30:591–596 (1991).

975. Sehgal SS, Ewing CK, Richards T, Taeusch HW. Modified bovine surfactant (Survanta) vs a protein-free surfactant (Exosurf) in the treatment of respiratory distress syndrome in preterm infants: A pilot study. J Natl Med Assoc 86:46–52 (1994).

976. Sempowski G, Chess P, Padilla J, Moretti A, Phipps R. CD40 mediated activation of gingival and periodontal ligament fibroblasts. J Periodontol 68:284–292 (1997).

977. Sempowski G, Chess P, Phipps R. CD40 is a functional activation antigen and B7-independent T cell costimulatory molecule in normal human lung fibroblasts. J Immunol 158:4670–1477 (1997).

978. Sen A, Hui S, Mosgrober-Anthony M, Holm BA, Egan EA. Localization of lipid exchange sites between bulk lung surfactants and surface monolayer: A freeze fracture study. J Colloid Interface Sci 126:355–360 (1988).

979. Serrallach EN, de Haas GH, Shipley GG. Structure and thermotropic properties of mixed-chain phosphatidylcholine bilayers. Biochemistry 23:713–720 (1984).

980. Sessler C, Bloomfield G, Fowler A. Current concepts of sepsis and acute lung injury. Clin Chest Med 17:213–235 (1996).

981. Shaffer T, Tran N, Bhutani V, Sivieri E. Cardiopulmonary function in very premature lambs during liquid ventilation. Pediatr Res 17:680–684 (1983).

982. Shaffer T, Wolfson M, Clark M. State of the art review: Liquid ventilation. Pediatr Pulmonol 14:102–109 (1992).

983. Shaffer TH, Douglas PR, Lowe CA, Bhutani VK. The effects of liquid ventilation on cardiopulmonary function in preterm lambs. Pediatr Res 17:303–306 (1983).

984. Shaffer TH, Lowe CA, Bhutani VK, Douglas PR. Liquid ventilation: Effects of pulmonary function in distressed meconium-stained lambs. Pediatr Res 18:47–52 (1984).

985. Shah DO. Surface chemistry of lipids. Adv Lipid Res 8:347–431 (1970).

986. Shah DO. The world of surface tension. Chem Eng Educ Winter: 14–23 (1977).

987. Shah NS, Nakayama DK, Jacob TD, Nishio I, Imai T, Billiar TR, Exler R, Yousem SA, Motoyama EK, Peitzman AB. Efficacy of inhaled nitric oxide in a porcine model of adult respiratory distress syndrome. Arch Surg 129:158–164 (1994).

988. Shah NS, Nakayama DK, Jacob TD, Nishio I, Imai T, Billiar TR, Exler R, Yousem SA, Motoyama EK, Peitzman AB. Efficacy of inhaled nitric oxide in oleic acid–induced acute lung injury. Crit Care Med 25:153–158 (1997).

989. Shapiro D, Notter R, Eds. Surfactant Replacement Therapy. New York: AR Liss (1989).

990. Shapiro DL, Notter RH, Morin F, Deluga KS, Golub LM, Sinkin RA, Weiss KI, Cox C. A double blind randomized trial of calf lung surfactant extract administered at birth to very premature infants for prevention of respiratory distress syndrome. Pediatrics 76:593–599 (1985).

991. Shaw DJ. Introduction to Colloid and Surface Chemistry. London: Buttersworth (1980).

992. Sheehan PM, Stokes DC, Yeh Y, Hughes WT. Surfactant phospholipids and lavage phospholipase A_2 in experimental *Pneumocystis carinii* pneumonia. Am Rev Respir Dis 134:526–531 (1986).

993. Shelley SA, Balis JU, Paciga JE, Espinoza CG, Richman AV. Biochemical composition of adult human lung surfactant. Lung 160:195–206 (1982).

994. Sheridan BC, McIntyre RC, Meldrum DR, Fullerton DA. Pentoxifylline treatment attenuates pulmonary vasomotor dysfunction in acute lung injury. J Surg Res 71:150–154 (1997).

995. Sherman MP, Campbell LA, Merritt TA, Long WA, Gunkel JH, Curstedt T, Robertson B. Effect of different surfactants on pulmonary group B streptococcal infection in premature rabbits. J Pediatr 125:939–947 (1994).

996. Shiffer K, Hawgood S, Haagsman HP, Benson B, Clements JA, Goerke J. Lung surfactant proteins, SP-B and SP-C, alter the thermodynamic properties of phospholipid membranes: A differential calorimetry study. Biochemistry 32:590–597 (1993).

997. Shimizu H, Hosoda K, Mizumoto M, Kuroki Y, Sato H, Kataoka K, Hagisawa M, Fujiwara S, Akino T. Improved immunoassay for the determination of surfactant protein A (SP-A). Tohoku J Exp Med 157:269–278 (1989).

998. Shin Y. Spectrophotometric ultramicrodetermination of inorganic phosphorus and lipid phosphorus in serum. Anal Chem 34:1164–1166 (1962).

999. Shuster D. ARDS: Clinical lessons from the oleic acid model of acute lung injury. Am J Respir Crit Care Med 149:245–260 (1994).

1000. Sies H. Oxidative Stress. London: Academic Press (1991).

1001. Silverman WA. Dunham's Premature Infants. 3rd ed. New York: Harper & Row (1961).

1002. Simatos GA, Forward KB, Morrow MR, Keough KMW. Interaction between perdeuterated dimyristoyl phosphatidylcholine and low molecular weight surfactant protein (SP)-C. Biochemistry 29:5807–5814 (1990).

1003. Sims B, Zografti G. Dynamic properties of fatty acid molecular films. Chem Phys Lipids 6:109–120 (1971).

1004. Sims B, Zografti G. Time-dependent behavior of insoluble monomolecular films. J Colloid Interface Sci 41:35–46 (1972).

1005. Sitrin RG, Ansfield MK, Kaltreider HB. The effect of pulmonary surface-active material on the generation and expression of murine B- and T-lymphocyte effector functions in vitro. Exp Lung Res 9:85–97 (1985).

1006. Skita V, Chester DW, Oliver CJ, Turcotte JG, Notter RH. Bilayer characteristics of a diether phosphonolipid analog of the major lung surfactant glycerophospholipid dipalmitoyl phosphatidylcholine. J Lipid Res 36:1116–1127 (1995).

1007. Small DM. The Physical Chemistry of Lipids: From Alkanes to Phospholipids. New York: Plenum Press (1986).

1008. Smith GB, Tauesch HW, Phelps DS, Keough KM. Mixtures of low molecular weight surfactant proteins and dipalmitoyl phosphatidylcholine duplicate effects of pulmonary surfactant in vitro and in vivo. Pediatr Res 23:484–490 (1988).

1009. Smith JC, Stamenovic D. Surface forces in lungs. I. Alveolar surface tension–lung volume relationships. J Appl Physiol 60:1341–1350 (1986).

1010. Smith R, Tanford C. The critical micelle concentration of L-α-dipalmitoyl phosphatidylcholine in water and water-methanol solutions. J Mol Biol 67:75–83 (1972).

1011. Smith RD, Berg JC. The collapse of surfactant monolayers at the air-water interface. J Colloid Interface Sci 74:273–286 (1980).

1012. Smyth J, Metcalfe I, Duffy P, Possmayer F, Bryan M, Enhorning G. Hyaline membrane disease treated with bovine surfactant. Pediatrics 71:913–917 (1983).

1013. Snik A, Booman P, Gieles P, Egberts J. Viscoelastic properties of lung surfactant. Prog Respir Res 18:24–28 (1984).

1014. Snyder JM, Mendelson CR. Insulin inhibits the accumulation of the major lung surfactant apoprotein in human fetal lung explants maintained in vitro. Endocrinology 120:1250–1257 (1987).

1015. Snyder JM, Ridgers HF, Neilson HC, O'Brien JA. Uptake of the 35 kDa major surfactant apoprotein (SP-A) by neonatal rabbit lung tissue. Biochim Biophys Acta 1002:1–7 (1988).

1016. Soll RF. Surfactant therapy in the USA: Trials and current routines. Biol Neonate 71:1–7 (1997).

1017. Soll RF, Hoekstra RE, Fangmann JJ, Corbet AJ, Adams JM, James LS, Schulze K, Oh W, Roberts JD, Dorst JP, Kramer SS, Gold AJ, Zola EM, Horbar JD, McAuliffe

TL, et al. Multicenter trial of single-dose modified bovine surfactant extract (Survanta) for prevention of respiratory distress syndrome. Pediatrics 85:1092–1102 (1990).

1018. Soll RF, Merritt TA, Hallman M. Surfactant in the prevention and treatment of respiratory distress syndrome. In: Boynton BR, Carlo WA, Jobe AH, Eds. New Therapies for Neonatal Respiratory Failure. New York: Cambridge University Press, pp. 49–80 (1994).

1019. Soutar AK, Pownall HJ, Hu AS, Smith LC. Phase transitions in bilamellar vesicles. Measurements by pyrene excimer fluorescence and effect on transacylation by lecithin:cholesterol acyltransferase. Biochemistry 13:2828–2836 (1974).

1020. Spalding JW, Orgner MJ, Tombropoulos EG, Gilmore LB, Hook GER. Isolation and characterization of rabbit lung lamellar bodies. Exp Lung Res 4:171–190 (1983).

1021. Speer CP, Gefeller O, Groneck P, Laufkotter E, Roll C, Hanssler L, Harms K, Herting E, Boenisch H, Windeler J, Robertson B. Randomised clinical trial of two treatment regimens of natural surfactant preparations in neonatal respiratory distress syndrome. Arch Dis Child 72:F8–F13 (1995).

1022. Spragg R. Surfactant therapy in acute respiratory distress syndrome. Biol Neonate 74(Suppl):15–20 (1998).

1023. Spragg R, Gilliard N, Richman P, Smith R, Hite R, Pappert D, Robertson B, Curstedt T, Strayer D. Acute effects of a single dose of porcine surfactant on patients with the adult respiratory distress syndrome. Chest 105:195–202 (1994).

1024. Stafford RE, Fanni T, Dennis EA. Interfacial properties and critical micelle concentrations of lysophospholipids. Biochemistry 28:5113–5120 (1989).

1025. Stamenovic D. Micromechanical foundations of pulmonary elasticity. Physiol Rev 70:1117–1134 (1990).

1026. Stamenovic D, Smith JC. Surface forces in lungs. II. Microstructural mechanics and lung stability. J Appl Physiol 60:1351–1357 (1986).

1027. Stamenovic D, Smith JC. Surface forces in lungs. III. Alveolar surface tension and elastic properties of lung parenchyma. J Appl Physiol 60:1358–1362 (1986).

1028. Stamenovic D, Yager D. Elastic properties of air- and liquid-filled lung parenchyma. J Appl Physiol 65:2565–2570 (1988).

1029. Staubach K-H, Schroder J, Stuber F, Gehrke K, Traumann E, Zabel P. Effect of pentoxifylline in severe sepsis. Arch Surg 133:94–100 (1998).

1030. Staudinger T, Presterl E, Graninger W, Locker G, Knapp S, Laczika K, Klappacher G, Stoiser B, Wagner A, Tesinsky P, Kordova H, Frass M. Influence of pentoxifylline on cytokine levels and inflammatory parameters in septic shock. Intensive Care Med 22:888–893 (1996).

1031. Stefanovitch V. Effect of pentoxifylline on energy rich phosphate in rat erythrocytes. Res Commun Chem Pathol Pharmacol 10:745–750 (1975).

1032. Steim J, Redding R, Hauck C, Stein M. Isolation and characterization of lung surfactant. Biochem Biophys Res Commun 34:434–440 (1969).

1033. Stine K. Investigations of monolayers by fluorescence microscopy. Microsc Res Tech 27:439–450 (1995).

1034. Stratton C. Morphology of surfactant producing cells and of the alveolar lining layer. In: Robertson B, van Golde L, Batenburg J, Eds. Pulmonary Surfactant. Amsterdam: Elsevier, pp. 67–118 (1984).

1035. Stratton C, Zasadzinski J, Elkins D. Lung lamellar body amphiphilic topography: A morphological evaluation using the continuum theory of liquid crystals. II. Disclinations, edge dislocations, and irregular defects. Anat Rec 221:520–532 (1988).

1036. Stratton CJ. The periodicity and architecture of lipid retained in extracted lung surfactant and its origin from multilamellar bodies. Tissue Cell 9:301–316 (1977).

1037. Stratton CJ. Three dimensional aspect of the mammalian lung surfactant myelin figure. Tissue Cell 9:285–300 (1977).

1038. Stratton CJ, Zasadzinski JAN, Elkins D. Lung lamellar body amphiphilic topography: A morphological evaluation using the continuum theory of liquid crystals. I. Closed surfaces: Closed spheres, concentric tori, and dupin cyclides. Anat Rec 221:503–519 (1988).

1039. Strayer D, Herting E, Sun B, Robertson B. Antibody to surfactant protein A increases sensitivity of pulmonary surfactant to inactivation by fibrinogen. Am J Respir Crit Care Med 153:1116–1122 (1996).

1040. Strayer DS. Immunogenicity of pulmonary surfactant preparations: Implications for the therapy of respiratory distress syndrome. Clin Immunother 1:441–448 (1994).

1041. Strayer DS, Merritt TA, Lwebuga-Mukasa J, Hallman M. Surfactant-antisurfactant immune complexes in infants with respiratory distress syndrome. Am J Pathol 122:353–362 (1986).

1042. Strayer DS, Robertson B. Surfactant as an immunogen: Implications for therapy of respiratory distress syndrome. Acta Paediatr 81:446–447 (1992).

1043. Stryer L. Biochemistry. 4th ed. New York: WH Freeman (1995).

1044. Stuart M, Wegh R, Kroon J, Sudholter E. Design and testing of a low-cost and compact Brewster-angle microscope. Langmuir 12:2863–2865 (1996).

1045. Stults JT, Griffin PR, Lesikar DD, Naidu A, Moffat B, Benson BJ. Lung surfactant protein SP-C from human, bovine, and canine sources contains palmityl cysteine thioester linkages. Am J Physiol 261:L118–L125 (1991).

1046. Sueishi K, Benson BJ. Isolation of a major apolipoprotein of canine and murine pulmonary surfactant. Biochemical and immunochemical characteristics. Biochim Biophys Acta 665:442–453 (1981).

1047. Sun B, Curstedt T, Robertson B. Exogenous surfactant improves ventilation efficiency and alveolar expansion in rats with meconium aspiration. Am J Respir Crit Care Med 154:764–770 (1996).

1048. Sun B, Curstedt T, Robertson B. Surfactant inhibition in experimental meconium aspiration. Acta Paediatr 82:182–189 (1993).

1049. Sun B, Curstedt T, Song GW, Robertson B. Surfactant improves lung function and morphology in newborn rabbits with meconium aspiration. Biol Neonate 63:96–104 (1993).

1050. Sundell H, Garrot J, Blankenship WJ. Studies on infants with type II respiratory distress syndrome. J Pediatr 78:754–764 (1971).

1051. Suter PM, Domenighetti G, Schaller M-D, Laverriere M-C, Ritz R, Perret C. N-Acetyl-cysteine enhances recovery from acute lung injury in man, a randomized, double-blind, placebo-controlled clinical study. Chest 105:190–194 (1994).

1052. Suzuki Y, Curstedt T, Grossman G, Kobayashi T, Nilsson R, Nohara K, Robertson B. The role of the low-molecular weight (≤15,000 daltons) apoproteins of pulmonary surfactant. Eur J Respir Dis 69:336–345 (1986).

1053. Suzuki Y, Fujita Y, Kogishi K. Reconstitution of tubular myelin from synthetic lipids and proteins associated with pig lung surfactant. Am Rev Respir Dis 140:75–81 (1989).

1054. Suzuki Y, Robertson B, Fujita Y, Grossman G. Respiratory failure in mice caused by a hybridoma making antibodies to the 15k Da surfactant apoprotein. Acta Anaesthesiol Scand 32:283–289 (1988).

1055. Sweet M, Hume D. Endotoxin signal transduction in macrophages. J Leukoc Biol 60:8–26 (1996).

1056. Szoka F, Papahadjopoulos D. Comparative properties and methods of preparation of lipid vesicles. Annu Rev Biophys Bioeng 9:467–508 (1980).

1057. Tabak SA, Notter RH. Effect of plasma proteins on the dynamic π-A characteristics of saturated phospholipid films. J Colloid Interface Sci 59:293–300 (1977).

1058. Tabak SA, Notter RH. A modified technique for dynamic surface pressure and relaxation measurements at the air-water interface. Rev Sci Instrum 48:1196–1201 (1977).

1059. Tabak SA, Notter RH, Ultman JS, Dinh S. Relaxation effects in the surface pressure behavior of dipalmitoyl lecithin. J Colloid Interface Sci 60:117–125 (1977).

1060. Tabor B, Ikegami M, Yamada T, Jobe A. Rapid clearance of surfactant-associated palmitic acid from the lungs of developing and adult animals. Pediatr Res 27:268–273 (1990).

1061. Taeusch HW, Ballard RA. Avery's Diseases of the Newborn. 7th ed. Philadelphia: WB Saunders (1998).

1062. Taeusch HW, Keough KMW, Williams W, Slavin R, Steele E, Lee AS, Phelps D, Kariel N, Floros J, Avery ME. Characterization of bovine surfactant for infants with respiratory distress syndrome. Pediatrics 77:572–581 (1986).

1063. Tagan M, Markert M, Schaller M, Feihl F, Chiolero R, Perret C. Oxidative metabolism of circulating granulocytes in adult respiratory distress syndrome. Am J Med 91(Suppl 3C):72–78 (1991).

1064. Takahashi A, Waring A, Amirkhanian J, Fan R, Taeusch HW. Structure-function relationships of bovine pulmonary surfactant proteins SP-B and SP-C. Biochim Biophys Acta 1044:43–49 (1990).

1065. Tanaka Y, Takei T. Lung surfactants : II. Effects of fatty acids, triacylglycerols, and protein on the activity of lung surfactant. Chem Pharm Bull 31:4100–4109 (1983).

1066. Tanaka Y, Takei T, Aiba T, Masuda K, Kiuchi A, Fujiwara T. Development of synthetic lung surfactants. J Lipid Res 27:475–485 (1986).

1067. Taneva S, Keough KM. Pulmonary surfactant proteins SP-B and SP-C in spread monolayers at the air-water interface: I. Monolayers of pulmonary surfactant protein SP-B and phospholipids. Biophys J 66:1137–1148 (1994).

1068. Taneva S, Keough KM. Pulmonary surfactant proteins SP-B and SP-C in spread monolayers at the air-water interface: II. Monolayers of pulmonary surfactant protein SP-C and phospholipids. Biophys J 66:1149–1157 (1994).

1069. Taneva S, Keough KM. Pulmonary surfactant proteins SP-B and SP-C in spread monolayers at the air-water interface: III. Proteins SP-B plus SP-C with phospholipids in spread monolayers. Biophys J 66:1158–1166 (1994).

1070. Taneva S, McEachren T, Stewart J, Keough KM. Pulmonary surfactant protein SP-A with phospholipids in spread monolayers at the air-water interface. Biochemistry 34:10279–10289 (1995).

1071. Taneva S, Stewart J, Taylor L, Keough K. Method of purification affects some interfacial properties of pulmonary surfactant proteins B and C and their mixtures with dipalmitoylphosphatidylcholine. Biochim Biophys Acta 1370:138–150 (1998).

1072. Taneva SG, Keough KM. Dynamic surface properties of pulmonary surfactant proteins SP-B and SP-C and their mixtures with dipalmitoylphosphatidylcholine. Biochemistry 33:14660–14670 (1994).

1073. Tanswell A, Freeman B. Liposome-entrapped antioxidant enzymes prevent lethal O_2 toxicity in the newborn rat. J Appl Physiol 63:347–352 (1987).

1074. Tashiro K, Li W-Z, Yamada K, Matsumoto Y, Kobayashi T. Surfactant replacement reverses respiratory failure induced by intratracheal endotoxin in rats. Crit Care Med 23:149–156 (1995).

1075. Taussig LM, Landau LI, Le Souef PN, Morgan WJ, Martinez FD, Sly PDE. Pediatric Respiratory Medicine. St. Louis: Mosby (1999).

1076. Temmesfeld-Wollbruck B, Walmrath D, Grimminger F, Seeger W. Prevention and therapy of the adult respiratory distress syndrome. Lung 173:139–164 (1995).

1077. Ten Centre Study Group. Ten centre trial of artificial surfactant (artificial lung expanding compound ALEC) in very premature babies. Br Med J 294:991–996 (1987).

1078. Tenner AJ, Robinson SL, Borchelt J, Wright JR. Human pulmonary surfactant protein (SP-A), a protein structurally homologous to C1q, can enhance FcR- and CR1-mediated phagocytosis. J Biol Chem 264:13923–13928 (1989).

1079. Teraska D, Clark D, Singh B, Rokahr J. Free fatty acids in meconium. Biol Neonate 50:16–20 (1986).

1080. The American College of Chest Physicians/Society of Critical Care Medicine Consensus Conference. Definitions for sepsis and organ failure and guidelines for the use of innovative therapies for sepsis. Crit Care Med 20:864–874 (1992).

1081. The Diagram Group. Rules of the Game. New York: St. Martin's Press (1990).

1082. Thompson W. Positional distribution of fatty acids in brain polyphosphoinositides. Biochim Biophys Acta 187:150–153 (1969).

1083. Tierney DF, Johnson RP. Altered surface tension of lung extracts and lung mechanics. J Appl Physiol 20:1253–1260 (1965).

1084. Timmons OD, Havens PL, Fackler JC (Pediatric Critical Care Study Group and the Extracorporeal Life Support Organization). Predicting death in pediatric patients with acute respiratory failure. Chest 108:789–797 (1995).

1085. Touchstone JC, Chen JC, Beaver KM. Improved separation of phospholipids in thin-layer chromatography. Lipids 15:61–62 (1980).

1086. Tracey K, Lowry S, Cerami A. Cachectin/TNF-alpha in septic shock and septic adult respiratory distress syndrome. Am Rev Respir Dis 138:1377–1379 (1988).

1087. Tracey KJ, Fong Y, Hesse DG, Manogue KR, Lee AT, Kuo GC, Lowry SF, Cerami

A. Anti-cachectin/TNF monoclonal antibodies prevent septic shock during lethal bacteremia. Nature 330:662–664 (1987).

1088.   Troncy E, Blaise G. Phenylephrine and inhaled nitric oxide (letter). Anethesiology 89:538–539 (1998).

1089.   Troncy E, Collet JP, Shapiro S, Guimond J-G, Blair L, Ducruet T, Francceur M, Charbonneau M, Blaise G. Inhaled nitric oxide in acute respiratory distress syndrome. A pilot randomized controlled study. Am J Respir Crit Care Med 157:1483–1488 (1998).

1090.   Troncy E, Francoeur M, Blaise G. Inhaled nitric oxide: Clinical applications, indications, and toxicology. Can J Anaesth 44:973–988 (1997).

1091.   Turcotte JG, Lin WH, Pivarnik PE, Motola NC, Bhongle NN, Heyman HR, Notter RH. Chemical synthesis and surface activity of lung surfactant phospholipid analogs. III. Chiral N-substituted ether-amide phosphonolipids. Chem Phys Lipids 58:81–95 (1991).

1092.   Turcotte JG, Lin WH, Pivarnik PE, Sacco AM, Bermel MS, Lu Z, Notter RH. Chemical synthesis and surface activity of lung surfactant phospholipid analogs. II. Racemic N-substituted diether phosphonolipids. Biochim Biophys Acta 1084:1–12 (1991).

1093.   Turcotte JG, Sacco AM, Steim JM, Tabak SA, Notter RH. Chemical synthesis and surface properties of an analog of the pulmonary surfactant dipalmitoyl phosphatidylcholine analog. Biochim Biophys Acta 488:235–248 (1977).

1094.   Turner SR, Litt M, Lynn WS. Permeation of water vapor through lipid monolayers. J Colloid Interface Sci 50:181–193 (1975).

1095.   Tutuncu A, Faithful N, Lachmann B. Intratracheal perfluorocarbon administration combined with mechanical ventilation in experimental respiratory distress syndrome: Dose-dependent improvement in gas exchange. Crit Care Med 21:962–969 (1993).

1096.   Tutuncu AS, Faithful NS, Lachmann B. Comparison of ventilatory support with intratracheal perfluorocarbon administration and conventional mechanical ventilation in animals with acute respiratory failure. Am Rev Respir Dis 148:785–792 (1993).

1097.   Uchida T, Nakazawa K, Yokoyama K, Makita K, Amaha K. The combination of partial liquid ventilation and inhaled nitric oxide in the severe oleic acid lung injury model. Chest 113:1658–1666 (1998).

1098.   Ueda T, Ikegami M. Change in properties of exogenous surfactant in injured rabbit lung. Am J Respir Crit Care Med 153:1844–1849 (1996).

1099.   Ueda T, Ikegami M, Henry M, Jobe AH. Clearance of surfactant protein B from rabbit lungs. Am J Physiol 268:L636–L641 (1995).

1100.   Vadas P, Pruzanski W. Biology of disease: Role of secretory phospholipases A_2 in the pathobiology of disease. Lab Invest 55:391–404 (1986).

1101.   van Daal GJ, Bos JAH, Eijking EP, Gommers D, Hannappel E, Lachmann B. Surfactant replacement therapy improves pulmonary mechanics in end-stage influenza A pneumonia in mice. Am Rev Respir Dis 145:859–863 (1992).

1102.   van Daal GJ, Eijking EP, So KL, Fievez RB, Sprenger MJW, van Dam DW, Erdmann W, Lachmann B. Acute respiratory failure during pneumonia indeced by Sendai virus. Adv Exp Med Biol 316:319–326 (1992).

1103. van Daal GJ, So KL, Gommers D, Eijking EP, Fievez RB, Sprenger MJ, van Dam DW, Lachmann B. Intratracheal surfactant administration restores gas exchange in experimental adult respiratory distress syndrome associated with viral pneumonia. Anesth Analg 72:589–595 (1991).

1104. Van Der Beek J, Plotz F, van Overbeek F, Heikamp A, Beekhuis H, Wildevuur C, Oklen A, Bambang OS. Distribution of exogenous surfactant in rabbits with severe respiratory failure: The effect of volume. Pediatr Res 34:154–158 (1993).

1105. van Golde LMG, Batenburg JJ, Robertson B. The pulmonary surfactant system: Biochemical aspects and functional significance. Physiol Rev 68:374–455 (1988).

1106. van Golde LMG, Casals CC. Metabolism of lipids. In: Crystal RG, West JB, Weibel ER, Barnes PJ, Eds. The Lung: Scientific Foundations. 2nd ed. Philadelphia: Lippincott-Raven, pp. 9–18 (1997).

1107. van Iwaarden F, van Strijp JAG, Ebskamp MJM, Welmers AC, Verhoef J, van Golde LMG. Surfactant protein A is opsonin in phagocytosis of herpes simplex virus type I by rat alveolar macrophages. Am J Physiol 261:L204–L209 (1991).

1108. van Iwaarden F, Welmers B, Verhoef J, Haagsman HP, van Golde LMG. Pulmonary surfactant protein A enhances the host-defense mechanism of rat alveolar macrophages. Am J Respir Cell Mol Biol 2:91–98 (1990).

1109. van Iwaarden JF, Shimizu H, van Golde PHM, Voelker DR, van Golde LMG. Rat surfactant protein D enhances the production of oxygen radicals by rat alveolar macrophages. Biochem J 286:5–8 (1992).

1110. van Schaik SM, Vargas I, Welliver RC, Enhorning G. Surfactant dysfunction develops in BALB/c mice infected with respiratory syncytial virus. Pediatr Res 42:169–173 (1997).

1111. Vandenbussche G, Clercx A, Clercx M, Curstedt T, Johansson J, Jörnvall H, Ruysschaert JM. Secondary structure and orientation of the surfactant protein SP-B in a lipid environment. A Fourier transform infrared spectroscopy study. Biochemistry 31:9169–9176 (1992).

1112. Vandenbussche G, Clercx A, Curstedt T, Johansson J, Jörnvall H, Ruysschaert JM. Structure and orientation of the surfactant-associated protein SP-C in a lipid bilayer. Eur J Biochem 203:201–209 (1992).

1113. Veldhuizen EJA, Batenberg JJ, Vandenbussche G, Putz G, van Golde LMG, Haagsman HP. Production of surfactant protein C in the baculovirus expression system: The information required for correct folding and palmitoylation of SP-C is contained within the mature sequence. Biochim Biophys Acta 1416:295–308 (1999).

1114. Veldhuizen R, McCaig L, Akino T, Lewis J. Pulmonary surfactant subfractions in patients with the acute respiratory distress syndrome. Am J Respir Crit Care Med 152:1867–1871 (1995).

1115. Veldhuizen RAW, Hearn SA, Lewis JF, Possmayer F. Surface-area cycling of different surfactant preparations: SP-A and SP-B are essential for large aggregate integrity. Biochem J 300:519–524 (1994).

1116. Venkitaraman A, Hall S, Whitsett J, Notter R. Enhancement of biophysical activity of lung surfactant extracts and phospholipid-apoprotein admixtures by surfactant protein A. Chem Phys Lipids 56:185–194 (1990).

1117. Venkitaraman AR, Baatz JE, Whitsett JA, Hall SB, Notter RH. Biophysical inhibition of synthetic phospholipid-surfactant protein admixtures by plasma proteins. Chem Phys Lipids 57:49–57 (1991).

1118. Venkitaraman AR, Hall SB, Notter RH. Hydrophobic homopolymeric peptides enhance the biophysical activity of synthetic lung phospholipids. Chem Phys Lipids 53:157–164 (1990).

1119. Vermont-Oxford Neonatal Network. A multicenter randomized trial comparing synthetic surfactant with modified bovine surfactant extract in the treatment of neonatal respiratory distress syndrome. Pediatrics 97:1–6 (1996).

1120. Vilallonga F. Surface chemistry of L-α-dipalmitoyl lecithin at the air-water interface. Biochim Biophys Acta 163:290–300 (1968).

1121. Villar J, Slutsky AS. The incidence of the adult respiratory distress syndrome. Am Rev Respir Dis 140:814–816 (1989).

1122. Vincent JS, Revak SD, Cochrane CD, Levin IW. Interactions of model human pulmonary surfactants with a mixed phospholipid bilayer assembly: Raman spectroscopic studies. Biochemistry 32:8228–8238 (1993).

1123. von Nahmen A, Schenk M, Sieber M, Amrein M. The structure of a model pulmonary surfactant as revealed by scanning force microscopy. Biophys J 72:463–469 (1997).

1124. von Neergaard K. Neue auffassungen uber einen grundbegriff der atemmechanik. Dieretraktionskraft der lunge, abhangig von der oberflachenspannung in den alveolen. Z Ges Exp Med 66:373–394 (1929).

1125. Voorhout WF, Veenendaal T, Haagsman HP, Verkkleij AJ, van Golde LMG, Geuze HJ. Surfactant protein A is localized at the corners of the pulmonary tubular myelin lattice. J Histochem Cytochem 39:1331–1336 (1991).

1126. Voorhout WF, Veenendaal T, Haagsman HP, Weaver TE, Whitsett JA, van Golde LMG, Geuze HJ. Intracellular processing of pulmonary surfactant protein B in an endosomal/lysosomal compartment. Am J Physiol 263:L479–L486 (1992).

1127. Voorhout WF, Weaver TE, Haagsman HP, Geuze HJ, Van Golde LM. Biosynthetic routing of pulmonary surfactant proteins in alveolar type II cells. Microsc Res Tech 26:366–373 (1993).

1128. Vorbroker DK, Profitt SA, Nogee LM, Whitsett JA. Aberrant processing of surfactant protein C in hereditary SP-B deficiency. Am J Physiol 268:L647–L656 (1995).

1129. Vorhout WF, Veenendaal T, Kuroki Y, Ogasawara Y, van Golde LMG, Geuze HJ. Immunocytochemical localization of surfactant protein D (SP-D) in type II cell, Clara cells, and alveolar macrophages of rat lung. J Histochem Cytochem 40:1589–1597 (1992).

1130. Voss T, Eistetter H, Schafer KP, Engel J. Macromolecular organization of natural and recombinant lung surfactant protein SP28-36. Structural homology with the complement factor C1q. J Mol Biol 201:219–227 (1988).

1131. Voss T, Melchers K, Scheirle G, Schafer KP. Structural comparison of recombinant pulmonary surfactant protein SP-A derived from two human coding sequences: Implications for the chain composition of natural human SP-A. Am J Respir Cell Mol Biol 4:88–94 (1991).

1132. Wagner PD, Mathieu-Costello O, Bebaut DE, Gray AT, Natterson PD, Glennow C.

Protection against pulmonary O$_2$ toxicity by *N*-acetylcysteine. Eur Respir J 2:116–126 (1989).

1133. Walker SA, Kennedy MT, Zasadzinski JA. Encapsulation of bilayer vesicles by self-assembly. Nature 387:61–64 (1997).
1134. Walker SR, Williams MC, Benson B. Immunocytochemical localization of the major surfactant proteins in type II cells, Clara cells, and alveolar macrophages of rat lungs. J Histochem Cytochem 34:1137–1148 (1986).
1135. Walmrath D, Gunther A, Ghofrani HA, Schermuly R, Schnedier T, Grimminger F, Seeger W. Bronchoscopic surfactant administration in patients with severe adult respiratory distress syndrome and sepsis. Am J Respir Crit Care Med 154:57–62 (1996).
1136. Walmrath D, Schneider T, Schermuly R, Olschewski H, Grimminger F, Seeger W. Direct comparison of inhaled nitric oxide and aerosolized prostacyclin in acute respiratory distress syndrome. Am J Respir Crit Care Med 153:991–996 (1996).
1137. Walther F, Hernandez-Juviel J, Bruni R, Waring AJ. Protein composition of synthetic surfactant affects gas exchange in surfactant-deficient rats. Pediatr Res 43:666–673 (1998).
1138. Walther FJ, David-Cu R, Leung C, Bruni R, Hernandez-Juviel J, Gordon LM, Waring AJ. A synthetic segment of surfactant protein A—Structure, in vitro surface activity, and in vivo efficacy. Pediatr Res 39:938–946 (1996).
1139. Walther FJ, David-Cu R, Lopez SL. Antioxidant-surfactant liposomes mitigate hyperoxic lung injury in premature rabbits. Am J Physiol 269:L613–L617 (1995).
1140. Walther FJ, Hernandez-Juviel J, Bruni R, Waring A. Spiking Survanta with synthetic surfactant peptides improves oxygenation in surfactant-deficient rats. Am J Respir Crit Care Med 156:855–861 (1997).
1141. Walther FJ, Nunex Fl, David-Cu R, Hill KE. Mitigation of pulmonary oxygen toxicity in rats by intratracheal instillation of polyethylene glycol–conjugated antioxidant enzymes. Pediatr Res 33:332–335 (1993).
1142. Wang Z, Gurel O, Baatz JE, Notter RH. Acylation of pulmonary surfactant protein-C is required for its optimal surface-active interactions with phospholipids. J Biol Chem 271:19104–19109 (1996).
1143. Wang Z, Gurel O, Baatz JE, Notter RH. Differential activity and lack of synergy of lung surfactant proteins SP-B and SP-C in surface-active interactions with phospholipids. J Lipid Res 37:1749–1760 (1996).
1144. Wang Z, Gurel O, Weinbach S, Notter RH. Primary importance of zwitterionic over amnionic phospholipids in the surface active function of calf lung surfactant extract. Am J Respir Crit Care Med 156:1049–1057 (1997).
1145. Wang Z, Hall SB, Notter RH. Dynamic surface activity of films of lung surfactant phospholipids, hydrophobic proteins, and neutral lipids. J Lipid Res 36:1283–1293 (1995).
1146. Wang Z, Hall SB, Notter RH. Roles of different hydrophobic constituents in the adsorption of pulmonary surfactant. J Lipid Res 37:790–798 (1996).
1147. Wang Z, Notter RH. Additivity of protein and non-protein inhibitors of lung surfactant activity. Am J Respir Crit Care Med 158:28–35 (1998).
1148. Warburton D, Buckley S, Cosico L. P1 and P2 purinergic receptor signal transduction in rat type II pneumocytes. J Appl Physiol 66:901–905 (1989).

1149. Ward A, Clissold SP. Pentoxifylline. A review of its pharmacodynamic and pharmacokinetic properties, and its therapeutic efficacy. Drugs 34:50–97 (1987).

1150. Ward JA, Roberts RJ. Effect of hyperoxia on phosphatidylcholine synthesis, secretion, uptake and stability in the newborn rabbit lung. Biochim Biophys Acta 796:42–50 (1984).

1151. Ward JA, Roberts RJ. Hyperoxia effects on pulmonary pressure:volume characteristics and lavage surfactant phospholipid in the newborn rabbit. Biol Neonate 46:139–148 (1984).

1152. Ward JA, Roberts RJ. Vitamin E inhibition of the effects of hyperoxia on the pulmonary surfactant system of the newborn rabbit. Pediatr Res 18:329–334 (1984).

1153. Ward PA, Hunninghake GW. Lung inflammation and fibrosis. Am J Respir Crit Care Med 157(Suppl):S123–S129 (1998).

1154. Waring A, Faull L, Leung C, Chang-Chien A, Mercado P, Taeusch HW, Gordon L. Synthesis, secondary structure and folding of the bend region of lung surfactant protein B. Peptide Res 9:28–31 (1996).

1155. Waring A, Taeusch W, Bruni R, Amirkhanian J, Fan B, Stevens R, Young J. Synthetic amphipathic sequences of SP-B mimic several physicochemical and in vitro properties of native pulmonary surfactant proteins. Peptide Res 2:308–313 (1989).

1156. Warr RG, Hawgood S, Buckely DI, Crisp TM, Schilling J, Benson BJ, Ballard PL, Clements JA, White RT. Low molecular weight human pulmonary surfactant protein (SP5): Isolation, characterization and cDNA amino acid sequences. Proc Natl Acad Sci USA 84:7915–7919 (1987).

1157. Watkins JC. The surface properties of pure phospholipids in relation to those of lung extracts. Biochim Biophys Acta 152:293–306 (1968).

1158. Weast RC, Astle MJ, Beyer WH. CRC Handbook of Chemistry and Physics. 69th ed. (1988).

1159. Weaver TE. Surfactant proteins and SP-D. Am J Respir Cell Mol Biol 5:4–5 (1991).

1160. Weaver TE, Lin S, Bogucki B, Dey C. Processing of surfactant protein B proteolipid by a cathepsin D–like protease. Am J Physiol 263:L95–L103 (1992).

1161. Weaver TE, Whitsett JA. Antigenicity of surfactant protein A, B and C. In: Robertson B, Taeusch HW, Eds. Surfactant Therapy for Lung Disease. New York: Marcel Dekker, pp. 293–306 (1995).

1162. Weaver TE, Whitsett JA. Function and regulation of expression of pulmonary surfactant–associated proteins. Biochem J 273:249–264 (1991).

1163. Weaver TE, Whitsett JA. Processing of hydrophobic pulmonary surfactant protein B in rat type II cells. Am J Physiol 257:L100–L108 (1989).

1164. Weg JG, Balk RA, Tharratt RS, Jenkinson S, Shah JB, Zaccardelli D, Horton J, Pattishall EN. Safety and potential efficacy of an aerosolized surfactant in human sepsis-induced adult respiratory distress syndrome. JAMA 272:1433–1438 (1994).

1165. Weibel ER. Morphometry of the Human Lung. Berlin: Springer (1963).

1166. Weibel ER, Bachofen H. How to stabilize the pulmonary alveoli: Pulmonary surfactant or fibers? News Physiol Sci 2:72–75 (1987).

1167. Weibel ER, Bachofen H. Structural design of the alveolar septum and gas exchange. In: Fishman AP, Renkin EM, Eds. Pulmonary Edema. Bethesda, MD: American Physiological Society Press, pp. 1–20 (1979).

1168. Weibel ER, Gil J. Electron microscopic demonstration of an extra-cellular duplex lining layer of alveoli. Respir Physiol 4:42–57 (1968).

1169. Weibel ER, Gil J. Structure-function relationship at the alveolar level. In: West JB, Ed. Bioengineering Aspects of the Lung. New York: Marcel Dekker, pp. 1–81 (1977).

1170. Weikert LF, Bernard GR. Pharmacology of sepsis. Clin Chest Med 17:289–305 (1996).

1171. West JB. Bioengineering Aspects of the Lung. New York: Marcel Dekker (1977).

1172. White CW, Jackson JH, Abuchowski A, Kazo GM, Mimmack RF, Berger EM, Freeman BA, McCord JM, Repine JE. Polyethylene glycol–attached antioxidant enzyme decrease pulmonary oxygen toxicity in rats. J Appl Physiol 66:584–590 (1989).

1173. White RT, Damm D, Miller J, Spratt K, Schilling J, Hawgood S, Benson B, Cordell B. Isolation and characterization of the human pulmonary surfactant apoprotein gene. Nature 317:361–363 (1985).

1174. Whitsett JA, Baatz JE. Hydrophobic surfactant proteins SP-B and SP-C: Molecular biology, structure and function. In: Robertson B, van Golde LMG, Batenburg JJ, Eds. Pulmonary Surfactant: from Molecular Biology to Clinical Practice. Amsterdam: Elsevier Science Publishers, pp. 55–75 (1992).

1175. Whitsett JA, Hull WM, Ross G, Weaver T. Characteristics of human surfactant-associated glycoproteins A. Pediatr Res 19:501–508 (1985).

1176. Whitsett JA, Nogee LM, Weaver TE, Horowitz AD. Human surfactant protein B structure, function, regulation, and genetic disease. Physiol Rev 75:749–757 (1995).

1177. Whitsett JA, Ohning BL, Ross G, Meuth T, Weaver T, Holm BA, Shapiro DL, Notter RH. Hydrophopic surfactant-associated protein in whole lung surfactant and its imporance for biophysical activity in lung surfactant extracts used for replacement therapy. Pediatr Res 20:460–467 (1986).

1178. Whitsett JA, Weaver TE, Clark JC, Sawtell N, Glasser SW, Korfhagen TR, Hull WM. Glucocorticoid enhances surfactant proteolipid Phe and pVal synthesis and RNA in fetal lung. J Biol Chem 262:15618–15623 (1987).

1179. Wilcox DT, Glick PL, Karamanoukian HL, Leach C, Morin FC, Fuhrman BP. Perfluorocarbon-associated gas exchange improves pulmonary mechanics, oxygenation, ventilation, and allows nitric oxide delivery in the hypoplastic lung congenital diaphragmatic hernia lamb model. Crit Care Med 23:1858–1863 (1995).

1180. Wilhelmy L. Ueber die abhangigkeit der capillaritats-constanten des alkohols von substanz und gestalt des benetzten festen korpers. Ann Phys Chem 119:117–217 (1863).

1181. Wilkinson A, Jenkins PA, Jeffrey JA. Two controlled trials of dry artificial surfactant: Early effects and later outcome in babies with surfactant deficiency. Lancet 2:287–291 (1985).

1182. Williams MC. Conversion of lamellar body membranes into tubular myelin in alveoli of fetal rat lungs. J Cell Biol 72:260–277 (1977).

1183. Williams MC. Ultrastructure of tubular myelin and lamellar bodies in fast-frozen rat lung. Exp Lung Res 4:37–46 (1982).
1184. Williams MC, Hawgood S, Hamilton RL. Changes in lipid structure produced by surfactant proteins SP-A, SP-B, and SP-C. Am J Respir Cell Mol Biol 5:41–50 (1991).
1185. Willson DF, Bauman LA, Zaritsky A, Dockery K, James RL, Stat M, Conrad D, Craft H, Novotny WE, Egan EA, Dalton H. Instillation of calf lung surfactant extract (calfactant) is beneficial in pediatric acute hypoxemic respiratory failure. Crit Care Med 27:188–195 (1999).
1186. Willson DF, Jiao JH, Bauman LA, Zaritsky A, Craft H, Dockery K, Conrad D, Dalton H. Calf lung surfactant extract in acute hypoxemic respiratory failure in children. Crit Care Med 24:1316–1322 (1996).
1187. Wilson TA. A continuum analysis of a two-dimensional mechanical model of the lung parenchyma. J Appl Physiol 33:472–478 (1972).
1188. Wilson TA. Relations among recoil pressure, surface area, and surface tension in the lung. J Appl Physiol 50:921–926 (1981).
1189. Wilson TA. Surface tension–surface area curves calculated from pressure-volume loops. J Appl Physiol 53:1512–1520 (1982).
1190. Wilson TA, Bachofen H. A model for mechanical structure of the alveolar duct. J Appl Physiol 52:1064–1070 (1982).
1191. Windsor AC, Walsh CJ, Mullen PG, Cook DJ, Fisher BJ, Blocher CR, Leeper-Woodford SK, Sugerman HJ, Fowler AA. Tumor necrosis factor-alpha blockade prevents neutrophil CD18 receptor upregulation and attenuates acute lung injury in porcine sepsis without inhibition of neutrophil oxygen radical generation. J Clin Invest 91:1459–1468 (1993).
1192. Windsor ACJ. Acute lung injury: What have we learned from animal models? Am J Med Sci 306:111–116 (1993).
1193. Wirtz HRW, Dobbs LG. Calcium mobilization and exocytosis after one mechanical stretch of lung epithelial cells. Science 250:1266–1269 (1990).
1194. Witschi H. Proliferation of alveolar type II cells: A review of common responses in toxic lung injury. Toxicology 5:267–277 (1976).
1195. Witschi H. Responses of the lung to toxic injury. Environ Health Perspect 85:5–13 (1990).
1196. Wojciak JF, Notter RH, Oberdoerster G. Size stability of phosphatidylcholine-phosphatidylglycerol aerosols and a dynamic film compression state from their interfacial impaction. J. Colloid Interface Sci. 106:547–557 (1985).
1197. Wolfson MR, Greenspan JS, Deoras KS, Rubenstein SD, Shaffer TH. Comparison of gas and liquid ventilation: Clinical, physiological and histological correlates. J Appl Physiol 72:1024–1031 (1992).
1198. Wong CJ, Akiyama J, Allen L, Hawgood S. Localization and development expression of surfactant proteins D and A in the respiratory tract of the mouse. Pediatr Res 39:930–937 (1996).
1199. Wright JR. Clearance and recycling of pulmonary surfactant. Am J Physiol 259:L1–L12 (1990).

1200. Wright JR, Benson BJ, Williams MC, Goerke J, Clements JA. Protein composition of rabbit alveolar surfactant subfractions. Biochim Biophys Acta 791:320–332 (1984).

1201. Wright JR, Clements JA. Metabolism and turnover of lung surfactant. Am Rev Respir Dis 135:426–444 (1987).

1202. Wright JR, Wager RE, Hamilton RL, Huang M, Clements JA. Uptake of lung surfactant subfractions into lamellar bodies of adult rabbit lungs. J Appl Physiol 60:817–825 (1986).

1203. Wright JR, Wager RE, Hawgood S, Dobbs LG, Clements JA. Surfactant apoprotein Mr 26,000–30,000 enhances uptake of liposomes by type II cells. J Biol Chem 262:2888–2894 (1987).

1204. Wysocki M, Roupie E, Langeron O, Liu N, Herman B, Lemaire F, Brochard L. Additive effect on gas exchange of inhaled nitric oxide and intravenous almitrine bismesylate in the adult respiratory distress syndrome. Intensive Care Med 20:254–259 (1994).

1205. Xu J, Richardson C, Ford C, Spencer T, Li-juan Y, Mackie G, Hammond G, Possmayer F. Isolation and characterization of the cDNA clone for pulmonary surfactant–associated protein-B (SP-B) in the rabbit. Biochem Biophys Res Commun 160:325–332 (1989).

1206. Yager D, Cloutier T, Feldman H, Bastacky J, Drazen JM, Kamm RD. Airway surface liquid thickness as a function of lung volume in small airways of the guinea pig. J Appl Physiol 77:2333–2340 (1994).

1207. Yamada T, Ikegami M, Tabor BL, Jobe AH. Effects of surfactant protein-A on surfactant function in preterm ventilated rabbits. Am Rev Respir Dis 142:754–757 (1990).

1208. Yao L-J, Richardson C, Ford C, Mathialagan N, Mackie G, Hammond GL, Harding PG, Possmayer F. Expression of mature pulmonary surfactant–associated protein B (SP-B) in *Escherichia coli* using truncated human SP-B cDNAs. Biochem Cell Biol 68:559–566 (1990).

1209. Yokoi K, Mukaida N, Harada A, Watanabe Y, Matsushima K. Prevention of endotoxemia-induced acute respiratory distress syndrome–like lung injury in rabbits by a monoclonal antibody to IL-8. Lab Invest 76:375–384 (1997).

1210. Yoshida K, Mohsenin V. Inhibition of protein kinase C of human neutrophils by phosphatidylcholines. Life Sci 54:515–524 (1994).

1211. Young JD, Brampton WJ, Knighton JD, Finfer SR. Inhaled nitric oxide in acute respiratory failure in adults. Br J Anaesth 73:499–502 (1994).

1212. Young SL, Wright JR, Clements JA. Cellular uptake and processing of surfactant lipids and apoprotein SP-A by rat lung. J Appl Physiol 66:1336–1342 (1989).

1213. Yu SH, Harding PGR, Possmayer F. Artificial pulmonary surfactant: Potential role for hexagonal HII phase in the formation of a surface active monolayer. Biochim Biophys Acta 776:37–47 (1984).

1214. Yu SH, Harding PGR, Smith N, Possmayer P. Bovine pulmonary surfactant: Chemical composition and physical properties. Lipids 18:522–529 (1983).

1215. Yu SH, Possmayer F. Adsorption, compression, and stability of surface films of natural, lipid extract, and reconstituted pulmonary surfactants. Biochim Biophys Acta 1167:264–271 (1993).

1216. Yu SH, Possmayer F. Comparative studies on the biophysical activities of the low-molecular-weight hydrophobic proteins purified from bovine pulmonary surfactant. Biochim Biophys Acta 961:337–350 (1988).

1217. Yu SH, Possmayer F. Reconstitution of surfactant activity by using the 6kDa apoprotein associated with pulmonary surfactant. Biochem J 236:85–89 (1986).

1218. Yu SH, Wallace D, Bhavnani B, Enhorning G, Harding PG, Possmayer F. Effect of reconstituted pulmonary surfactant containing the 6000 dalton hydrophobic protein on lung compliance of prematurely delivered rabbit fetuses. Pediatr Res 23:23–30 (1988).

1219. Yu X-G, Feet BA, Moen A, Curstedt T, Saugstad OD. Nitric oxide contributes to surfactant-induced vasodilatation in surfactant-depleted newborn piglets. Pediatr Res 42:151–156 (1997).

1220. Yukitake K, Brown CL, Schlueter MA, Clements JA, Hawgood S. Surfactant apoprotein A modifies the inhibitory effect of plasma proteins on surfactant activity in vivo. Pediatr Res 37:21–25 (1995).

1221. Zapol WM. Nitric oxide inhalation in acute respiratory distress syndrome: It works, but can we prove it? Crit Care Med 26:2–3 (1998).

1222. Zelter M, Escudier BJ, Hoeffel JM, Murray JF. Effects of aerosolized artificial surfactant on repeated oleic acid injury in sheep. Am Rev Respir Dis 141:1014–1019 (1990).

1223. Zeni F, Pain P, Vindimian M, Gay J-G, Gery P, Bertrand M, Page Y, Page D, Vermesch R, Bertrand J-C. Effects of pentoxifylline on circulating cytokine concentrations and hemodynamics in patients with septic shock: Results from a double-blind, randomized, placebo-controlled study. Crit Care Med 24:207–214 (1996).

1224. Zhang K, Phan SH. Cytokines and pulmonary fibrosis. Biol Signals 5:232–239 (1996).

1225. Zhang Y-P, Lewis RNAH, Hodges RS, McElhaney RN. Interaction of a peptide model of a hydrophobic transmembrane α-helical segment of a membrane protein with phosphatidylcholine bilayers: Differential scanning calorimetric and FTIR spectroscopic studies. Biochemistry 31:11579–11588 (1992).

1226. Zheng H, Crowley JJ, Chan JC, Hoffman H, Hatherill JR, Ishizaka A, Raffin TA. Attenuation of tumor necrosis factor–induced endothelial cell cytotoxicity and neutrophil chemiluminescence. Am Rev Respir Dis 142:1073–1078 (1990).

1227. Zhu GF, Sun B, Niu S, Cai YY, Lin K, Lindwall R, Robertson B. Combined surfactant therapy and inhaled nitric oxide in rabbits with oleic acid–induced acute respiratory distress syndrome. Am J Respir Crit Care Med 158:437–443 (1998).

1228. Ziegler EJ, Fisher CJ, Sprung CL, Straube RC, Sadoff JC, Foulke GE, Wortel CH, Fink MP, Dellinger RP, Teng N, Allen IE, Berger HJ, Knatterud GL, LoBuglio AF, Smith CR. Treatment of gram negative bacteremia and septic shock with HA-1A human monoclonal antibody against endotoxin: A randomized, double-blind, placebo-controlled trial. N Engl J Med 324:429–436 (1991).

1229. Zlatkis A, Zak B. Study of a new cholesterol reagent. Anal Biochem 29:143–148 (1969).
1230. Zucker A, Holm BA, Wood LDH, Crawford G, Ridge K, Sznajder IA. Exogenous surfactant with PEEP reduces pulmonary edema and improves lung function in canine aspiration pneumonitis. J Appl Physiol 73:679–686 (1992).
1231. Zwissler B, Gregor K, Habler O, Kleen M, Merkel M, Haller M, Brigel J, Welte M, Peter K. Inhaled prostacyclin (PGI_2) versus inhaled nitric oxide in adult respiratory distress syndrome. Am J Respir Crit Care Med 154:1671–1677 (1996).

INDEX

A

Absorption spectroscopy, 92
Acetic acid, surface tension of, 75
Acetone, surface tension of, 75
N-Acetylcysteine (NAC), 309, 310, 314, 315–316
Active lung surfactant, biophysical properties of, 146–147
Acute lung injury (ALI), 1, 233
  ARDS and, 239–247
  See also ARDS-related acute lung injury
Acute respiratory distress syndrome (ARDS), 1, 2, 3, 233
  in adults, surfactant therapy for, 305–308
  anti-inflammatory agents and antioxidants for use in, 313–316
  clinical diagnosis of, 241
  INO and other vasoactive agents for use in, 311–313
  modes of ventilation for treatment of, 316–317
  multimodal therapy for, 309–311
  pathophysiology of, 299–300
  -related acute lung injury, 239–247
    surfactant dysfunction in animal models of, 265–269

[Acute respiratory distress syndrome (ARDS)]
  surfactant-related events in, 242, 246
  surfactant replacement in animal models of, 269–278
  -related respiratory failure
  surfactant therapy in children with, 303–305
  surfactant therapy in full-term infants with, 303
  surfactant therapy for, 4, 300–301, 308–309
  factors complicating assessment, 301–302
"Adult" respiratory distress syndrome:
  see Acute respiratory distress syndrome (ARDS)
Adsorption:
  behavior of surfactants suspended in liquid phase and, 34–39
  definition of, 35–37
  with minimized diffusion resistance, 79, 82–84
  of phospholipids, 68–70
Adsorption of lung surfactant, 7
  modeling of, 163–165
  tubular myelin and, 143–144